The Urban Book Series

The Urban Book Series is a resource for urban studies and geography research worldwide. It provides a unique and innovative resource for the latest developments in the field, nurturing a comprehensive and encompassing publication venue for urban studies, urban geography, planning and regional development.

The series publishes peer-reviewed volumes related to urbanization, sustainability, urban environments, sustainable urbanism, governance, globalization, urban and sustainable development, spatial and area studies, urban management, transport systems, urban infrastructure, urban dynamics, green cities and urban landscapes. It also invites research which documents urbanization processes and urban dynamics on a national, regional and local level, welcoming case studies, as well as comparative and applied research.

The series will appeal to urbanists, geographers, planners, engineers, architects, policy makers, and to all of those interested in a wide-ranging overview of contemporary urban studies and innovations in the field. It accepts monographs, edited volumes and textbooks.

Indexed by Scopus.

More information about this series at https://link.springer.com/bookseries/14773

Kwi-Gon Kim · Massamba Thioye
Editors

Planning Climate Smart and Wise Cities

A Multidisciplinary Approach

Springer

Editors
Kwi-Gon Kim
Seoul National University
Seoul, Korea (Republic of)

Massamba Thioye
Mitigation Programme
UNFCCC Secretariat
Bonn, Germany

ISSN 2365-757X ISSN 2365-7588 (electronic)
The Urban Book Series
ISBN 978-3-030-80164-9 ISBN 978-3-030-80165-6 (eBook)
https://doi.org/10.1007/978-3-030-80165-6

This Springer imprint is published by the registered company Springer Nature Switzerland AG
The registered company address is: Gewerbestrasse 11, 6330 Cham, Switzerland

Preface

Climate change, natural disasters, and dramatically increasing urbanization-associated challenges raise disruptive issues that have come to the forefront of global discussion on the agenda of sustainable development and sustainable "human friendly" living. The recent "UN 2019 Climate Action Summit" in New York has highlighted the urgent need for and recognized the criticality of immediate action to deliver a sustainable and livable planet and world to the future generations. To be effective, the actions at the city level should be multidimensional, with the integration of disruptive and transformational platform solutions leveraging disruptive technologies, policies, business models, financing instruments, but also leadership for a response from all levels, from the city government level to the level of individual citizens.

Climate actions must be mainstreamed in integrated sustainable development goals (SDGs) actions. The main challenge in the pursuit of the SDGs is that actions targeting simultaneously the achievement of several of them are rare. Most of the time, the SDGs are addressed separately, without seeking synergy. Yet all the goals are interconnected. The big challenge now is how to integrate different actions for SDGs.

Rationale for This Book

The post-COVID-19 pandemic recovery provides an opportunity for enhanced climate action mainstreamed in the recovery measures planned by major countries all over the world. Shaping a low GHG emission and resilient post-COVID-19 urban era and charting path forward on the Green Deal have become new agendas for urban planning and management. The European Union has announced plans to set aside 25% of a proposed $826 billion coronavirus-related temporary recovery instrument NextGenerationEU for climate-friendly measures such as building retrofits, clean energy, electric vehicles, and sustainable land use. If such a recovery plan goes through, it could be the most significant green investment in history. Bonus points: it will create 1 million new jobs (Joel Jager, WRI DIGEST, June 10, 2020).

Climate smart city planning and the digital, carbon and ecological technologies behind it hold to the cities' stakeholders much promise and other interesting opportunities for new markets, services, and practices for the Green Deal and post-COVID-19 recovery plans. Though cities and communities around the world have elaborated a wide variety of climate change actions and plans, their development and implementation often follow a project-based approach with no network connections between projects. The increasing number of urban responses to climate change, Green New Deal, and COVID-19 highlights the need for carbon-centered comprehensive (3Cs) smart planning model which incorporates climate change mitigation and adaptation and Green Deal policy goals, and post-COVID-19 recovery plans at each stage of the planning process.

Such a model aims for digitalizing platform solutions for individual actions and viability decisions in a connected manner, and, therefore, the model needs to combine a large number of different ICT uses that are relevant to climate and COVID stimulus measures. Lots of institutional capacity are needed in execution of smart ICT networks.

In urban systems modeling, existing urban models are not sufficient to take into account mechanisms relevant for climate, Green Deal, and post-COVID-19 recovery issues which introduce complexities with high level of uncertainties, volatilities, ambiguities, and interactions across the issues. This requires developing a new integrated dynamic urban model to address more parameters with the smart, resilient, and resource wise grid platform systems on the concept of artificial neural network.

We can learn lessons from the coronavirus lockdown and use big data, AI program, and machine learning to tackle urgent urban planning challenges relevant to the global crisis. Cutting edge data, creative uses of technology, and forward-thinking partnerships can help find solutions to many urban challenges.

Another important challenge is the one related to incentivizing climate action in cities. Indeed, there is a mismatch between the GHG accounting system of cities and the type of climate actions their potential climate actors can undertake, what they are accountable for or can claim credit for. While a production-based approach is widely used for cities GHG accounting, cities are less and less producers of carbon-intensive products and more and more consumption centers. As the GHG accounting system is the basis for incentivizing action, it should address the types of climate actions the cities' climate actors are able to conduct. These are mainly need-based and solution-oriented climate actions aiming at satisfying the needs of the citizens at the lowest level of GHG emissions. Hence the need to use Consumption Based Approaches for GHG accounting. However, this approach to the GHG accounting for cities raises challenges including complexities and availability of data. One chapter of this book is dedicated to addressing this important issue and presents a new GHG accounting system for cities, leveraging the frontier technologies of the fourth industrial revolution such as distributed ledger technologies.

Access to low cost finance is key for cities to implement green action plans. The book opens a dialogue on how green bonds, the most widely used instruments for the mobilization of finance for green activities, can serve cities and the long-term climate goals.

Methodologies: A Multidimensional Approach

This book adopts a multidimensional but integrated approach which combines different sectors and themes of cities, leveraging combination of climate solutions such as digital systems, policies and planning instruments, financing mechanisms and disruptive leadership and ensuring their prompt but inclusive deployment. This approach would leave no one behind.

In terms of planning models, the book focuses on the state-of-the-art spatial planning practices to regulate development and land and resource use in a coordinated manner. The new spatial planning model seeks to streamline climate, Green Deal, and COVID-19 issues into coordinated land and resource uses in a sustainable, smart, resilient, and resource wise manner. Land use planning may incorporate mechanisms to coordinate other sector policies. However, spatial planning is more than the land use planning and is more centrally concerned with the coordination or integration of the spatial dimension of sectoral policies through a territorially based strategy. The strategy acts as a framework for the formulation and implementation of sectoral policy. One of the sectors will be land use planning. In this sense, spatial planning seeks to identify and address the contradictory effects of sectoral policies, and the opportunities for synergy through the territorial strategy.

Climate and urbanization-based global challenges are increasing exponentially. Our individual disciplines are unable to grasp the magnitude of climate and environmental challenges ahead. For that, we need a multidisciplinary approach to work collaboratively, calling on the knowledge of climatologists, engineers, sociologists, economists, ecologists, urban planners, designers, architects, AI experts, community organizers, and more, using an integrated holistic approach. For the scale and pace of the response to be commensurate to the challenge, transformative solutions are expected from these experts and tools to measure their impacts are needed.

Based on decades of teaching, research, exploration, advice, guidance, implementation, and brainstorming with scientists, politicians, practitioners, civil societies, researchers, and private sectors on various occasions, authors are now proposing challenging, innovative, and stimulating solutions as a response to the climate emergency. Subsequently, the action proposals to support the Green Deal and climate pact are addressed in the relevant chapters. Each chapter of the book is an open door to further exploration and development, and cross-referenced carefully to avoid overlapping and conflicts between chapters.

All together, the elements of these various chapters—with the help of ICT—propose a fully integrated, disruptive, and transformative approach to support city's climate goals.

The key point with regard to the multidimensional approach is that digital technology-based approach for climate action is a useful tool for addressing a need-based and solution-oriented approach to climate action than expand the climate action space for cities with possibilities of disrupting markets as well as entire value chains.

Structure of the Book

As the table of contents shows, this book is divided into four parts and eleven chapters with an appendix. Following this foreword, the first part contains the overview of climate smart city plans and other planning instruments, and climate big data, IoT, and AI for better decision. It focuses on theory and practices of integration platform cities.

Part 2 discusses the nature-based approach from urban nature, vulnerability and risk assessment, and mangrove forest perspective. Part 3 addresses several new developments to foster climate action in cities. It includes (1) possible climate actions currently available in cities, (2) market mechanisms from project to mitigation activities, (3) how global carbon markets can promote low-carbon cities in developing countries, (4) climate change mitigation action enablers for climate smart cities, (5) market mechanisms in the building sector from an urban perspective, and (6) new advanced GHG accounting system for cities enabled by distributed ledger technologies.

Finally, an appendix offers the readers a practical guide for preparing the carbon neutral city plan 2050 within the broad framework of climate smart cities described in this book. It is intended to assist cities to develop strategies, measures, and project activities for carbon neutrality in response to global requirements to keep the global temperature to 1.5 degree Celcius above preindustrial era levels. This appendix is oriented to some principles, practical methodologies, procedures, implementation, co-benefits, indicators, and suggestions for writing carbon neutral city plans, and alleviating common criticisms. The guide provided herein can be improved by the appropriate pilot test and validation. We have great expectations that this guide will help turn the ideas and proposals contained in this book into action in a consistent and coordinated manner.

Seoul, Korea (Republic of) Kwi-Gon Kim
Bonn, Germany Massamba Thioye

Acknowledgements

As editors of this book, we are grateful to all contributors for their willingness to participate in this challenging and interestingly demonstrated project and for meeting the stringent deadlines we imposed upon them.

As the book evolved individual chapters to the edited book, all of the papers presented in this book are cross-referenced with common style or structure. A special word of thanks in this respect must go to all the authors for never-ending efforts to support our team work. Many of the manuscripts are completely new research results to pay attention to new developments at the city level. We owe thanks to Springer which enabled us to execute this research initiative.

We owe special thanks to reviewers who provide us challenging feedback on the project proposal and final manuscripts in the preparation of the book. Finally, we are most grateful to our publisher Springer for their kind professional and efficient support in the publication of this book.

A complete list of contributors and their affiliation is provided following the table of contents.

March 2021

Kwi-Gon Kim
Massamba Thioye

Disclaimer

This book has no relationship whatsoever with the UNFCCC process or the work of the secretariat. Its findings, interpretations and conclusions do not reflect the views of the secretariat. The secretariat does not vouch for the accuracy, currency or completeness of the data or information contained in the book and excludes all liability for any loss or damage arising (including through negligence) in connection with and as a result of any reliance on the data and information contained in the book and/or its use. Questions concerning the analysis and views expressed in the different chapters shall not be addressed to the secretariat.

Contents

Contributors

Hee-Sun Choi Korea Environment Institute, Sejong, South Korea

Edoardo Croci GREEN Research Center, Bocconi University, Milan, Italy

Gajanana Hegde UNFCCC secretariat, Bonn,, Germany

Eddly Joel National Capital District Commission, National Capital District, Moresby, Papua New Guinea

Regina Kiele School of Natural and Physical Sciences, University of PNG, National Capital District, Port Moresby, Papua New Guinea

Kwi-Gon Kim Seoul National University, Seoul, South Korea

Benedetta Lucchitta GREEN Research Center, Bocconi University, Milan, Italy

Axel Michaelowa Institute of Political Science, University of Zurich, Zurich, Switzerland;
Perspectives Climate Research, Freiburg, Germany

Gertrude Miria National Capital District Commission, National Capital District, Moresby, Papua New Guinea

Harald Rauter EIT Climate-KIC, Zurich, Switzerland

Maike Sippel University of Applied Science, Konstanz, Germany

Massamba Thioye UN Climate Change Secretariat, Haus Carstanjen, Bonn, Germany

Abbreviations

AFOLU	Agriculture, Forestry, and Other Land Use
BID	Business Improvement District
BIT	Bus information Terminal
BRT	Bus Rapid Transit
C40	Network of the World's Megacities
CBA	Consumption-Based Approach
CCDA	Climate Change and Development Authority
CCP	Climate Protection Campaign
CDM EB	CDM Executive Board
CDM	Clean Development Mechanism
CDR	Carbon Dioxide Removal
CEPA-JICA	Central Provinces
CERs	Certified Emission Reductions
CI	Coverage Index
CICES	Common International Classification of Ecosystem Services
CIFA	Community-wide Infrastructure Footprint Approach
CPA	Component of Project Activities
CPAs	CDM Project Activities
CPLC	Carbon Pricing Leadership Coalition
DAOs	Decentralized Autonomous Organizations
dApps	Decentralized applications
DEM	Digital Elevation Model
DER	Distributed Energy Resources
DLT	Distributed Ledger Technologies
EBA	Ecosystem-based Adaptation
EcoDRR	Ecosystem-based Disaster Risk Reduction
EEIO	Environmentally Extended Input–Output
EFP	Emission Footprint
ER	Emission Right
ES	Ecosystem Services
ESCO	Energy Service Companies
ESCO	Energy Service Company

ESS	Energy Storage System
EU	European Commission
EV	Electric Vehicle
FEMA	Environment and Sustainable Development Fund
GHG	Green House Gas
GHG	Greenhouse Gas
GIS	Geographical Information System
GIZ	Deutsche Gesellschaft für Internationale Zusammenarbeit (GIZ) GmbH
GPS	Global Positioning System
HVAC	Heating, Ventilation, and Air Conditioning
IASS	International Advanced Sustainability Studies
ICT	Information and Communication Technology
IEA	International Energy Agency
IET	International Emission Trading
InVEST	Integrated Valuation of Environmental Services and Trade-offs
IoT	Internet of Things
IPCC	Intergovernmental Panel on Climate Change
ITS	Intelligent Transport System
JI	Joint Implementation
KP	Kyoto Protocol
LCA	Life Cycle Assessment
LDC	Less Developed Country
LED	Light Emitting Diode
LFG	Landfill Gas
LGGFA	Local Government Green Funding Agency
LID	Low Impact Development
LoA	Letter of Agreement
LPA	Local Planning Area
MOOCs	.
MOS	Mitigation Outcome Security
MRV	Monitoring, Reporting, and Verification
NBS	Nature-based Solution
NBS	Nature-Based Solutions
NCD	National Capital District
NCDC	National Capital District
NDCs	Nationally Determined Contributions
NSCs	Need Satisfaction Companies
NWRM	National Water Retention Measures
PA	Paris Agreement
PAYT	Pay As you Throw
PBA	Production-Based Approach
PDD	Project Design Document
PES	Payments for Ecosystem Services
PNG	Papua New Guinea

PoA	Program of Activities
RFID	Radio Frequency Identification
SC	Smart City
SCADA	Supervisory Control and Data Acquisition
SDGs	Sustainable Development Goals
SEA	Strategic Environmental Assessment
SID	Small Islands Developing States
SLR	Sea Level Rise
TEEB	The Economics of Ecosystems and Biodiversity
TOC	Theory of Change
UEA	Urban Environment Accords
UERA	Urban European Research Alliance
UN SDSN	UN Sustainable Development Solutions Network
UNCCD	United Nations Convention to Combat Desertification
UoP	Use of Proceeds
VUCA	Volatile, Uncertain, Complex, and ambiguous world

Part I
Climate Smart and Wise Cities

Chapter 1
Planning Instruments for Climate Smart and Wise Cities: A Spatial, Green and Digital Deal Approach

Kwi-Gon Kim and Hee-Sun Choi

Abstract The Smart City Council defines a smart city as one that has simple, easy, and practical digital technology embedded across all city functions. Though cities and communities around the world have elaborated a variety of climate actions and plans, their development and implementation often follow a project-based approach with no network connections between projects. Climate smart and wise cities extends the concept and approach of smart cities to include the diverse dimensions of climate actions in a holistic manner.

This chapter describes how climate smart and wise cities can support climate actions with a variant of the New Urbanism called the climate smart and wise urbanism., which is intended to be more climate-oriented and nature-based. The chapter argues that our task now is to mainstream climate city planning into urban planning and development policy. The contents of this chapter include new issues and solutions for climate emergency which need a new tenet of planning, a conceptual framework, and key components of the planning system for integrated climate smart and wise cities, planning procedure, and planning methodology. The models include holistic planning models, sub-system planning models, and sectoral dynamic platform models with the use of automation and Internet control system technology development.

Keywords Climate smart and wise cities · Climate emergency · Sustainable development goals · Environmental integrity · Carbon footprint · Urban normal · Digital space mapping · Energy transition · COVID-19 deal · Green transition · Energy transition · State of pandemic · Hydrologic simulation models · Flood risk assessment · Land use platform · Smart growth scenarios · Digital city · Spatial planning · Transparency framework · Katowice rulebook · Digital transformation

---

K.-G. Kim (✉)
Seoul National University, 1 Gwanak-ro, Gwanak-gu, Seoul, South Korea
e-mail: kwigon@snu.ac.kr

H.-S. Choi
Korea Environment Institute, Bldg B, 370 Sichenong-daero, Sejong, South Korea
e-mail: choihs@kei.re.kr

K.-G. Kim and M. Thioye (eds.), *Planning Climate Smart and Wise Cities*,
The Urban Book Series, https://doi.org/10.1007/978-3-030-80165-6_1

1.1 Introduction: Issues and Solutions for Climate Emergency

Climate change is our common enemy and became a global multidimensional issue. The next coming generation is watching the present generation. We should not underestimate the climate change threat to them.

We are at a pivotal moment in urban planning history where the way in which we integrate the Green Deal movement offers the possibility to reorient urban planning toward sustainable development goals (SDGs) and climate.

The Green Deal movement is growing all over the world. In particular, the Green Deal is an integral part of the European Commission's strategy to implement the United Nation's 2030 Agenda and SDGs. As part of the Green Deal, the Commission will review the regulatory framework which should foster the deployment of innovative technologies and infrastructure, such as smart grids, hydrogen networks or carbon capture, storage and utilization, energy storage, also enabling sector integration. The European Commission's Green Deal policy and strategy have many implications for the development of climate smart and wise city (European Commission 2019). Article 6 of the Paris Agreement (PA) offers opportunities for Parties to voluntarily cooperate to promote Nationally Determined Contributions (NDCs) implementation and sustainable development. Capacity building is crucial for enhancing the transparency framework, thus requires multiple approaches and actors.

In this context, at the UNFCCC COP 25, groups assembled for side events entitled "The role of CDM in decarbonizing cities and its co-benefits," "How digital technologies can support climate action,", "Nature-based solutions: Coastal ecosystems in 2020 NDCs," and many other climate-friendly themes and topics (Photos. 1.1 and 1.2).

There are increasing interests in the potential of cities to contribute to climate change and COVID-19 recovery. Multiple assessments have evaluated the scale and composition of urban GHG emission, while others have evaluated some aspects of urban mitigation potential. Climate smart and wise cities need smart innovative solutions. There have been newly emerging solutions for climate emergency which can be summarized as follows:

Planning-based solutions

Future cities need to be more consciously planned if they are to address sustainability and climate issues properly. These issues cannot be left to spontaneous mechanisms or to market forces. Urban planning practices also need to be changed to reflect a new awareness and international movements, and to integrate environmental, climate, economic and social concerns in the twenty-first century. The relevant planning policy guides have to be developed for climate smart and wise cities.

For example, these solutions need to explore the role of national planning laws for local flood resilience and the interplay between adaptation planning laws and disaster risk reduction and management. The linkages between urban planning law and environmental planning law have to be strengthened to integrate green spaces

Photo 1.1 Side event on "The role of CDM in decarbonizing cities and its co-benefits," at the UNFCCC COP 25, Madrid, December 2019

and river systems with wind passages in cities to cope with fine dust (particles) and for climate change adaptation.

Cities are on the frontlines of COVID-19. Reopening our cities in the midst of COVID-19 in the era of climate change calls for different approaches to urban planning and implementation.

Nature-based solutions (NBS)

A renewed focus should be on the need for the city to make a committed transition toward a 'green', environmental integrity and carbon neutrality 2050.

Nature-based Solutions (NBS) are defined by IUCN as actions to protect, sustainably manage and restore natural or modified ecosystems, which address societal challenges (e.g., climate change, food and water security, or natural disasters) effectively and adaptively, while simultaneously providing human wellbeing and biodiversity.

If nature-based action is to provide part of the climate solution, it must do it by reducing emissions, integrating biodiversity, protecting the ecosystem and supporting the livelihoods.

These solutions aim to integrate urban and coastal ecosystems in 2020 NDCs. The approaches include protection and wise use of ecosystems with multiple co-benefits and IPCC guidelines for measuring mitigation value. The solutions will serve as contributions to the broader discussion on the role of climate smart and wise cities.

Photo 1.2 Side event on "How digital technologies can support climate action," at the UNFCCC COP 25, Madrid, December 2019

Nature-based solutions enable cities to learn best practices of design and implement policies and programs to better equip and manage urban spaces under a changing climate.

Wetlands, peatlands, mangroves, forests, and other primary ecosystems retain massive carbon stocks and improve resilience. Biodiversity underpins carbon storage and sequestration. Harnessing and mapping linkages between nature-rich and carbon-rich areas will catalyze higher ambition for NDCs. These solutions provide policy and practice sharing on mangrove conservation and zero deforestation, agricultural and forestry supply chain in time of climate change.

AFOLU (agriculture, forestry and other land use) sector presents great opportunities to increase NDC ambitions. The strategic alignment of all urban and regional plans and initiatives in the AFOLU sector, enhances its urban and regional mitigation and adaptation ambition, toward carbon neutrality in the sector, by 2050. Contributing up to 8.4% of GHG globally, new approaches to agriculture have a key role to play in slowing down the effects of climate change by adopting innovations and evidence-based good practice. These solutions explore options for farmers to address the biggest challenge of the twenty-first century. The solutions take a look at earth, air, fire and water elements for sustainability.

Climate change is inextricably linked to water for adaptation and mitigation, and there are significant co-benefits to managing climate and water in a more coordinated and sustainable manner. We have to manage water for climate change resilience. Water security is linked to ecosystem services. Therefore, the recent calls to action demand a new narrative on water management with ecosystem services to water. Nature-based solutions help catalyze the revolution by highlighting how climate-resilient water management delivers critical adaptation solutions for food systems, cities, finance infrastructure and local action.

Nature-based adaptation to climate change in cities explores how to measure the impact of NBS not only on climate change adaptation but also on its multiple benefits of health and wellbeing, social cohesion, and sustainable economic development. Nature-based approaches also contribute to disaster risk reduction. Carbon dioxide removal can be obtained from nature and technology-based solutions. These solutions present their co-benefits, adverse effects, cost, technical potential and enabling conditions and implementation of landscape restoration.

Ecosystem-based adaptation in the water sector mobilizes "no-regret" nature-based solutions including resilient land and forest management that best deliver at the scale of the basins of lakes, rivers and aquifers based on bioregional units. The ecological approaches can be used for landslide hazard mitigation and management.

There is potential for coastal blue carbon to support countries' NDCs. The effects of climate change on oceans continue to adversely impact economies, societies and biodiversity. Yet, the potential of coastal and marine NBS to support global mitigation and adaptation efforts remains underutilized. New science and successful NBS projects, a global cooperation and fundraising will be necessary to develop the NBS tools and methods.

Faced with the need for greater NDC ambition, we need to show how land is a key to nature-based solutions with a mitigation potential of 1/3 of the gap. We can share experiences of protecting and restoring diverse biomes, like mountains and forests.

Nature-based approaches that help people adapt to climate change and simultaneously reduce poverty and protect or restore biodiversity and ecosystems are particularly promising in mountain regions. Forest-based solutions in climate change mitigation and adaptation incentivize mechanisms to promote global green supply chains. Sustainable supply of goods and ecosystem services provided by forests are essential to economic development and to the environment.

Tree planting with "Plant-for- Planet App" is a good example. It's a solution put forward by the UN Environment to solve the climate crisis as a part of the UN Decade of Ecosystem Restoration (2021–2030). The App redefines transparency of planting projects. Forest, water, and carbon storage create synergies and balancing trade-offs in view of climate change.

There is an urgency for all countries to develop more ambitious commitments in order to achieve the Paris goals. Renewables- solar, wind, biomass, hydro & geothermal-working together can strengthen national and local commitments to achieve a just and equitable transformation to a renewable energy future. An integrated bioregional approach managing from hilltops to oceans is needed to take full advantage of the benefit that nature offers- be they ecological, economic, social, or cultural.

Market-based solutions: market-ready technological solutions for climate emergency

These solutions explore approaches to guide an international carbon market based on experiences with the Kyoto mechanisms and regional ETS to guard the environmental integrity of the market mechanism under the Paris Agreement (PA). These solutions include the impact of using offsets on the economy. Significant challenges remain to account for the transfer of emission reductions under Article 6 of the Paris Agreement. New analysis has to be made on the risk of spreading hot air throughout the Paris agreement, as well as suggested ways to ensure that all emission reductions are properly accounted for.

Developing countries are now earning credits from REDD+ activities, verified by the UNFCCC.

Future carbon markets can be designed and blended with climate finance in order to mobilize finance for NDC implementation in developing countries and LDCs.

Digital technology-based solutions: innovative technical solutions

It is important to modernize, simplify and transform IT with solutions from climate smart and wise city perspectives.

For global CO_2 emission reductions, accelerating innovation that benefits each country by adapting the advanced decarbonizing technologies respectively are crucial not only for climate action mitigation but also for multiple SDGs.

Digital technologies can revolutionize climate action. The combination of IoT, AI and DLTs/Blockchain is critical to the mobilization of the full potential of climate policy making. It enables the measurement and attribution of scope 3 (indirect emissions)) climate contribution as well as climate contribution arising from enhancing the circularity of a city.

Data-driven approaches lay out a new strategy for equal access to climate action, "Climate Action for All, not just the Privileged." Digitalizing traffic and parking data in cities for traffic and park equity is a good example experienced Chandler, Arizona, USA.

With the Katowich rulebook in place, transparency will inform the global stocktake. This will require data/ information and analyzes of countries' policies and pledges. Other digital data techniques are digital geophysical mapping and digital twin (3D virtual simulation).

Best available science requires high-quality monitoring of GHGs. High-precision, sustained, interoperable and freely accessible data on climate impacts are a prerequisite for relevant climate policies. Research infrastructures have to cooperate globally to provide standardized observations and contribute to enhanced monitoring, reporting and verification (MRV).

Financing fit for Paris (3fP)-based solutions: effective carbon pricing to reach global carbon mitigation goals

Aligning financing systems with the Paris Agreement—The Pathway to 1.5C convened by UNEP FI and partners, leading banks and investors are targeting net zero emissions by 2050, striving for a 1.5C world and driving GHG emission reductions in the real economy.

For the implementation of the commitments made by countries through the NDCs under the Paris Agreement, one of the major challenges is the mobilization of related financial resources despite the existence of several climate funds.

2021 is a critical year for scaling up climate finance to respond to climate priorities and the needs of developing countries. Carbon financing issues include carbon pricing in emerging economies, policy choices between CDM, ETS and carbon taxes, options to address competitiveness concerns, and how to create domestic support for carbon pricing. Climate finance and project development is another finance infrastructure issue for bridging the gaps between climate finance access and building resilience on the ground. These solutions can demonstrate that ambitious climate action can be a driver for inclusive green growth, job creation, rights and social justice.

Many developing countries that are vulnerable to climate change are unsuccessful in accessing the magnitude of finance required to turn their climate targets into action. Financing solutions help support improved access to climate finance.

Alignment of the financial sector and climate finance obligations constitute challenges for both countries with large financial centers and recipient countries. Financing-based solutions cover regulatory approaches to shift and mobilize finance but also strategies to overcome unintended side effects. Financial regulation can support the Paris goals by shifting billions toward a low-carbon and climate-resilient economy. The 3fP-Tracker tool provides transparency across jurisdictions and helps define the global knowledge frontier, including for non-state actors.

Banks, insurers, and investors have a key role in ensuring the implementation of the PA, especially in article 2.1c. The financing solutions explore which measures are necessary to divert funds from the fossil industry to the sustainable energy transition and low-carbon development.

To shift from incremental improvements to transformational changes in climate actions, financiers should be incentivized to conduct green investments. This requires the measurement and attribution of their scope 3 climate contribution, a complex task that raises several new and challenging issues. Business opportunities in climate action accelerate investments that are needed to close the gap in adaptation and mitigation financing.

Climate governance-based solutions: local sustainable solutions

Growing international momentum around net-zero targets can drive more ambitious NDCs and share experience in setting the domestic legislation and governance arrangements to support this. The role of people on the frontlines, such as children, youth, women, workers, including those in the informal economy, and indigenous peoples is to drive ambitious climate action that ensures ecosystem integrity and their empowerment through the fulfillment of the human right to participate. Creative, visual, and digital methods can be used for empowering communication to connect personal stories to collective issues.

New city master plans, new neighborhood plans and climate action plans can be guided by a set of community values and a new community vision involving the citizens. Private actors engagement and community mobilization promote low-carbon water and climate change: the engagement of private actors, water partnership, promoting low-carbon and resilient development through community mobilization advocacy and equity for enhanced climate action. Transboundary water governance could be a good solution for adaptation to climate change.

Ambition, innovation and leadership by non-party stakeholders are crucial for accelerating the transition to a low-carbon economy. These governance-based solutions could lead to a new Movement of Climate Change through New Climate and Practice Establishments.

Corona virus and green new deal-based solutions

The world is changing at an incredible rate and cities are at the heart of it all.

The Corona virus has been sweeping the globe and many organizations have issued emergency advisories. Across the globe, COVID-19 is threatening cities and communities, endangering not only public health but also the economy and fabric of urban society. COVID-19 is already having a deep, multidimensional impact and is expected to create the worst economic contraction in decades, with job losses already at an all-time high.

Given these shocks, the World Bank estimates that about 100 million people will likely fall into poverty due to the impact of the pandemic, with as many as 49 million falling into extreme poverty.

Residents in slums and informal settlements are particularly vulnerable. The one billion people who live in slums and informal settlements lack infrastructure and basic services, including water, sanitation, waste collection and access to basic health care. Many rely on communal water and toilet facilities, which accelerates infection rates (World Bank Blogs, "Cities are on the front lines of COVID-19", May 13, 2020).

Times are uncertain as city officials respond to a health pandemic and find ways to build urban resilience. Every day, across the globe, new and innovative approaches are emerging. These ideas will lead us into an urban normal.

Petersburg climate dialogue took place via video conference. Officials from 30 countries took part in the dialogue. The talks focused on making a green economic recovery after the COVID-19 pandemic. Guteres says that we have an opportunity to build our world for the better. The Green New Deal could offer us many opportunities for post COVID 19.

In light of steep declines in transit ridership and municipal budgets as a result of COVID-19, transit providers are rapidly becoming more reactive and nimble in their offerings. For example, to support cities and businesses for their efforts to provide transportation for their riders, Ford Mobility subsidiaries TransLoc, Ride Systems and DoubleMap are providing free transit consulting and demand response software for transit agencies in an effort to help them quickly deploy a responsive service that can both support evolving rider demand, and adhere to quickly-changing health guidelines (Meeting of the Minds Webinar, May 13, 2020).

It has been suggested in the recent World Health Assembly (WHA), delivered via video conference that corona virus vaccines and treatments should be public goods for humankind. The principles of transparency obligation and early warning system have to be adopted to prevent COVID-19. President Moon Jae-in proposed in the WHA Monday (May 18, 2020) that the international community cooperate beyond borders to develop vaccines and treatments for COVID-19 and other pandemics that may emerge in the future (The Korea Times, Tuesday, May 19, 2020).

We need to create transformative urban solutions to address global pandemics within the broad framework of healthy cities. The decision tree can be used for predicting severe causes of COVID-19 with help from AI and machine learning. AI experts can analyze historical patterns of corona virus disease cases with Big Data. Several actions to be taken can be summarized as follows:

- Assessment of impact on ICT industries.
- Analysis of present ICT policies and future tasks to cope with COVID-19.
- Government's countermeasures: Building of digital infra, digitalizing of social overhead capital (SOC), promoting of untact industries, etc.
- Tasks to prepare for post COVID-19.
- New digital standards for new group facility development and operation (e.g., virtual shops).
- Empowering communication: The use of creative, visual and digital methods to connect personal stories to collective issues.
- Driving economic growth and recovery, helping to advance productive and sustainable industries, and building long-term resilience in cities and communities.
- Matching a new economic vision with the legislation.

Endemics and pandemics are both standalone city risk as well as an amplifier of existing trends and vulnerabilities. The sustainable urban agenda must not lose momentum, and climate smart and wise city should be at the forefront of urban planning facing years of climate and health uncertainty. As urban climate planners adapt to a period of extreme volatility, there are many questions to be answered as follows:

- What are the pandemic implications for the cities?
- The impacts of COVID on the green urban development: challenge or opportunity?
- The difficulties of mobilizing finance to reach a 1.5 degree target in the midst of a pandemic.
- Forecasting the short and long-term impacts.
- Preparation of the city digital management method to minimize urban functional failures such as city lockdown or blockade as a long-term countermeasure (e.g., digital early warning system).
- Considerations for the economically and socially deprived people to solve the polarization of the digital economy at the urban policy, planning and design level.
- Updating utility grids, automating systems and services, and meeting increased connectivity need to create flexible and resilient cities that can outlast and grow beyond whatever comes their way are just a few of the critical upgrades to be considered.

- Dynamic street management, city airspace, and capturing the value of municipal infrastructure including transport:
  - Whether streets temporarily closed to vehicle traffic during COVID-19 should be permanently closed or not.
  - Before we take the next step with drone delivery, cities and FAA need to come to a conclusion on who controls, and in what manner, the airspace above cities.
  - How can we leverage public/private partnerships to create new opportunities and experiences for constituents?
  - How can we move from ad hoc solutions to a well-planned path to recovery?

Stimulus packages, including COVID-19 economic recovery funding and digital services, that address climate change and prioritize efficiency, sustainability, circularity, connectivity, equality, and resiliency are likely to have higher returns than those that merely shore up past approaches. As with pandemics, early planning action on climate reduces risks and slashes costs for the journey toward the "next normal".

We have to deal with the root causes of the COVID-19 pandemic for our 'New Normal.' Shaping a resilient post-COVID-19 urban era and charting a path forward on the Green New Deal have become new agendas for urban modeling. As an example, the European Union has announced plans to set aside 25% of a proposed $826 billion corona virus proposal for climate-friendly measures such as building retrofits, clean energy, electric vehicles, and sustainable land use. If a recovery plan like the proposal goes through, it could be the most significant green investment in history. Bonus points: it will create 1 million new jobs (Joel Jager, WRI DIGEST, June 10, 2020).

Holistic thinking-based solutions

We have to take our climate action to the next level with multi-sector solutions to deal with complex climate problems.

Many urban planners realize that the environmental, social and economic factors that promote wellbeing are complex. The holistic approach, advocated by planning pioneers, looks at the interrelationships between the whole person and his or her environment. This generalist approach is not opposed to the specialized thought and detailed work on health carried out by the medical profession. These two approaches are not only complementary but also mutually indispensable.

Our task now is to integrate Green New Deal, Digital Deal, COVID-19 and climate change issues into urban planning policy. To achieve these goals, this book suggests four deals as follows:

- Land and resource, and climate deal aim for ecological transition to avoid climate impacts and contribute to sustainable use of land and resources;
- Green deal is designed for green transition to implement the United Nation's 2030 Agenda and the Sustainable Development Goals;
- Digital deal addresses for digital transition to transform cities into digitalized, automated, connected cities;
- COVID-19 deal is geared up for recovery transition to improve human health and reopen cities in the midst of COVID-19.

1.2 Climate Smart and Wise City Planning Framework

This section describes an integrated, whole of urban planning response to climate emergency in the digital era. It includes planning system, focusing on a hierarchy of physical and digital space planning, and planning procedure targeting spatial planning process at the city level.

1.2.1 Planning System: A Hierarchy of Physical and Digital Space Planning

(1) **Definition and Principles of Climate Smart and Wise Cities**

Cities can be defined according to a multiplicity of perspectives on emerging fields of urban development and tools. The definition of the city sought by eco-city, sustainable city, smart city and resource wise city can be injected into the new climate smart and wise cities. In terms of planning principles, in addition to efficiency, sustainability, connectivity, circularity, resiliency principles of cities, environmental integrity and resource wisdom principles have become increasingly influential in the emergence of climate smart and resource wise urbanism.

Plunz and Sutto (2010) argued that sustainable cities will be sustainably integrated into local and global ecological processes, using energy efficiency and low-carbon technologies, emphasizing the quality of life in local communities and building healthy systems for transport, recreation, and employment. According to them, ecology and economics will merge, rather than being distant and often conflictive modes of thinking and action. Design in its broadest sense- integrating technology, spatial analysis, community participation, and a deep appreciation of ecology- will provide an integrative framework for action.

This is the digital age. The Smart City Council defines a smart city as one that has simple, easy, and practical digital technology embedded across all city functions. On the other hand, climate-resilient and low-carbon smart cities can be defined as ones that have digitalized connections of all sectors and functions, in which everything is connected, supporting sustainability, resiliency, circularity, efficiency, and connectivity of the city to combat climate change by promoting climate actions. It incorporates climate change mitigation and adaptation policy goals at each stage of the planning process and in urban policies.

A new concept and approach called climate smart and wise cities aim to reform all aspects of urban planning, in terms of 'environmental integrity' and 'carbon neutrality' rather than in narrow terms considered sufficient in the nineteenth and twentieth centuries. The principle of environmental integrity expresses a complex set of concepts that describes a healthy natural system that can support essential processes. The principles of sustainability include the environmental integrity which refers to maintain the state of the environment. This means that human activities should not unduly disrupt the ecosystems and human communities located in the area. Furthermore, care should be taken so that the surrounding landscape is not drastically impacted by human activities.

The Paris Agreement introduces a quite different definition: environmental integrity here means not a natural system, but the transfer of units. The term 'environmental integrity' is used several times in Paris Agreement, but is not defined clearly. In the context of international market mechanisms, what we mean by environmental integrity is that the transfer of units does not result in higher global emissions than if the NDCs had been achieved only through domestic action. Environmental integrity is central to both the effectiveness and credibility of mark mechanisms.

Difficulties arise with the distinction between the definitions of natural system and international market mechanisms for maintaining environmental integrity. Climate smart and wise city planning systems provide an important mechanism for connecting these two definitions within an integrated framework. Smart networking and innovative cyber- physical solutions are vital for the utilization of the entire city's resource potentiality. Innovative solutions help individuals, communities, and the whole city to renew and manage resources wisely.

The principles of climate smart and wise cities are summarized as follows:

Principles of green new deal: idealogical revolution

Europe now has its own green deal. The European Commission President Ursula von der Leyen presented the European Green Deal on Dec. 11, 2019, which aims to make the EU climate neutral by 2050 as follows (European Commission 2019):

- A zero pollution ambition.
- Preserving and restoring ecosystems and biodiversity.
- From 'Farm to Fork': a fair, healthy and environmentally friendly food system.

Figure 1.1 shows the European Green Deal which contains 10 strategies.

Resource Wisdom Principles

As summarized in Table 1.1, resource wisdom principles have been developed by the Finish Environment Institute. A list of principles of resource wisdom is as follows:

- Zero climate emissions.
- A resource wise city aims to minimize its climate emissions or greenhouse gas emissions. The unit used for this indicator is the carbon dioxide equivalent per resident.
- Zero waste.
- Zero overconsumption.
- Sustainable wellbeing.

In Table 1.1, the indicators have been selected to correspond to the principles of resource wisdom.

Connectivity principles

- Entailing a greater scale of information gathering for big data, more integration of information and greater cooperation among information providers, both amateur and expert.

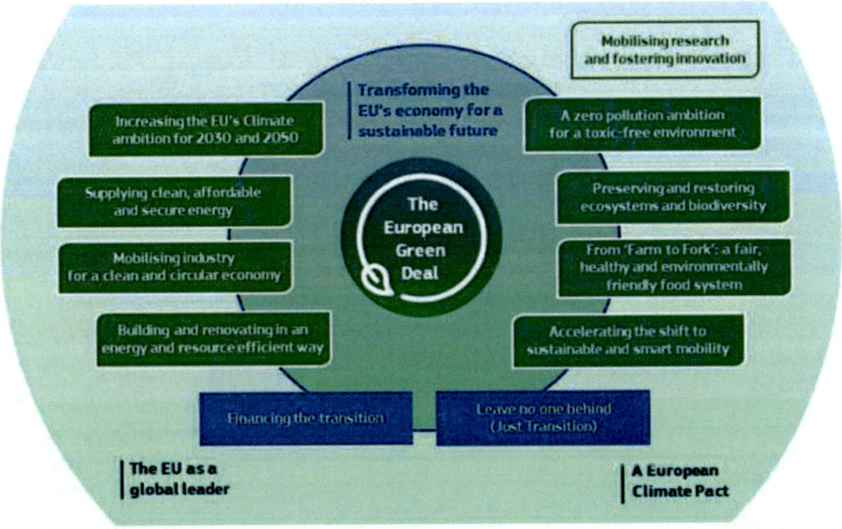

Fig. 1.1 The European Green Deal *Source* European Commission, the European green deal, Brussels, 11.12.2019, COM (2019) 640 final, p. 3

Table 1.1 Summary of main resource wisdom principles and indicators

Principle	Indicator	Unit	Accounting instructions
Zero climate emissions	Consumption-based greenhouse gas emissions	tCO_2 eqv/res	KASVENER or CO_2 report
Zero waste	Material losses	Tonne	Flow into landfills + incineration of non-renewables + export of waste for final disposal
Zero overconsumption	Ecological footprint	gha/res	Global Footprint Network's instructions and presumptions
Sustainable wellbeing	Share of residents who perceive their quality of life to be at least good	% residents	The World Health Organization's quality of life indicator, WHOQOL-8

Source Finnish environment institute, "the indicators of the governance model of resource wisdom", updated (PDF)

- Developing independent enabling factors which are all crucial elements in facilitating the rapid connection.
- Encouraging a zero-carbon lifestyle that is connected with the connect-tech.
- Interconnecting street grid network which disperses traffic and eases walking and use blue, green, and white networks.

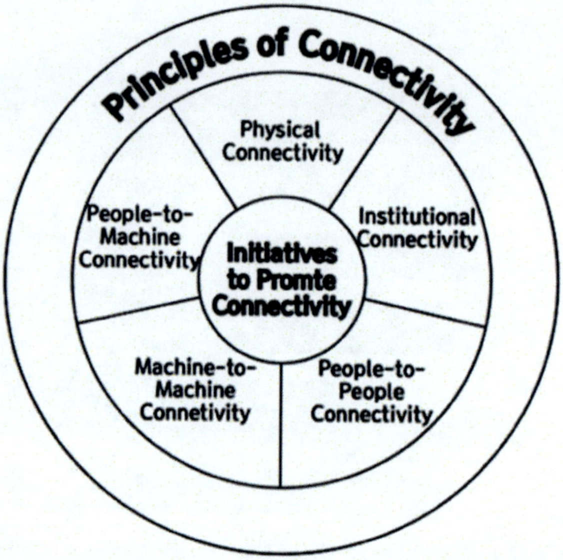

Fig. 1.2 Urban principles of connectivity

- Building the capacity of institutions to embed new connect-tech into the local institutional and social context.
- Establishing proper relationships or connections between man and machine (Fig. 1.2).

Circularity principles

- Developing the circular model which stirs the economy and has the ability to create the job as well.
- Setting a goal to increase the water diversion rate.
- Facilitating public–private collaborations that promote efficiency and restorative use of natural resources.
- Supporting improvements in material effectiveness and transition to renewable energy and a circular economy.
- Blend ancient approaches with modern digital technologies to drive a new urban metabolism (Fig. 1.3).

Sustainability principles

- Managing cities' resources sustainably for demonstrating long-term environmental sustainability.
- Designing with nature and adopting new climate planning and design approaches (Fig. 1.4).

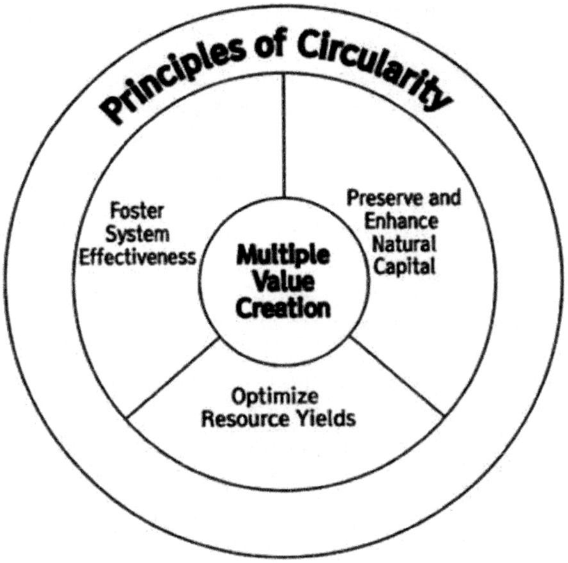

Fig. 1.3 Urban principles of circularity

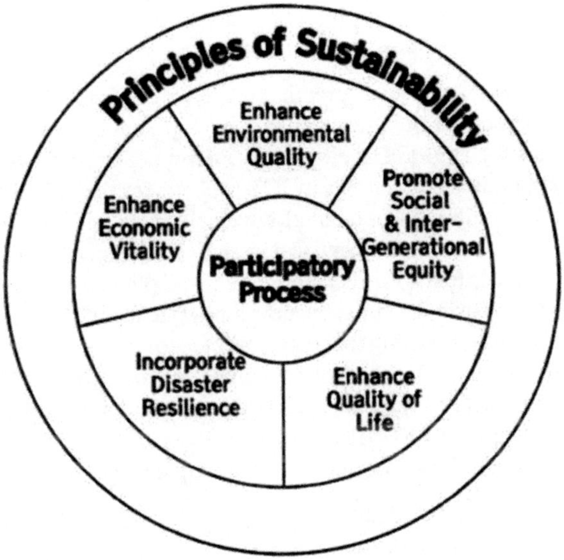

Fig. 1.4 Urban principles of sustainability

Resiliency principles

- Adopt a precautionary but positive approach to development that aims not just to avoid further climate change but also to reduce risks and enhance the resiliency of ecosystems and communities.

- Make our cities more resilient to climate change, while accommodating higher population densities.
- Require planning to protect, maintain, and enhance the multiple and interdependent co-benefits and ecosystem services offered by the total greenhouse neutrality.
- Develop a strategy which is built around diverse water resources and a mixed water infrastructure with smart water grid systems. It will allow cities the flexibility to access a portfolio of sources at least cost, in terms of environmental impacts and other externalities.
- Develop uniform standards and engineering procedures for designing climate adaptation measures.

Efficiency principles

- Create energy-efficient city and transportation planning in the climate context.
- Adopt transit-oriented development.
- Demonstrate how urban smart energy grid systems can play an important role in addressing climate issues.
- Try to have a net effect on our carbon footprint which requires an interdisciplinary approach.

Digital development principles

Based on their rich practical experiences from ICT interventions, a broad range of members of the international development community have developed nine Principles for Digital Development (Energypedia 2019). These principles are intended to strengthen the benefits of ICT projects for the affected individuals and communities while mitigating potential adverse effects and risks. Among others, they highlight the need for a user-centric design of ICT projects which pays attention to the specific local contexts. Furthermore, the principles underscore the importance of issues like data security and privacy and emphasize that ICT systems should be built in a sustainable way, meaning that they continue to function reliably and provide benefits to their users even after the official end of the intervention. Furthermore, the Principles for Digital Development promote the use of open software and the sharing of experiences and data (IASS 2019).

The 9 digital development principles include the following:

1. Design with the user.
2. Build for sustainability.
3. Reuse and improve.
4. Understand the existing ecosystem.
5. Be data driven.
6. Address privacy & security.
7. Design for scale.
8. Open source.
9. Be collaborative.

Data quality principles

As data quality principles, the 7 Cs of Data Q are essential:

1. Certified Accuracy.
2. Confidence.
3. Cost-Savings.
4. Campaign Intelligence.
5. Consolidated.
6. Completed.
7. Compliant.

Source: ITPro Today, 14 May 2020.

(2) **Planning System Framework**

We have to make "Climate Smart and Wise City" more mainstream. A new way to describe the relationship between technology, climate and resource management is the climate smart wise city planning.

The climate smart and wise city framework and components intend to lay the groundwork for the development plan of cities to transform it from a twentieth-century city into an integrated twenty-first-century city. This framework is based on the Post-2015 Sustainable Development Goals (SDGs) and Paris Climate Agreement, and more specifically, the low emission development strategy (LEDS) and nationally determined contributions (NDCs).

The integrated climate smart and wise city plan can be prepared based on this framework and key components and used as a sustainable tool for green urban development with particular emphasis on the green deal movement as referred to in the SDGs and Paris Agreement. ICT such as IoT and blockchain technologies is helping to achieve the SDGs and implement the Paris Agreement in human settlements.

Our task now is to mainstream climate smart and wise city planning into urban planning and development policy. This is only possible by creating the planning conditions necessary to integrate the full effects of climate change into the urban planning and development process, which is a new tenet of planning.

National planning policy framework

The framework of planning instruments, and the procedures which are followed in their creation and operation, provide an important starting point for understanding how the planning system works. The plan system is intended to produce a comprehensive and systematic hierarchy of national and regional guidance, and development plans (Cullingworth, Nardin 2002).

The hierarchy of urban planning is shown in England in Fig. 1.5. Traditional urban planning is mainly about land use planning. Planning decisions are affected by the national planning policy framework and by the development plan for the city, which includes the local plan and sustainable community strategic plan component. As an example, the possibilities for utilizing renewable energy can be influenced in practice by urban planning at the municipality level.

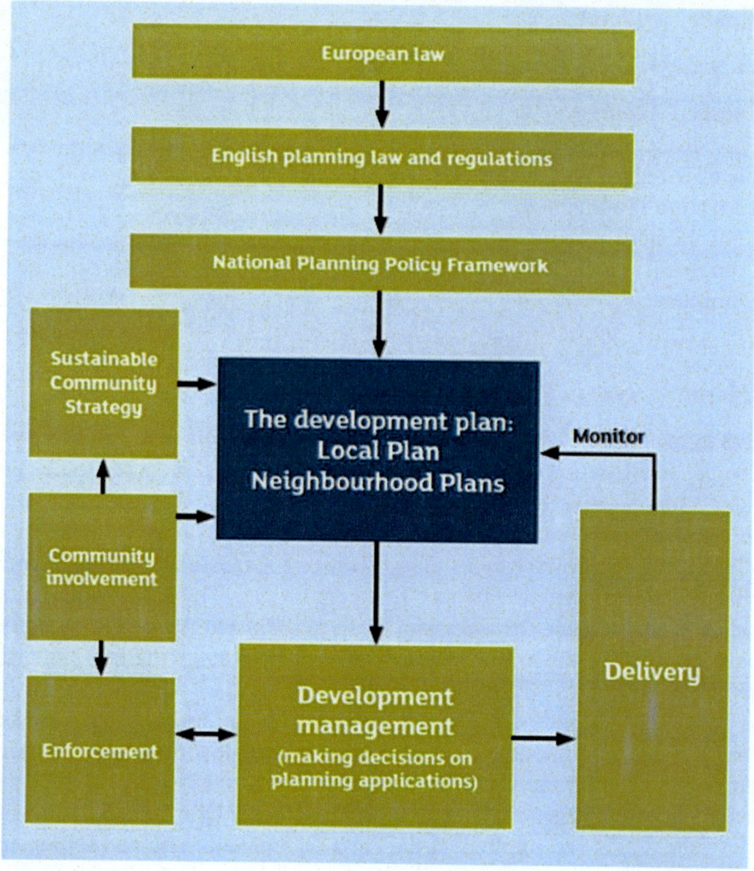

Fig. 1.5 Key components of the planning system: the case of England. *Source* Planning help

State of art spatial planning framework

The new spatial planning framework seeks to streamline the use and management of land and resources and to coordinate the spatial dimension of sectoral policies through the territorial strategy and action in a sustainable, smart, resilient and resource wise manner to achieve sustainability, connectivity, resiliency and, resource wisdom goals. It describes government action to regulate development and land use in pursuit of agreed objectives.

Figure 1.6 is a framework and components for state of art spatial planning practices which provide the basis for three types of spaces which are physical, digital and socio-financial systems. Figure 1.7 shows a transformative illustration of the system layers to put this framework into practice. The main features of three layers of spatial patterns in the pursuit of green deal, post-COVID-19 recovery and digital transition are as follows:

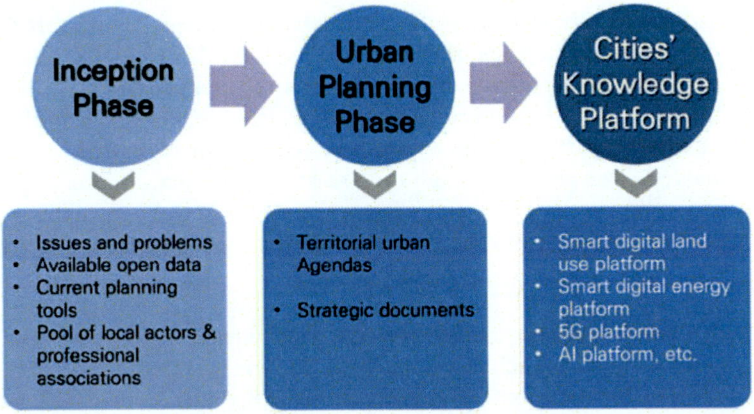

Fig. 1.6 Framework and components for state of art spatial planning practices

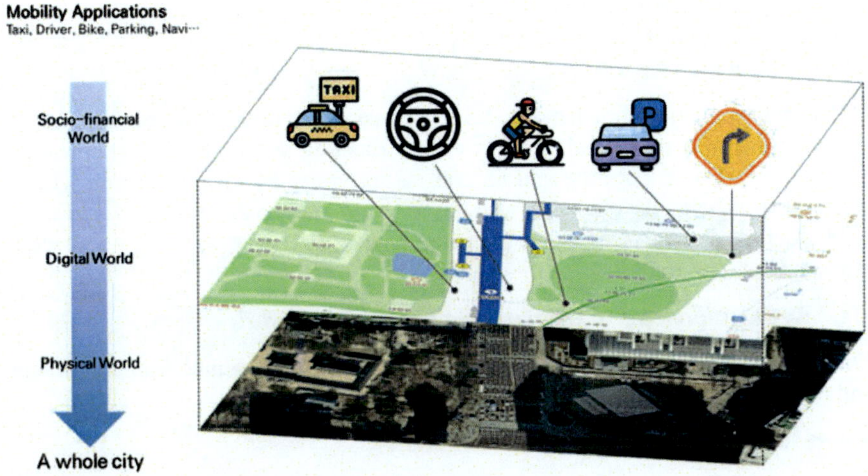

Fig. 1.7 Layers of physical, digital and socio-financial world

- The physical/territorial spatial system provides a tool that serves to guide land and other resource use patterns, mobility patterns, forms of urbanization and community, and green and park systems.
- The digital space system is a deployment of disruptive smart technologies for cities' knowledge platforms of different sectors of cities. In the COVID-19 era, the system is a virtual methodology for untact activities such as remote working home space.
- A virtuous circle heads toward sustainability through the deployment of disruptive leadership and governance, policies, business models and financial products.

- 'International SD Knowledge Platform' has been developed by the UN Department of Economics and Social Affairs, Division of SDGs. Cities' knowledge platform could aim to double energy and resource productivity by 2030 whilst transforming to a low-carbon coronomics. The Urban SDG Knowledge Platform is intended to support local action for the implementation of the 2030 Agenda for Sustainable Development, through up-take and replication of successful initiatives and good practices, by providing a repository of policies, initiatives and good practices at the city level by municipal governments and other stakeholders.

  Planning system framework can be further elaborated as follows:

- Growth and urban development framework at a strategic level.
- Physical planning approach and technology (IT strategy) framework.
- Physical planning describes government action to regulate development and land use in pursuit of agreed objectives. In a generic sense, the term is often used to describe spatial planning system.
- Master plan framework at the local level.
- Sector-specific planning framework.
- Planning framework for selected projects for implementation.
- Implementation framework.

The master plan, as shown in Fig. 1.5, is a general plan to direct the social structure and land use of a municipality in its entity. Existing social structures, economic and ecological sustainability, and natural values are to be given attention when a master plan is being developed. It also has to secure the inhabitants' access to social infrastructure and services, such as water supply and sewage, energy and waste management, and roads.

(3) Key Components of the planning system framework

The new approach to make urban planning more effective, climate friendly, smarter and resource wiser has the following components:

Smart infrastructure component

The transition to climate smart and wise cities requires smart infrastructure. This component should foster the deployment of innovative technologies and infrastructure, such as smart grids, hydrogen networks, or carbon capture, storage and utilization, also enabling sector integration.

Some existing infrastructure and assets should require upgrading to remain fit for purpose and climate resilience.

Global climate governance component

There are many methods that can be utilized to satisfy the feed forward and feedback features of a climate governance participation program: (1) keeping pace with the rate of technical change, (2) creating a need for a new consortium to reimagine and supercharge climate governance, and (3) public–private partnerships using blockchains to target our action and investments.

ICT component

Digital Technology can be a tool for the mobilization of the full potential of climate policy making, enabling measurement, and enhancing the circularity of a city. The Technology Framework (TF) adopted in Katowice provides overarching guidance to the Technology Mechanism in promoting enhanced action on technology development and transfer.

Nowadays we have effective ways to both decrease our carbon footprint and adapt to climate change through smart digital tech. Digital transformation can be accelerated through the use of IoT, AI and Big Data and cloud computing for climate action.

A new circular city model can be developed to strategically link urban sectors based on circulatory principles through intelligent digital connections (IDC). This model could be based on the holistic thinking of "Everything is connected" approach. It can be argued that the only way we can make a real difference in our impact on the global environment and climate is through a holistic approach to policy, planning, technology and governance where in all work collaboratively toward the common goal of climate smart and wise urbanism. This type of model has been illustrated by Cisco, IBM and AI Smart Open Platform in the form of IoT of connected SMART WORLD powered by AiSOP.

Big Data, IoT, AI and Blockchain technology that every circular climate smart and wise city should have, according to the circularity principles described Sect. 1.2.1,. are summarized below:

Firstly, IoT-based high-tech waste management systems are suggested to increase the waste diversion rate. The systems include:

- Smart PAY AS You Throw (PAYT) garbage proposal that would encourage people to recycle more and waste less.
- IoT solar-powered trash compactors that reduce volume of waste and travel distance to collect and dispose waste.
- IoT and AI-based eco-industrial park or smart factory that uses the waste of one factory as the raw material for another factory through smart tech facilitation.
- Smart graywater recycling is the filtration and utilization of graywater in place of portable water for non-portable functions such as irrigation and toilet flushing.

Secondly, Big Data, deep machine learning and AI are other technologies. They could be tools for CDM baseline and project scenario development through their prediction function. More and more advanced algorithms emerge every day to help us understand our city and predict the future. Big Data can be extensively used to support climate actions as follows:

- Satellite images overlaid with planning guidelines data are used to monitor the growth of "urban sprawl" and ensure that building development is taking place in line with strategy and regulations (e.g., Milton Keynes, England).
- Data are collected from refuse disposal sites, to ensure they are being emptied in an optimal way-cutting down on wasted journeys and unnecessary CO_2 emissions.

- Sensors monitoring the flow of traffic on roads through the city, which is used to alert motorists to congestion and plan future infrastructure development.
- Smart street lighting technology gathers data on when and where people are walking at night-time, to ensure illumination where necessary for safety and energy conservation were not.
- Data on water and energy usage is gathered to better understand demand and to help plan for supply.

Thirdly, there is a possibility of using DLT/Blockchain to drive energy prosumer behavior. Addressing the role of prosumers for climate change and renewable energy will contribute to the development of new carbon market mechanisms. Blockchain technology can play a role of an intermediary between producers and consumers in trading energy and carbon emissions.

Fourthly, various smart sensors can be deployed to measure GHG emissions before climate actions and the impacts after the climate actions are taken at various steps for MRV. With the intelligent digital connections with support from connected sensor networks, AI, IoT and DLTs enable the CDM to be expanded to include city-wide CDM (Urban CDM) approaches. The latest smartphones have sensors to tell where we are (GPS), how fast we are moving (accelerometer), what the weather is like around us (barometer), what force we are using to press the touch screen (touch sensor) and much more. Smart lights can incorporate a variety of sensors, including motion sensors. This enables them to turn on only when they detect a pedestrian or vehicle in the vicinity. This can save substantial amounts of energy by avoiding wasting energy-producing light that no one will benefit from.

Fifthly, the technologies for describing a solar energy-connected IoT microgrid system model equipped with ESS (off/on) provide electricity for solar water heaters, solar EV charging stations, solar LED park and street lighting, solar bus stop and bus information terminal (BIT), etc.

Sixthly, the rapid expansion of Internet communications gave rise to the concept of "smart city". This involves applying the UN Principles for digital development to every area of civic life, from waste management to public transport, with the aim of improving the quality of life for the people living there. Climate smart and wise cities use digital technologies to reduce the amount of carbon emissions released into the air, in order to mitigate the effects of climate change and to find the way to structure a program that could benefit populations vulnerable to the effects of climate change.

Seventhly and finally, geophysical information/data are needed to integrate the spatial dimension of sectoral policies and strategies of climate smart and wise cities through a territorially based strategy.

Here a question remains. How to integrate available smart digital technologies into GHG emission tracking and accounting in terms of institutional policy, law, regulations and guidelines on a mandatory or voluntary basis.

The author sincerely hopes that the UNFCCC and its partners will play an important role in facilitating the mobilization of the full potential of smart digital technologies for co-benefits of climate change mitigation, adaptation, biodiversity and social cohesion as well. Emergence of climate smart and wise cities provides the potential for using smart digital technologies to promote planning and technological cooperation to fight climate change.

We need to think about the many ways digital transformation will improve human life. In resource management and sustainability, resources will be matched with need, waste minimized, and constraints alleviated. Digital transformation even has the potential to completely decouple emissions and resource use from economic growth (Siebel 2019).

Resource wisdom component

Resource wisdom can be monitored through four key indicators selected by Sitra in cooperation with the Finnish Environmental Institute and the National Institute for Health and Welfare. These indicators are used to promote the realization of the resource wisdom principles described in the previous section. Naturally, in addition to these four indicators and the monitoring indicators supporting them (energy efficiency, bio-capacity, and material balance sheet), it is important to monitor economic and employment indicators as a component of resource wisdom indicators.

Layers of different components

The increasing number of urban responses to climate change highlights the need for carbon-centered comprehensive (3Cs) smart and resource wise planning model which incorporates climate change mitigation and adaptation policy goals at each stage of the process. Such a model aims for digitalizing solutions for individual actions and viability decisions in a holistic manner, and therefore, the model needs to combine a very high number of different ICT uses that are relevant to climate change mitigations and adaptation measures. Lots of institutional capacity is needed in the installment of execution of smart ICT uses networks.

climate smart and wise city is not just about technology. It is also about policy, planning, and governance. Climate smart and wise city can help transform cities into a low-carbon knowledge–based economy by nurturing innovation and creativity in all citizens without boundaries of technologies and expertise. The future of climate smart and wise city planning can be defined by continuing the emergency of new challenges such as climate challenges. The climate smart and wise city may be a "more connected city." It connects us to central nodes in the network. Building connectivity or relationships in cities is the key. More investment is needed in the research and development of new core technologies for the climate-land–water-energy-waste-food nexus.

Figure 1.8 shows technologies for the digitalizing connections of smart cities and low-carbon cities which can be extended to include resource wise city components described above.

Techniques for climate smart and wise cities are simply organized in terms of knowledge basis, connect-tech, sustainable smart technologies and technology facilitation mechanism (Fig. 1.9).

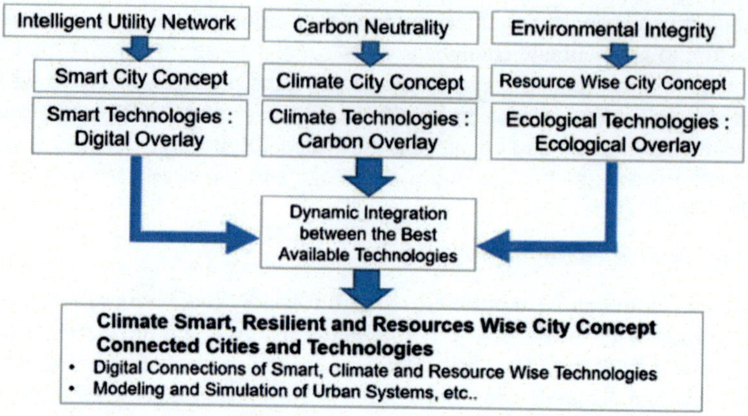

Fig. 1.8 Integration of technologies for smart city, low-carbon city and resource wide city components toward climate smart and wise cities: human and technologies work together

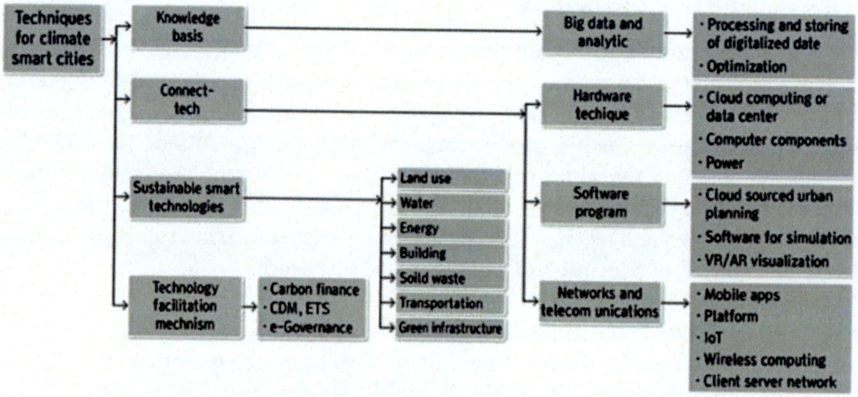

Fig. 1.9 Simplified organization of techniques for climate smart and wise cities

1.2.2 Planning Procedure: Spatial Planning Process at the City Level

(1) Considerations for developing spatial planning process

A few new attempts have been made to reorienting urban planning toward new state of art spatial planning practices.

International commitments to move toward sustainable development, involving all interests in the process of policy making, provide a framework for a new approach. Effectively coordinated programs and/or projects are necessary to integrate land use and transport, economic, social, environmental and climatic issues and resource planning in a streamlined process.

Spatial planning process is more centrally concerned with the coordination or integration of the spatial dimension of sectoral policies through a territorially based strategy. The strategy acts as a framework for the formulation and implementation of sectoral policy.

One of the sectors will be land use planning. In this sense, spatial planning seeks to identify and address the contradictory effects of sectoral policies, and the opportunities for synergy through the territorial strategy (Cullingworth, Nadin 2002). Figure 1.6 shows how the spatial dimension of sectoral policies can be integrated into different ways in pursuit of agreed objectives of climate smart and wise cities.

The continuing evolvement of the urban planning model and planning process, along with the adoption of numerous regulations and guidelines on the Green Deal and post COVID-19, leads to the conclusion that the planning model and process will continue to be transformative.

For the inception phase of the state of art spatial planning practices shown in Fig. 1.6, practical considerations may include:

- Try to keep a balance between dynamic city growth and the preservation and restoration of natural ecosystems;
- Make sure spatial connectivity with ICT applications for post-COVID recovery;
    - Connectivity of actions and plans;
    - Technologies for the digitalizing connections;
- Encourage each person to fulfill its own responsibilities as a member of the community for neighboring people. Our lives are connected to each other;
- Design a healthy and green community with nature;
- Assure for all people safe, healthful, resource wise, and resilient surroundings;
- Preserve and restore the natural ecosystems which support our lives;
- Establish remote control system to promote untact activities in public areas due to social distancing;
- Promote local food production for food security with urban farming;
- Develop automated early warning systems of the crisis;
- Detecting COVID-19 in wastewater, sewage treatment and sludge;
- Attain the widest range of beneficial use of the environment without degradation, risk to health and safety, or other undesirable and unintended consequences;
- Achieve a good balance between population and resource use which will permit high standards of living;
- Enhance the quality of renewable resources and approach the maximum attainable recycling of depletable resources;
- Provide digitally connected urban form and public space to avoid face-to-face contacts;
- Provide free transit consulting, demand response software for transit agencies and intelligent transport system (ITS);
- Prepare 'connected health care framework';
- Promote untact market;
- Assure remote access to content for library users; and

- Balance life and working remotely.

(2) **Physical master planning process**

City climate planning process is less than a decade old and, as such, represents a developing method. The future of this process can be described according to the normative and problem-solving planning process. There have traditionally been two schools of planning thoughts. They are normative (on the right) and problem-solving (on the left) approaches to planning, as shown in Fig. 1.10. Normative planning process starts with goal setting while problem-solving planning process with the issue and problem identification. Figure 1.10 has been developed to integrate city climate planning issues with conventional planning process.

Figure 1.11 is an example of the problem-solving planning process model.

Figure 1.12 depicts how cap and trade and the climate change action work together while Table 1.2 shows how financial analysis and carbon marketing can be integrated with physical/spatial/and ICT planning processes.

After having reviewed the major concerns with various planning processes, the sustainable eco-smart tech/integration platform process suggests operational nine steps as follows:

- Step 1. Assessment of impacts of urbanization on the urban ecosystem and their services. Climate is a part of the system
- Step 2. Overview of planning responses to climate change
- Step 3. Identification of planning components and elements, and their connections with digital technologies: Planning system framework for climate smart and wise cities
- Step 4. Adoption of climate integrated planning process: Planning procedure
- Step 5. Preparation of an integrated physical/spatial climate plan: Integrated total system modeling

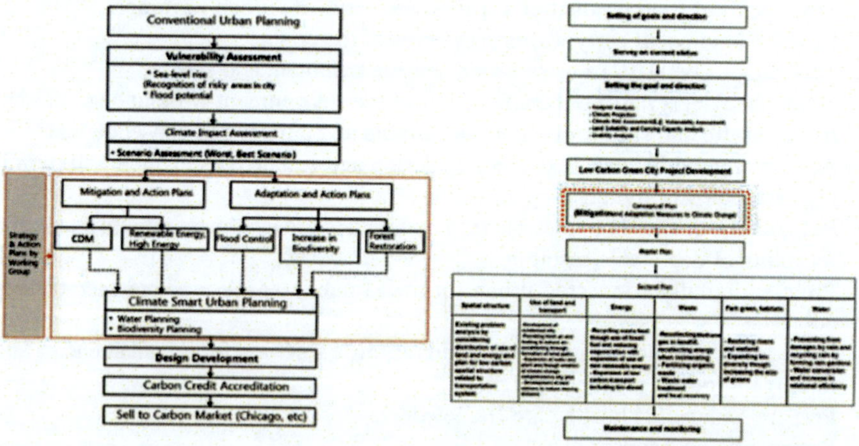

Fig. 1.10 Climate smart comprehensive urban planning processes for super-connected cities

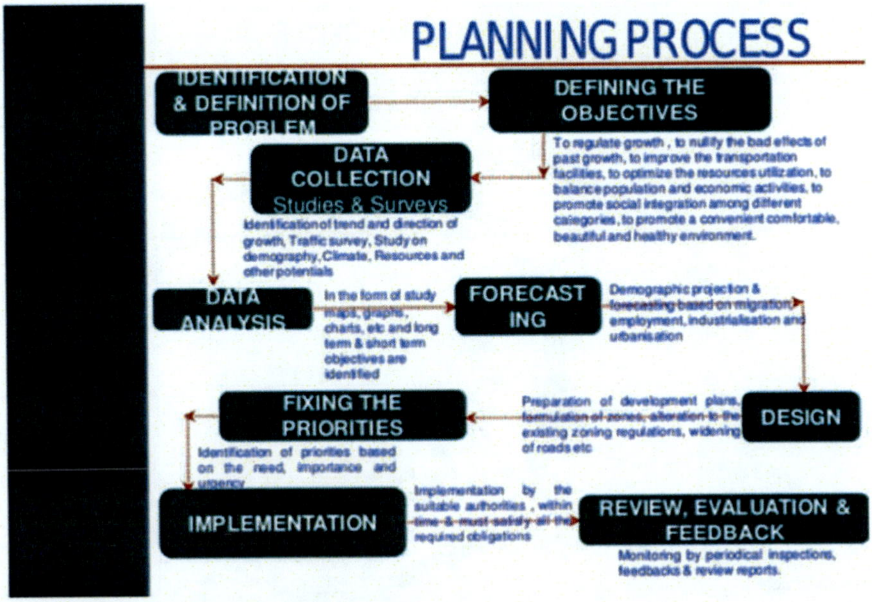

Fig. 1.11 An example of a problem-solving planning process. *Source* https://www.slideshare.net/ctlachu/introduction-to-town-planning

- Step 6. Exploration of methods and techniques to form a connected system of individual platform layers: sector integration platform modeling
- Step 7. Attraction of carbon financing and investment
- Step 8. Identification of further research needs
- Step 9. Establishment of implementation tools
- Step 10. Considerations for the future of low-carbon smart cities.

Finally, we need to look at a planning process that incorporates the carbon neutrality plan into revised physically spatial planning process above. The feedback process from carbon neutrality by 2050 must be used in climate smart and wise city planning or the purpose of the planning will not have been fully satisfied. Carbon neutrality plans can be useful in: (1) defining planning response to climate change, (2) identifying planning components and elements, and their connections, (3) adopting planning process, (4) exploring methods and techniques, (5) attracting carbon financing, and (6) establishing implementation tools.

Among the carbon neutral city planning procedures, two levels are suggested for incorporation of carbon neutrality information in a climate smart and wise city plan.

First, an overall organizational framework should be established in:

(1) Facilitating and coordinating interdepartmental links and multi-stakeholder partnerships for organizational action for city profile and action plan for climate neutrality.

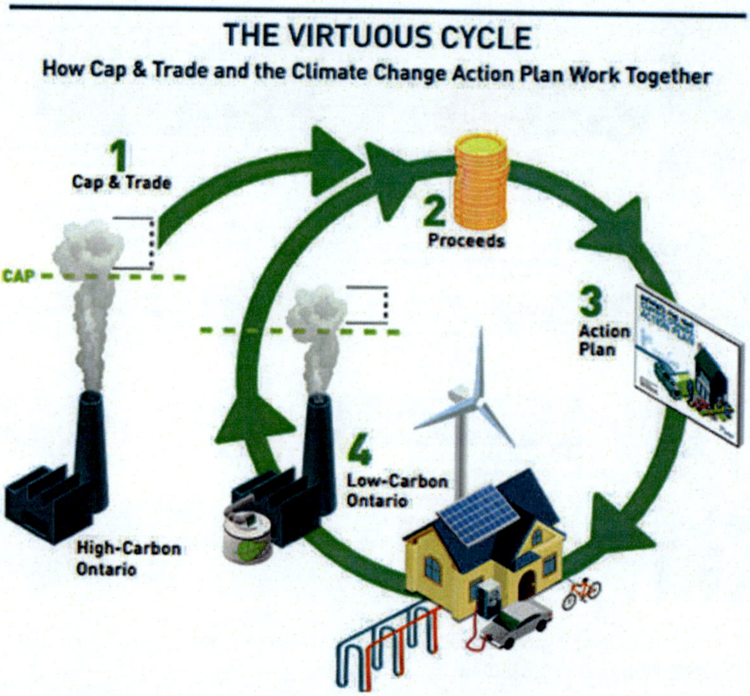

Fig. 1.12 The virtuous cycle—how cap and trade and the climate change action plan work together. *Source* https://www.ontario.ca/page/climate-change-action-plan

(2) Developing and implementing a city profile and action plan:

- To serve as an "evidence baseline" for activities supporting climate neutrality.
- To draft a city profile for climate and energy on an inventory and analysis of the city's existing energy balance, carbon balance sheet, climate footprint, vulnerabilities and opportunities.
- To develop an integrated action plan, which should include emission-reduction targets, climate change mitigation and adaptation measures, support mechanisms for most vulnerable groups, and time frames for the climate neutral transition.
- To include sectoral measures and aim at creating synergies and integration between the different sectors.
- To draw up procedures for continuous evaluation, reviewing and updating the action plans.
- To mobilize financial resources at national and international levels for implementing the action plan, pilot projects and other climate neutral initiatives, including public–private partnerships and the private sector.
- To implement the action plan.

Table 1.2 A protocol of step, activity, participating agency, budget and timeline: an investment business proposal

Step	Activity	Participating Agency	Budget	Timeline
1. Pre-planning	1. Formation of project team(Climate Smart and Wise City Consortium) 2. Capacity building training			
2. Project planning design	3. Identification of participating cities 4. Preparation of the master plan and site briefs for the selected climate smart and wise city 5. Preparation of the detailed plan for the action area 6. Selection and design of infrastructure projects			
3. Financial analysis	7. Cost-effective analysis 8. Additionality analysis • Technical additionality • Financial additionality			
4. Development management	9. Construction 10. Supervision			
5. Operation	11. Overall project operation/maintenance 12. ICT operation			
6. Monitoring	13. Monitoring of ICT operation 14. Monitoring of GHG reduction activities 15. Quantification or estimation of the cumulative volume of the reduced emissions, compared with the "business-as-usual(BAU)" baseline level of emissions 16. Reporting and validation of monitoring data/information			
7. Marketing	17. Selling and transfer of development products 18. Selling and trading of carbon emission rights(CERs)			
8. Evaluation	19. Periodic project evaluation 20. Feedback to project operation			
9. Project finalization	21. Final project reporting			

- To review public procurement practices and permission systems based on the action plan.
- To incorporate the plan into revised spatial planning: standard smart integrated platform models(physical space model, digital space model, virtual connection model).

(3) Educating the public, professional associations, businesses and industry about climate neutral solutions, establishing advice centers and knowledge-sharing platforms, addressing skills development needs and providing capacity building programs for various groups and job markets.

(4) Establishing a system of climate monitoring early warning and contingency measures at the city level.

(5) Monitoring and evaluating the implementation of the action plans and the city's climate neutral policies, including spatial planning, and participating in inter-urban networks and cooperation activities at the national and international levels for mutual learning and greater visibility.

Second, a framework for identifying priority sectors for action should be established as follows:

(1) To use spatial planning, zoning and other instruments to reduce and prevent urban sprawl, stimulate mixed-use compact communities and transit-oriented development, and implement protection and resilience measures for vulnerable zones.

(2) To assist in the development of green markets, green business start-ups and green jobs, and support public and private R&D.

(3) To increase the use of green energy, stimulate distributed renewable energy generation, as well as city comprehensive smart energy and water grid platform, use energy-efficient technologies for city facilities, and develop district heating–cooling systems and tri-generation.

(4) To stimulate and promote energy service companies, which will help finance and coordinate energy efficiency measures.

(5) To promote walking and cycling by integrating these into all spatial planning policies and infrastructure improvements, develop good public transport, provide disincentives for using cars, and install alternative fuel infrastructures such as hydrogen and electric vehicle charging stations.

(6) To improve energy efficiency and the physical condition of buildings via the promotion of zero-energy building design, climate-proof development and low energy retrofitting, strengthen the quality of property management and management systems, and integrate energy efficiency into the city's housing programs at different levels of scale.

(7) To preserve and expand green open spaces, and greenbelts, promote tree planting and green roofs and consider other measures to mitigate the urban heat island effect such as blue-green-white corridor infrastructure.

(8) To increase recycling infrastructure in the city, install waste-to-energy technologies, and promote sustainable material life cycle via platform design control.

(9) To carry out project demonstration projects, set up "champion" projects in the public sector and award champions in the private sectors, promote pilot projects of different kinds and in different sectors, and different smart digital technologies to test and verify the effectiveness and replicability of different neutral solutions.

(10) To designate areas as testing grounds for zero net-carbon neighborhood and city development, and consider expanding them over time.

1.3 Planning Methodology: Development of Smart Platform Models

The integration of key components of the spatial planning system in the planning process requires to use systematic and multidisciplinary platforms and to develop methods and techniques to ensure that emerging issues and agendas and instruments may be given appropriate consideration in planning climate smart and wise cities.

1.3.1 Definition and Types of Platform Models

The climate integration platform theory is a theory of climate action integration that suggests the multi-layers of individual platform models. A platform means a common knowledge-based which is combined with physical space, digital space and socio-economic space. The climate-integrated platforms include a common thematic integration platform, spatial integration platform, actor integration platform and sector integration platform.

A climate smart and wise city platform can be defined as a framework for sensing, communications, integration, and intelligent decision making (www.beechamresearch.com/files/Br Smart City EXEC Summ... 2015). At a minimum, the functionality of a smart city platform must include:

- Visualization: Digital twin.
- Application enablement.
- Data management.

The platform may also include:

- Cybersecurity.
- Device management.
- Network management.
- Application development.

- Analytics.
- GIS.
- Mobility.
- Reporting.
- Simulation.
- Back office, and more….

Climate-integrated smart city platforms attempt to integrate information from the disparate systems of climate actions such as intelligent grids, smart lighting systems, building automation systems, emergency management systems, security and access control systems, renewable power, water treatment and supply, transportation, and more into a unified whole to provide a holistic view of the overall performance and state of the city and its various functions.

In this chapter, climate smart and wise city platforms include AI smart city open platform, smart green blue grid platform, integrated rainwater drainage system platform, smart solar connected community microgrid platform, deployment of smart digital sensor platform, home IoT platform and smart wetland management platform. For example, some ICT solutions for climate smart and wise cities include the use of virtual models as a communication tool for smart infrastructure planning and development.

1.3.2 New Green Agendas as Instruments for Planning Models

COVID-19 is now a critical urban agenda because cities are on the front lines of COVID-19. There are many lessons learned from the lockdown in cities. A renewed focus should be on the need for the city to make a committed transition toward a 'green' for post-COVID-19 recovery. If Green Deal movements growing across the globe lead to green investment, it could be a significant momentum for implementing climate-friendly measures. This section discusses Green Deal movements and post-COVID-19 recovery plan as the basics for 'urban normal.' Climate smart and wise city planning can accelerate SDG action and the COVID-19 recovery.

(1) **Green Deal Movements: A carbon neutrality 2050 approach**

Different continents and countries have different principles and project activities for the Green Deal policy and strategy.

1. **European Commission(EU)-Style Green Deal**

Principles

The Green Deal is an integral part of the European Commission's strategy to implement the United Nation's 2030 Agenda and the sustainable development goals. It states that:

"To deliver the European Deal, there is a need to rethink policies for clean energy supply across the economy, industry, production and consumption, large-scale infrastructure, transport, food and agriculture, construction, taxation and social benefits.

To achieve these aims, it is essential to increase the value given to protecting and restoring natural ecosystems, to the sustainable use of resources and improving human health. This is where transformational change is most needed and potentially most beneficial for the EU economy, society and natural environment.

The EU should also promote and invest in the necessary digital transformation and tools as these are essential enablers of the changes (Commission Communication. 11/12/2019)".

Transformational project activities

The various elements of the EU Green Deal are illustrated below:

- Smart grids
- Hydrogen networks
- Carbon capture, storage and utilization
- Energy storage (ESS for example)
- Enabling sector integration
- Capacity building.

Technologies

One of the main features of the EU Green Deal Strategy is the use of Blockchain technology for the SDGs and climate in the digital era:

- Blockchains for the Green New Deal: a decentralized scheme to empower local authorities and people
- A Blockchain ecosystem for digital transformation.

2. **"South Korea Embraces EU-Style Green Deal For COVID-19 Recovery"**

With strong public backing, the Korean government set out enacting its Green New Deal, published in March 2020 as part of its manifesto. President Moon Jae-in has explicitly referred to the European Commission's European Green Deal as instrumental in the formulation of the Korean plan, and bilateral meetings between South Korea and EU officials have enabled the two regions to share policy ideas and form cooperation projects (Forbes 2020).

Potential project activities: digital, green new deal projects

- Carbon Tax
- Requirements to end investment in coal
- Investment in renewable energy source
- Hydrogen strategy
- Potential digital projects: Digital transformation and digital tools

- Digitalized, connected, automated cities: Digital transformation of cities
- Digital SOC (or smart infrastructure)
- Renewable mini-grids and digital technologies
- Vocational training centers for digital jobs in the fields of sensors, AI, IoT, cloud computing, Big Data, 5G, DLT/Blockchain technology, etc.
- Integrated building energy management system called 'Cloud BEMS' which connects cloud (virtual storage space) with BEMS
- Online digital untact business for informal work force including women
- Covid-19 Recovery Plan: A new instrument for urban planning and development for transitions to post-COVID WORLD.

1. **The approach**

Predictive methodologies such as the decision tree can be used for predicting severe causes of COVID-19 with help from AI and machine learning. AI experts can analyze historical patterns of corona virus disease cases with Big Data and help to make intelligent decisions.

2. **Potential project activities**

Generalized activities include:

- Empowering communication: The use of creative, visual (or virtual) and digital methods to find out the path forward on the Green New Deal and post COVID-19 with stakeholder meetings
- Post lockdown public space management
- Responsive transit services that can both support evolving rider demand and adhere to quickly-changing health guidelines.

1.3.3 Practical Considerations for the Green Deal and Post COVID-19 in Developing Urban Planning Models: Towards Ecological Health, Economic Health Versus Human Health

This chapter suggests the following strategies for model development:

- To implement the Green Deal and post- COVID-19 recovery at the city level;
  - The urban dimension of cohesion policy should be strengthened. For example, the proposed European Urban Initiative provides assistance to cities to help them make the best use of opportunities to develop sustainable urban development (European Commission, The European Green Deal, Brussels, 11.12.2019, COM(2019) 640 final);
- To develop climate smart and wise city platform models as an interface between mandatory planning and new smart technologies;

- To link the Green Deal and post-COVID-19 recovery with urban and regional spatial plans;
- To incorporate climate action goals, Green Deal policy goals, post-COVID-19 recovery goals and resource wisdom goals at each stage of the planning process;
- To include the platform models in Green Deal laws, and COVID-19 guidelines and plans;
- To predict future scenarios with help from Big Data, AI program, algorithm and machine learning.

We need to be open to changes for the city of the future.

1.3.4 *Types of Planning Models*

In cities, everything is directly or indirectly connected. In this regard, urban infrastructure and service networks are of growing importance and are in high demand to enhance the efficiency of urban connectivity. However, currently, there is no overarching model integrating these networks into city planning.

There is a need for an integrated neural modeling approach to solve the current limitations of isolated models and the deficiencies that could be addressed by the models based on the integration platform theory.

The platforms below could impact the level of smartness and wisdom of climate smart and wise cities. For each platform, the following should be discussed and evaluated to develop innovative models to address how isolated and fragmented networks could be improved to meet the requirements for the advancement of smart infrastructure and service planning and management.

- To develop Big Data and Analytics supported by IoT and AI for knowledge-based networking;
- To carry out social surveys to identify challenges and limitations of current isolated models, rising issues and needs for innovation networking;
- To explore various grid platform technology potentials;
- To suggest possible testbeds to demonstrate how alternate networks are different in practice;
- To organize a public–private partnership platform for better implementation of the networks; and
- To compare the carbon footprints between isolated and fragmented, and integrated networks.

Planning platform models can be divided into:

- Holistic planning model
  - Assembly of urban components (urban matrix): Physical/spatial, digital and socio-economic overlay
  - IoT and 5G virtual connection platform for all sectors of a city: Digital overlay

  - Spatially coordinated climate smart and wise city platform: A synthesis (3D digital map: Cities made by digital twin).

- Sub-system planning models
  - Digital city models
  - Low-carbon city models
  - Resource wise city models.
- Sectoral dynamic platform models
  - The standard platform model for land use
  - The standard platform model for solar energy-connected smart microgrid
  - The standard platform model for smart water grid
  - The standard platform model for smart rainwater and storm management
  - The standard platform model for green–blue (rivers) and white (wind passage) infrastructure
  - The standard platform model for wetland connectivity: An IoT/Blockchain–based approach
  - The standard platform model for sensor networks and IoT
  - The standard platform model for blockchain
  - The standard platform model for geophysical information.

The standardization of geospatial information systems would be more ambitious and futuristic climate smart and wise city innovation.

- The standard platform model for smart technology transfer
- The standard platform model for climate disasters
- The standard platform model for COVID-19 pandemic.

(1) **Holistic planning models: Integrated platform project models**

The due issue is how to connect the real physical world with the digital world with smart ambient media (applications) for the city as a whole. Carbon footprint overlay can be a product of connecting ecological mapping with digital connect-tech through ICT.

Table 1.3 shows components of ecological mapping overlay (physical world) and digital connect-tech overlay (digital world) which can be superimposed on the carbon footprint overlay.

The following is the key limitations of the current planning models according to the definition of climate smart and wise cities and ICT application:

- Lack of spatial analysis in terms of physical-digital integration of infrastructure
- Lack of acceptance and understanding of system dynamics for planning based on system thinking
- Not enough research, innovation and new thinking
- Lack of standards and skills
- Failure to meet the needs of people involved is often a problem

Table 1.3 Components of overlay approach to climate smart and wise cities

Ecological mapping overlay	Digital connect-tech overlay	Carbon overlay
• Land Development – Land use zoning – Protected areas – Flood-prone areas • Mitigation – Renewable energy – Carbon capture and utilization – Carbon storage – Carbon forests • Adaptation – Natural ground surface (natural wetlands) – Mangrove plantation – LID (Rain gardens) • Physical Infrastructure – Drinking water – Wastewater – Landfill site – Food waste treatment – Other solid waste infrastructure	• Mobile Techniques – iPhone – smartphone • Sensors – Led lighting sensor – CO_2 sensor – Car navigation system • Intelligent CCTV – GIS-based CCTV monitoring for crime • Safety – Disaster (natural & human-induced) monitoring – Automatic disaster warning system – Water level perceiving system – Flood risk management • Meters for metering utilities – Electricity consumption – Water consumption – City gas • Energy & heat density mapping • Monitoring for operation and maintenance (control tower)	• Smart housing carbon footprint • Shopping carbon footprint by social media (internet: behavioral change) • Smart connected factory carbon footprint • Smart parking carbon footprint • Smart hospital carbon footprint • Traffic flow organization carbon footprint • Intelligent digital signage carbon footprint
Integrated superimposed synthesized map	Integrated superimposed synthesized map	Total amount of carbon footprint & carbon reduction Risk reduction

Source Kim (2018)

- Lack of investment by the green and digital industry.

The holistic planning model of climate smart and wise cities is the digitalizing connections of the structure and function of cities with technologies that unite Smart and digital city models, Low-Carbon City models and Resource Wise City models for the Green Deal policies and post-COVID-19 recovery plans. It should be comprehensive, carbon centered (3Cs). This model can be practiced in the form of testbeds of new and regeneration projects as the cities of the future.

Testbeds for new digital technologies with the Green Deal and post-COVID-19 recovery plan applications aim to be "global center for urban innovation". They can be used for the following:

- R & D
- Living laboratory for smart and green technologies

- Cloud innovation center
- Center for smart data innovation
- International technology transfer and export market.

Two testbed illustrations at the different sectors of cities are shown in Figs. 1.13 and 1.14 which include grid technologies, smart technologies, climate, energy and trash technologies, eco-tech and transportation technologies.

To solve the above-mentioned limitations, each of the major urban components needs to be analyzed separately prior to coming together to form an interactive urban system. Sectoral urban components are shown in Fig. 1.15. The urban matrix (Fig. 1.16) is assembled by superimposing the different components. The aim is to integrate the components with each other so that the systems for land use, mobility, water, blue-green-white infrastructure, energy, digital technology and social distancing work well together in a holistic, sustainable, resilient, efficient, climate friendly and connected manner (Fig. 1.16). IoT virtual platforms have changed urban planning solutions (Fig. 1.17). AI is changing everything in cities.

Figure 1.18 uses interaction massing technique for integrating urban elements included in the physical matrix and digital overlay.

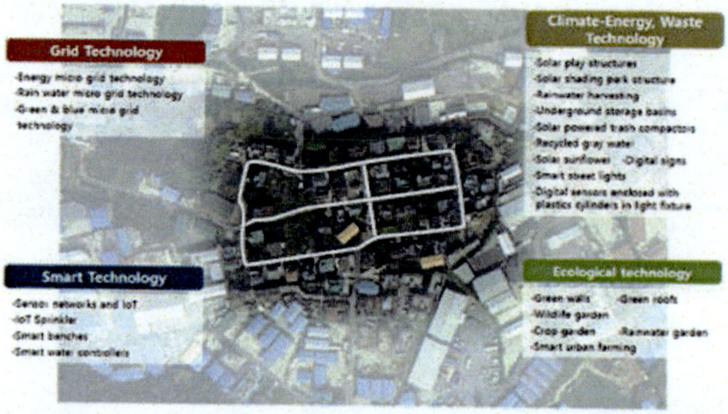

Fig. 1.13 Urban village testbed: an illustration

Fig. 1.14 Commercial area testbed: an illustration

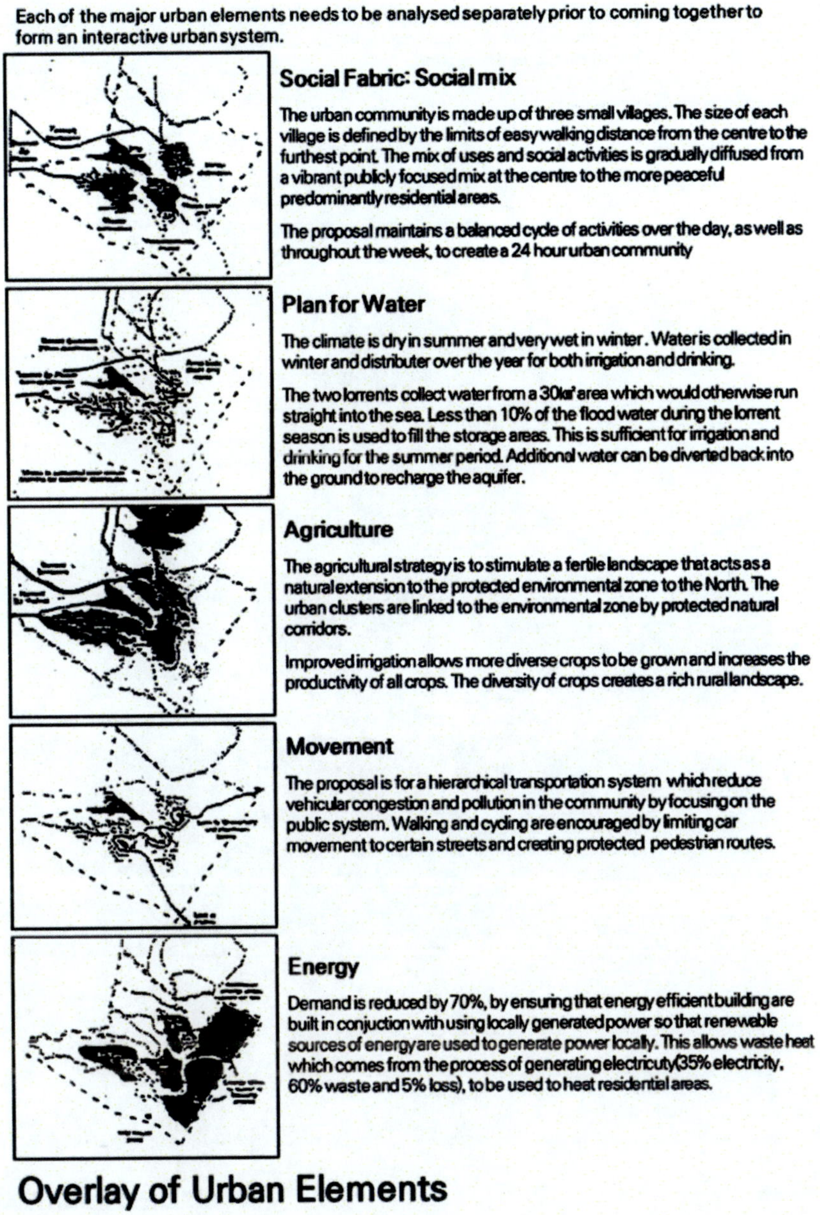

Fig. 1.15 Urban grid components with 3D digital mapping. *Source* Rogers (1996). (modified by the authors)

Fig. 1.16 Assembly of urban components: physical overlay. *Source* Rogers (1996). (modified by the authors)

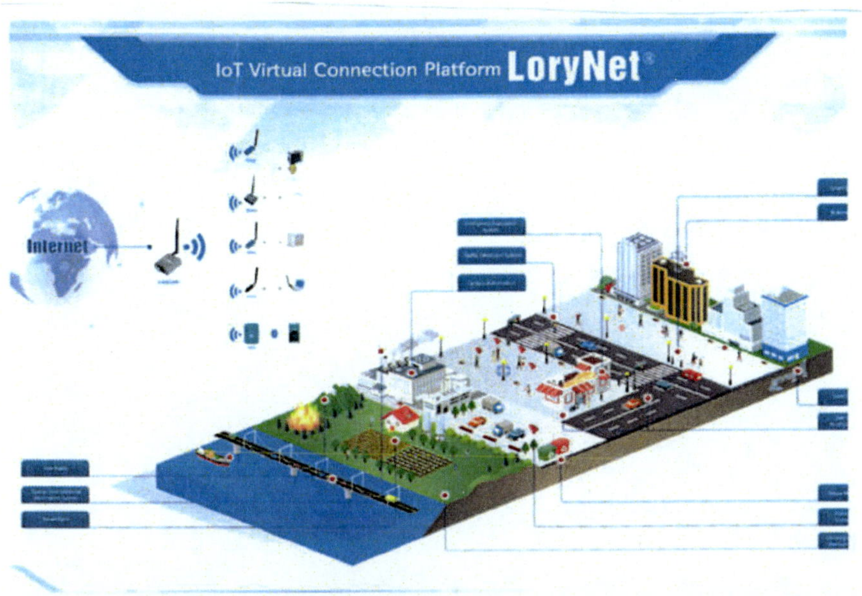

Fig. 1.17 IoT virtual connection platform model for all sectors of a city: digital overlay

One of the main features of this holistic planning model includes a digital overlay of different technology layers toward the smart urban transformation goals. It is time to invest in low-carbon, digital and ecological tech.

(2) **Sub-system planning models**

3Ds (digitalization, decarbonization, decentralization) are instrumental in operating the holistic planning model. They allow the holistic planning model to make its system functional through the sub-system models used in climate smart and wise city planning.

1. **Smart and digital city models: IoT transition to digitalization in the era of digital new deals**

Digital transformation of solution for climate emergency will rapidly change the climate action value chain. It can create quality jobs, save life cycle costs, secure transparency and reduce more GHG emissions by providing better structure and function of cities.

Definition and goals

The Digital City is a city connected to the net, equipped with technological platforms of information and communications management that can enable the Internet of Things. These platforms also make it possible for huge amounts of data and information to offer new services to the inhabitants of urban areas and new functionality for the management of the urban environment.

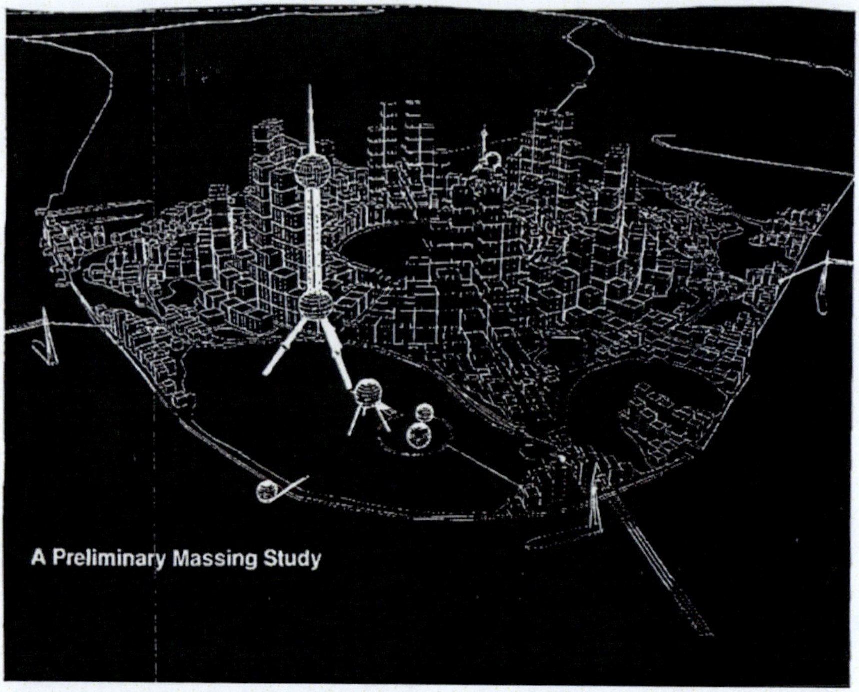

Fig. 1.18 Coordinated climate smart and wise city platform model: synthesis of physical and digital world using big data. *Source* Rogers (1996). (modified by the authors)

As the climate smart and wise cities have linked new digital technologies to the development of the built environment, one could expect that climate change and the growing demand for natural resources change the focus of strategic spatial planning toward the digital city model. Digitalization is the use of digital technologies to change a city model and to provide new revenue and value-producing opportunities; it is the process of moving to a digital business.

A broad meaning of digitalization goes beyond business and refers to the ongoing adoption of digital technologies across all possible societal and human activities in cities. We have to rethink the scale and pace of 5G investment in the New Normal. Digital transformation arises from the intersection of cloud computing, big data, IoT, and AI, and it is vital to industries across the market today. Some describe it as the power of digital technology applied to every aspect of the organization. Some refer to it as using digital technologies and advanced analytics for economic value, agility, and speed (Siebel 2019). Originally digital transformation was used to describe the transformation of, for instance, paper into digital information. Here we can see where the confusions come from the difference between digitalization and digital information (https://www.i-scoop.eu/digital-transformation/digitization-digitalization-digital-transformation-disruption/).

The approach

Digitalization is a key factor in making cities more sustainable and untact, enabling the development of new forms of urban land use, buildings, energy, water, waste treatment, logistics and mobility, such as electric mobility and car sharing. Digitalization also encourages the introduction of new safety and renewable source technologies. It also promotes the introduction of a system for withstanding extreme meteorological events.

Digital city can be further defined as a community digital space, which is used to facilitate and augment the activities and functions, taking place within the physical space of the city (Smart City Research Larissa Romuldo 2013). In the future, the successful provision of services throughout the cities, which are rapidly increasing both in size and population, will depend on physical infrastructure which new digital technologies are able to interlink and operate. Telecommunications networks are going to be used for a large volume of machine-to-machine data transmission. Sensors and actuators will be essential components in the digital cities of the future. We can point out that to improve existing services and the creation of new ones, it is necessary to integrate the various component systems within a city. Hence, physical-digital integration of infrastructure will be of utmost importance for the success and survival of the next generation of cities. The sectoral digital space and platforms will be discussed more in detail in the next section.

The Models

A smart and digital city model entails the use of intelligent solutions for infrastructure, energy, housing, mobility, services, and security based on integrated sensor technology, connectivity, data analytics, and independently value-added processes.

2. **Digital shadow models**

The most important model for smart and digital city planning is the so-called digital shadow model. It is a prerequisite for any smart and digital city to create a raw input of data about the city, coming from an interconnected network of sensors, devices, and other digital services and products.

In Fig. 1.19, just imagine a city where thousands (or even hundreds of thousands) of sensors (Figs. 1.20 and 1.21) record information about air quality, temperature, traffic, energy consumption in buildings, smog, garbage production, or humidity. The data is continuously uploaded to the cloud, creating a digital image of the "behavior" of the city, available online at every moment. This digital shadow is possible because of the latest achievements in connectivity, especially the Internet of Things, a set of technologies connecting "intelligent" objects through the Internet, adding another layer of data.

The resulting corpus of data is the foundation for many other services and products, created by institutions and companies that might build second-layer solutions. For instance, apps could be created that allow citizens to improve their mobility with eco-friendly, connected, solutions or to help building owners improve energy efficiency through data and algorithms that control heating and cooling in real time or

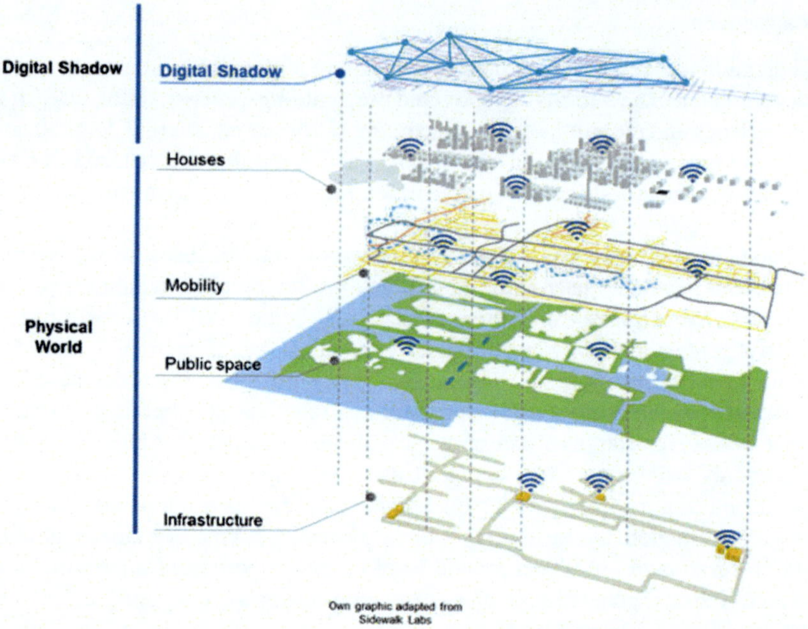

Fig. 1.19 Digital shadow model: this image shows how the digital shadow mirrors the city through a miriad of sensors working 24 h/day. *Source* BMI-Lab/blog/2019/9/13/smart-cities-and-digital-innovation-a-guide-to-the-future

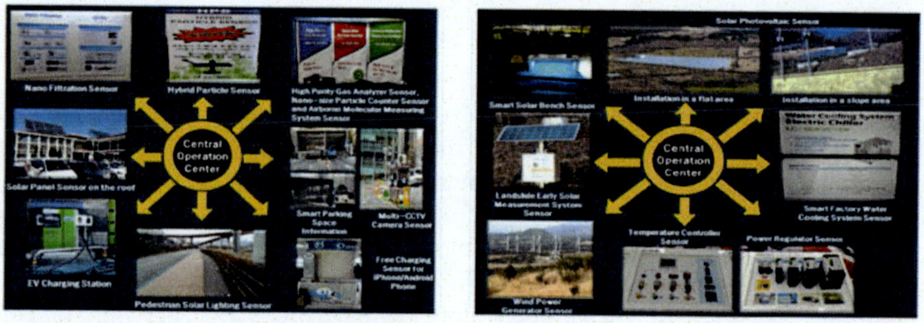

Fig. 1.20 Positioning of sensors in climate smart and wise cities (1) and Fig. 1.21 Positioning of sensors in climate smart and wise cities (2) Source Kim (2018)

perhaps a new service to connect government officials and citizens, improving their communication by digital means. Other options might be related to urban farming, telehealth, air and water quality, trash treatment, and data management.

The range of potential solutions built over the digital shadow is vast and continually growing. These platform solutions can be broken down into these twelve areas:

Fig. 1.21 Platform model for "Internet of things in connected smart world": physical-digital integration of infrastructure

- Smart digital land use—minimizing the ecological footprint of a city.
- Smart digital living—improving the quality of life in the city.
- Smart digital economy—increasing the city's competitiveness.
- Smart digital mobility—creating new, sustainable and untact mobility options in the city.
- Smart digital government and governance—promoting the participation of citizens and increasing transparency in the public sector.
- Smart digital people—preserving and increasing the human capital of a city, helping people develop their full potential, and fostering their participation in social life.
- Smart IoT home—making home more digitalized.
- Smart digital green–blue-white network—improving the city's pattern and functionality.
- Smart digital water—reducing the city's water consumption.
- Smart digital energy—enhancing the city's energy efficiency.
- Smart digital wetland—enhancing the connectivity of urban wetlands
- Smart digital disease pandemic—stimulating recovery package.

The above twelve areas of the digital shadow model are shown both in a pictorial form (Fig. 1.21) and in a cross-sectional diagram of a city (Fig. 1.22).

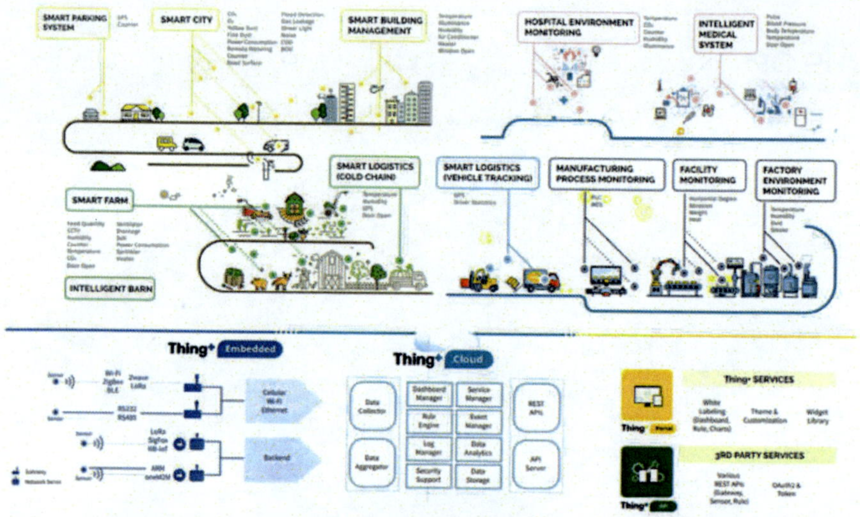

Fig. 1.22 Platform model for "How thing + Makes IoT REAL"

3. Low-carbon city models: Climate transition to decarbonization in the era of climate new deal

Definition and goals

Decarbonization means reducing greenhouse gas (GHG) emissions produced as a result of transport, buildings, water and waste, including emissions released directly during the sectors and emissions due to the production of transport- for example emissions from the production of electricity used to power a given mode.

Decarbonizing cities is a global agenda with huge significance for the future of cities. Cities offer some of the best opportunities for decarbonization. And a few sectors such as buildings, transport, water and waste have the greatest potential for high decarbonization impacts. Creating an enabling environment for cities to invest in these sectors is essential for meeting the less than 2 degrees Celcius target of the Paris Agreement in view of an urban population growing by approximately 1.4 million weekly.

The approach

In response to the above-mentioned challenges, there is increasing attention to the possibilities for 'cracking open' the transformation processes for decarbonizing urban living through the co-creation of future visions involving researchers, designers, professionals, and the community. The challenge of decarbonization and resilience involves no less than a transition from long-established systems (technologies, infrastructures, practices, lifestyles, values and policies) to very different ones.

There are framework models, tools and even digital games that can help explore futures and the complex dynamics of physical, technological and societal systems including financial systems (Fig. 1.7) in order to transform those systems and our cities, as described so far in this chapter.

As freely available approaches, they should become vehicles for expanding public interest in thinking beyond business-as-usual realities to the creative design of alternative cities and urban living.

The models

Models that reduce fossil-fuel use can achieve both global carbon mitigation and local health-protection goals. Now research shows the dual benefits of compact urban design and circular economy policies in cities. These models move beyond CO_2 mitigations and single sector approaches.

Growing awareness of the devastating impacts of climate change adds a new angle to urban infrastructure and resource- management for saving the planet. Emergence of climate smart and resource wise cities provide an opportunity for the introduction of innovative new market mechanisms from project to mitigation activities in decarbonizing cities and their benefits. An effectively coordinated carbon market mechanism is necessary to integrate land use and transport, energy generation, building energy, solid waste, and wastewater and water efficiency with new technologies in a holistic manner (Table 1.4).

Table 1.4 Comparison of conventional CDM and innovative CDM methodology (Kim 2018, p. 217)

	Conventional methodology	Innovation methodology
Theoretical underpinnings	Isolated standalone view	Holistic, integrated, smart view
Context	Singular project development framework	Urban planning and development framework
Scope	Technology and products-oriented system	Extension of CDM to include policy, planning and governance
Methods	Use of existing methods approved by UNFCCC	Combination of existing and new methods
Baseline scenario	Constant level of emission	Considerations for population and economic growth, etc
Measurement	Quantification	Quantification and estimation
Certification	Tradable certified emission reduction (CERs) at single CDM project level	Aggregated tradable certified emission reduction (CERs) at the city scale and between cities
Market	Carbon markets implemented at the national or regional level	Carbon markets to be implemented at the city level
Tool for urban planning	Tool for disconnected infrastructure	Tool for 3Cs (carbon centered comprehensive) planning, land use-based mechanism
Effectiveness analysis	Project cost–benefit analysis	Carbon balance sheet analysis

More holistic approach which integrates all sectors into one common approach will increase co-benefits of climate mitigation and adaptation. Cities can benefit from the new policy on blockchain-based urban market mechanisms as follows:

- Enabling cities and climate defenders to innovate by considering climate actions as essential to the urban planning and development
- Keeping cities' climate actions continuously through financial gains by adopting and practicing the urban market mechanism
- Supporting cities in increasing their access to carbon markets through the urban carbon market
- Encouraging cities to raise citizens' awareness about climate change issues by sharing
- Blockchain-based database on city climate status and future planning with regard to actions to be taken at the city and community level
- Contributing to new climate economy for climate smart and wise cities.

On the other hand, urban methodology for carbon markets suggested by Hegde (2018) includes urban planning, new renewable energy, transportation, energy generation, building energy, solid waste, wastewater, waste collection and transportation energy, and waste to energy. He has added new project activities involving transit-oriented development, infill/densification, urban forestry and green space programs, waste transport vehicle efficiency/fuel switching, waste transport vehicle routing/reduce km traveled, waste treatment site operational efficiency, and installment of waste to energy plant to current guidelines and practices (Fig. 1.23).

4. **Resource wise city models: Resource transition to decentralization in the era of COVID-19 recovery deal**

Fig. 1.23 An image of low-carbon city: Walking along the carbon free street

Definition and goals

Resource wisdom means being able to reduce emissions and the consumption of natural resources, while actually improving the perceived well being of the residents in a given region (Systems 2017). Spatial planning integrates the key aspects, transportation, housing, and food production which are, on many occasions, stated as the most significant consumption factors causing environmental impacts and climate effects.

Taking account of environmental considerations will lead us into making choices to act within the Earth's carrying capacity or ecological footprint focusing on environmental impacts. It will also allow future generations the opportunity to lead rich and fulfilling lives.

The Approach

To achieve these goals, smart networking, co-creation, and innovative cyber-physical solutions are vital for the utilization of the entire city's resource potentiality. We need land suitability and carrying capacity, and strategic environmental assessment (SEA) to wisely manage urban growth and urbanization. The integration of the growth framework should be based on the fundamental alignment of its land development (settlement), water and ecological components. Each of these three interdependent resource systems has a hierarchal structure with complementary elements (Fig. 1.24). Connected through land use zoning, buildings, mobility, energy, water, social infrastructures and smart technologies, these structures act as a model for resource wise city plans.

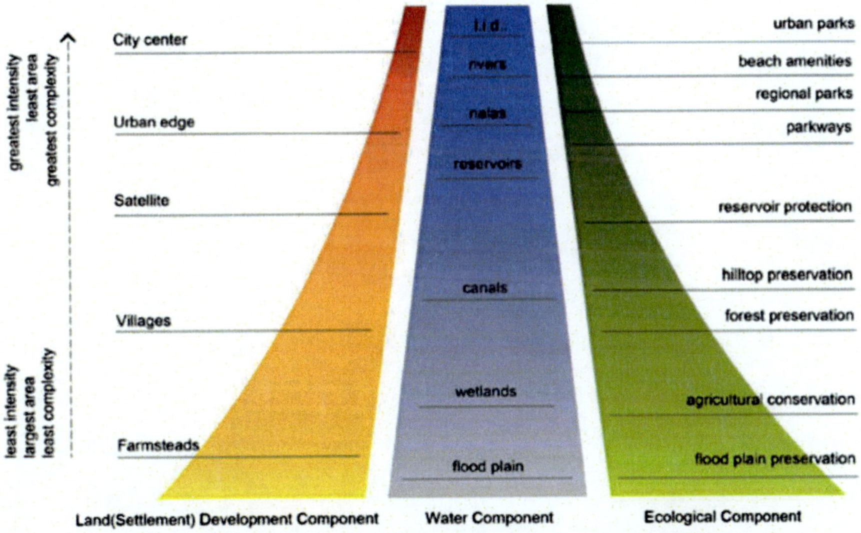

Fig. 1.24 Integrated framework for growth. *Source* AECOM, IBM and KPMG (2017)

The models: centralization versus decentralization

In the annals of population studies, COVID-19 will be the event that thrust demography into the global spotlight. The processes accompanying the pandemic, including the lockdowns, are intimately connected to demographic trends and patterns (World Bank 2020).

Since the mid-1960s, many cities have tried to cope with extraordinarily high rates of population growth. Some cities doubled their populations in less than a decade. This rapid growth is often accompanied by negative effects such as the loss of valuable open space, decreased air quality, and overloaded wastewater treatment plants. Often, rapidly growing communities also experience financial problems in trying to expand their public services to accommodate the new growth. The result is that traditional tools for managing growth such as comprehensive plans, zoning ordinances, and subdivision regulations have been supplemented in many cities by innovative strategies.

The following three scenarios for the growth pattern (Fig. 1.25) can be explored in innovative resource wise city models:

- Business-as-usual: Assume that existing urban development trends will continue. Also takes into account existing plans for local area development.
- Alternative 1: Future growth occurs primarily as infill development in existing urbanized areas, and incrementally within a number of smaller greenfield development areas adjacent to existing urbanized areas. Explicitly aims to achieve live/work balance at the city level and to locate development in areas with suitable slopes, soils/lower natural hazard risks. The natural hazard risk platform model is described in the next section.

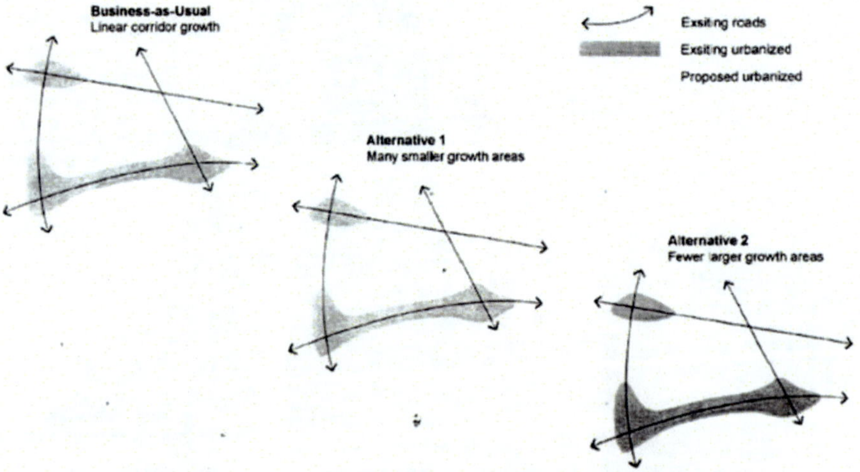

Fig. 1.25 Smart growth scenarios: business-as-usual, alternative1 and alternative 2

- Alternative 2: Future growth occurs primarily as infill development in existing urbanized areas, and incrementally within a few larger greenfield development areas. Explicitly aims to achieve live/ work balance and to locate development in areas with suitable slopes, soils/lower natural hazard risks.

A future image of resource wise cities is shown with various projects in Fig. 1.26.

(3) **Sectoral dynamic platform project models**

Each of the major urban platforms needs to be analyzed separately prior to coming together to form an interactive urban system.

The following platform models describe how to apply the requirements and functionality of the platform model for each sector of climate smart and wise cities with virtual diagrams. Each platform has the steps for sensoring with sensors and devices, communicating with 5G programming, storing with cloud computing, integrating with Big Data and making intelligent decisions with Big Data analytics and predictions with AI. The platforms below can be used to solve current global issues for various aspects of cities within a broad framework of integration platform theory described previously. The project models with Big Data can also activate the urban platform economy.

1. **The standard platform project model for land use**

Fig. 1.26 Future image of resource wise cities. *Source* Kim (2009)

This platform is for the urban pattern for sustainable urbanization. Urbanization alters a climate as the result of buildings occupying land, pavements covering streets, and smoke from factories pouring into the air. Even the seemingly most trivial of human activities can bring about vast environmental and climate changes. Urbanization is also a cause of disease pandemics across the world.

The land use plan is intended as an important means of reaching physical, economic, and social community goals. The plan, through its effects on public and private decisions and investments, can be a powerful influence on the growth rates, character, health (novel COVID-19 pandemic), quality, and pattern of the city's physical environment (Fig. 1.27).

The land use platform is for zoning to regulate the use of land. The zoning map delineates the cities and districts within which the provisions of the ordinance apply. The zoning plan expresses the basic policies which shape the community character, the general land use, circulation, and relationships among the variety of urban facilities.

The novel corona virus pandemic has cast a light on social and spatial disparities within cities like never before. A major concern is a city's economic geography: the interplay between its economic and physical setting.

Land use patterns- the spatial arrangement of the places in which we spend our daily lives- have a profound impact on the production of greenhouse gases and the corona virus pandemic disparities. We have to deal with the root causes of global warming and COVID-19 pandemic. We need an innovative ecological transition to land and resource use patterns together with conventional wisdom in a holistic way. As an example, the creation of efficient land use and other resource use patterns

Fig. 1.27 Land use plan at the district level

is an instrumental strategy for reducing three key sources of urban greenhouse gas emissions: road transportation, building energy use, embodied carbon, each of which is discussed later in this book by other authors.

Criteria for developing the land use platform: spatial multi-criteria approach

Spatial multi criteria should be developed for the following areas:

- Urban planning guidelines should be flexible to meet future possible needs (such as the smart grid layout of energy, water, blue-green-white corridor network, etc.)
- Fusion of form and function of cities should be required.
- Coordination or assembly of urban asset elements (land, water, vegetation, energy, etc.) should be encouraged for growth management and land use zoning (built-up area, natural protection area, industrial area, etc.) based on land use suitability and carrying capacity analysis.
- Existing natural ecosystems should be preserved and restored to enhance environmental integrity.
- The overall patterns of streets should be diverse for untact mobility.
- Edges of development should be defined with permanent open spaces or green belts for growth management and urbanization.
- Mixed-use development (e.g., day/night, commercial/public/industrial/residential, low/middle/high income, family/singles/seniors) should be encouraged and provided for social cohesion, energy efficient and resource wise planning.
- Social distancing should be taken into account for the urban pattern and sprawl.

A recent survey Harris Poll found that as a result of the pandemic, almost 40% of urban sites in the U.S. worked consider moving somewhere less densely populated (Meeting of the Minds 2020).

Components of the platforms

- Physical space: Physical world
  - Land use patterns (land use zoning), Growth patterns, Forms of urbanization and community, Mobility patterns, Natural patterns (green space, park, water, habitat, wetland, etc.), Public facility patterns, Building systems, Food systems (agriculture)
- Digital space: Digital world
  - AI and machine learning, IoT transformation, 3D printing, Digital twin, Cloud platform, Telecommunication, Machine-to-machine data transmission, Sensors and actuators, GIS/GPS, VR, AR, MR, AI controller, AI-based predictive algorithms and programming, Software, Land and resource use app.
- Big Data platform: Digital database
  - Sensors record information

Air quality
Temperature
Traffic
Energy consumption in buildings
Smog
Garbage production
Humidity.

- Planning and market factors

  Size of land ownership
  Present utility services.

- Ecological factors

  Vegetative cover
  Proximity to water surface features
  Air pollution
  Natural hazard
  Areas of high environmental sensitivity.

- Visual factors

  Cityscape
  Views.

- Geologic and soil factors

  Earthquake fault zone
  Landslides
  Natural slope stability
  Soil erosion potential.

- Hydrological factors

  Watershed
  Stream zones
  Flood-prone areas
  Surface water features.

- Infrastructure and services

Water supply
Energy supply
Sewage treatment
Waste treatment
Transportation
Health care
Education
Police station

Fire protection.

The above components of land use patterns can be used for ecological transition, climate transition, digital transition and post-COVID-19 recovery transition of conventional planning models to innovative ones (Table 1.5). Digital transition to land use patterns is to create untact public space and public facilities.

Virtual connection model for the land use patterns using land use big data

The virtual connection model for land use patterns is the relation between physical reality and digital image with Big Data for land use platform projects. It unites smart grid infrastructure, stormwater, circulation, housing and public open space as part of an integrated climate smart and wise city and plan. The following mappings are useful tools for integrating different components described in Table 1.5.

- Physical space mapping: Overlay of the physical world for land use transformation
- Digital space mapping: Overlay of the digital world for digital transformation
  - A digital space map shows the deployment of digital transformation technologies (AI, Big Data, Cloud, Blockchain, 5G, VR, AR, MR, logical and predictive algorithms, land use software, etc.) for land use transition, and land use software development.
- Virtual connection mapping: Assembly of physical and digital world using land use Big Data.

2. **The standard platform model for climate disasters and recovery**

A disaster is a sudden, calamitous event that seriously disrupts the functioning of a community or society and causes human, material, and economic or environmental losses that exceed the community's or society's ability to cope using its own resources. Though often caused by nature, disasters can have human origins. Disaster can be a function of hazard, vulnerability and capacity as shown in the equation.

$$\textbf{(VULNERABILITY} + \text{HAZARD)}/\textbf{CAPACITY} = \textbf{DISASTER}$$

In this equation, a disaster occurs when a hazard impacts vulnerable people. The combination of hazards, vulnerability and inability to reduce the potential negative consequences of risk results in disaster.

When disaster strikes, government and citizens are at risk, countries have protocols to respond, including shelters to protect displaced and affected populations and emergency plans for geographically vulnerable locations. Leaders and citizens know they must always be ready to respond and begin anew the process of recovery. This year, however, countries need a new plan because the economic recession and the COVID-19 pandemic are part of the equation (World Bank 2020).

In the field of disaster risk management, when a new risk appears in an already risky situation, we call it a compound risk. The COVID-19 pandemic means communities need to prepare even more thoroughly for disasters, as resources are limited or

Table 1.5 Components of land use patterns from different transition perspectives: an illustration

Components	Ecological transition	Climate transition	Digital transition	Post-COVID-19 recovery transition
Landuse zoning	• Densely populated neighborhoods	• Mixed land use	• Digitalized mixed development	• Sparsely, less densely populated neighborhoods
Growth pattern			• "Platform" value: new capability added through software	
Form of urbanization and community	• Car-free access to jobs	• Compact infill development		
Mobility patterns	• Coupled with strong pedestrian, bicycle and transit access	• Compact, walkable, and transit-rich communities	• Digitalized bus information terminal • Driverless autonomous cars	• Expansive of bike routes
Natural patterns	• Deep urban forest system	• Deep urban carbon forest system	• Real-time monitoring system	• Expansion of greenways • Recreation and amenity system in close proximity
Public facility patterns		• Car-free access to public facilities	• Expansion of untact services	
Building systems			• Smart home IoT	• Home near more open space, more bike paths
Infrastructure	• Less spread out, compact development which requires significantly less infrastructure			
Food systems	• Encouragement of farming near urban centers			

have been reallocated to respond to the health crisis. In simple terms, the economic and health impacts of COVID-19 are making already –disaster-prone countries more vulnerable than ever before.

Temperature and weather irregularity caused by climate change increases the possibility of flooding and droughts and impact the quality and availability of global freshwater supply. As the water supply is affected, farmers will need to irrigate crops.

The natural hazard might lead to disasters. A natural disaster is a major adverse event (damage and loss) resulting from natural processes of the Earth (hazards); examples are floods, hurricanes, tornadoes, volcanic eruptions, earthquakes, tsunamis, storms and other geologic processes (landslides, rock cracks, etc.).

Disasters can be categorized into geologic disasters, hydrologic disasters (floods, tsunami and limnic eruptions) and meteorological disasters. Risk assessment is a process to determine the probability of losses by analyzing potential hazards and evaluating existing conditions of vulnerability that could pose a threat or harm to property, people, livelihoods and the environment on which they depend.

Floods are considered one of the most frequently occurring natural hazards worldwide and are occurring more frequently in recent decades. In this section, indicator-based approach to urban flooding risk assessment is used as an example of natural hazard and disaster platform project models.

Criteria for developing the climate disaster platform: indicator-based approach

- Natural rainwater drainage and aquifer recharge patterns should be restored.
- Roadways, parking lots and public parks should incorporate retention pond marshes to naturally treat runoff.
- Water features such as ponds and underground tanks should be used to store water for fire protection and flood disaster emergencies.
- Creek valley, stream zones, riparian buffer, wetlands and flood-prone areas should be preserved and restored.
- The impacts of a natural hazard on human societies and ecosystems should be assessed by future climate scenarios with historical patterns and GIS/GPS, Big Data and AI-based prediction. The impacts can be compared with measurement-based hazard monitoring. Hydrometeorological hazards (HMHs) cataloging is an effective tool to evaluate their associated risks and plan appropriate remediation strategies and actions.
- Communities should be empowered to communicate on hazard impacts, remediation strategies and actions with the use of creative, visual and digital methods
- Investment in disaster preparedness should be ramped up. They can set humanity on a safe course for the next decades and beyond.

Components of the platforms

- Physical space: Physical world

- Spatial multi-criteria evaluation (MCE), Inventory map, Spatial overlay of hazard footprints and elements-at-risk location, Indicators, Flooding risk index map, Pairwise comparison and rank ordering weighting methods

- Digital space: Digital world

Table 1.6 illustrates digital technologies and equipments for climate smart project activities.

- Big Data platform: Digital database
  - Watershed information
  - Floodplain information
  - GIS mapping information
  - Hydrological factors
  - Sea level rise information
  - Flood disaster information
  - Mountain disaster information
  - Landsliding information
  - Coastal flooding and storm surge information
  - Disaster recovery information, etc.

Virtual connection model for the flood risk assessment using big data

- Physical space mapping: Overlay of physical world

If the proposed development site lies within a flood zone, a flood risk assessment is required. We have to look over the lifetime of the proposed development. The flooding risk assessment involves nine steps. Depending on the current situation, the following steps can be simplified to some extent.

Step 1: Delineation of the boundary of landscape district and place unit: Location map (1).

Bioregional mapping and approach analyze local landscape resources by setting a regional feature-based place unit as a standard unit and define natural and cultural resources forming the unique features of each place unit.

Step 2: Delineation of the boundary of catchment area: Location map (2).

A catchment area is a hydrological unit. Each drop of precipitation that falls into a catchment area eventually ends up in the same river going to the sea if it doesn't evaporate. However, it can take a very long time. Catchment areas are separated from each other by watersheds. A watershed is a natural division line along with the highest points in an area. Catchments are divided into sub catchments, also along the lines of elevation (Fig. 1.28).

Step 3: Identification of factors (indicators) to be included in the flood risk assessment exercise.

For example, potential flooding hazards, soil permeability, and distribution of people, property and ecosystems should be identified in any given catchment area.

Table 1.6 List and costing of equipments for project activities: an example

Project activity			Technology	Equipment					Remarks
Component	Subcomponent			Item	Number	Unit Price (US$)	Specification	Sub-total	
1. Slum Upgrading Settlement and Land Development	Testbed (New Village Development)—Affordable Housing	Land use	• GIS mapping	• Drone image embedded with a hyperspectral sensor	1set 1	50,000 8000	Drone + Hyperspectral camera ArcGIS (concurrent license)	58,000	
		Building	• Green building technology • Renewable energy—integrated buildings: Building-integrated photovoltaics	• Solar roof panels for heating, cooling and decentralized electricity • Municipal Wifi • Home IoT & Digital locker • Solar garden lighting • Digital meters for gas, water, electricity, etc		21.85/ft2 −150 (month) 150/set 40/set (gas) + 900/set(water)			
		Energy	• Community smart energy microgrids • Energy harvesting technology	• Smart solar collector field • Solar panel • Wind turbine • ESS (secondary cell battery)	1set	ESS: 100,000			
		Transport	• Integrated smart transport grid systems	• Smart bike count • Solar light					

(continued)

Table 1.6 (continued)

Project activity			Technology	Equipment					Remarks
Component	Subcomponent			Item	Number	Unit Price (US$)	Specification	Sub-total	
		Water	• Rainwater harvesting technology • Smart water grids: well (groundwater) • Water recycling technology	• Heat pump • Rainwater cistern • Smart sensors for smart water grid (1) Rainwater supply meters (2) Leakage monitoring meters (3) Billing meters • Solar water heater • Smart water controller					
		Waste	• Smart waste treatment system • Separate collection system	• Smart trash bin using smart green card • Collection bag • Collection vehicle					
		Green area network	• Ecological engineering technology • Smart community garden • Green–blue grid technology						

(continued)

Table 1.6 (continued)

Project activity			Technology	Equipment					Remarks
Component	Subcomponent			Item	Number	Unit Price (US$)	Specification	Sub-total	
2. Digital transformtion			• Microsoft Azure for blockchain technology • Smart digital technology • Big Data center for climate change, settlements and mangrove swamp eco-park • Climate Big Data platform • Digital climate change and training center • Database server and storage • Cloud hosting • Software integrated management system	• IoT cloud computing equipment					
				• Big Data and analytics equipment					
				• DLT/Blockchain equipment					
				• AI equipment					
				• Software	1set	500,000			
Total amount									

AI-based predictive algorithms, AI programming and software are important parts of flood risk management practice
Application exclusive for disaster management: Remotely monitoring environment information analyzed with a flood disaster management system allows producing a warning alarm in real time when a disaster occurs

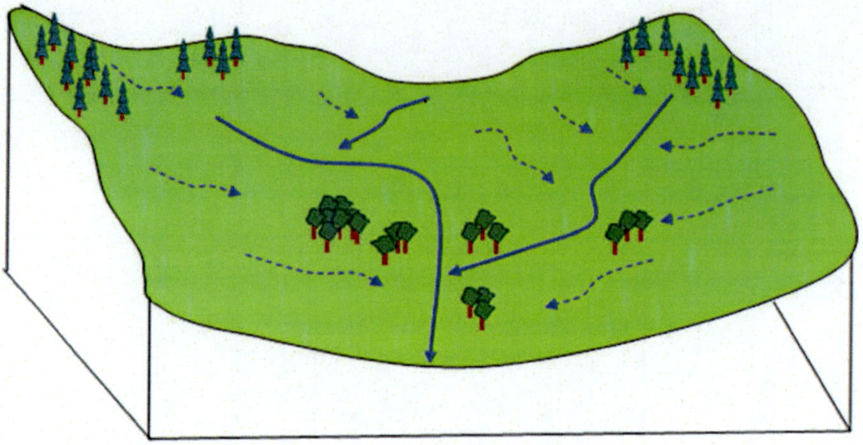

Fig. 1.28 A theoretical catchment area with flow paths of precipitation

Step 4: Preparation of an "inventory map" for each factor showing how it varies over the study area.

Data for each of these indicators should be collected at a particular spatial level.

Step 5: Creation of composite maps by overlaying two or more inventory maps.

The map overlay technique is a procedure for synthesizing the spatial data used in flooding risk assessment. It includes a spatial overlay of hazard footprints and elements-at-risk location.

Step 6: Analysis of the composite maps to make inferences relevant to flooding risk assessment.

The analysis could involve a synthesis of flooding hazard maps (historical patterns and future scenarios) and inventory maps for five factors as examples: slope, slope stability, stream zones and allocation maps of people, property and ecosystems.

A. Analyzing and predicting potential hazards: Hydrometeorological hazard (HMH) mapping.

   - Historical patterns: past episodes and events (past flooding levels)
   - Hazards based on future climate scenarios (future flooding level scenarios)
   - Use of AI programming and algorithm for predicting potential hazards

B. Analyzing vulnerability of people, property, livelihoods, and ecosystems.

Allocation maps of people, property, livelihoods, and ecosystems are superimposed on the hazard maps,

Step 7: Evaluating risk (probability of damage and losses) and risk mapping: The use of spatial multi-criteria evaluation.

Potential hazards and their associated risks can be evaluated by using HMHs cataloging (developed mathematical model exercise), and land use suitability and risk probability mapping below.

Flooding risk index mapping includes standardization, pair-wise comparison and rank ordering weighting method.

The selected indicators or factors are standardized (e.g., by reclassifying them between 0 and 1), weighted internally within sub-objectives and then the various sub-objectives are also weighted amongst themselves. If the scores for flooding potential, and soil type and infiltration are approximately equal for all grid segments, these factors can be dropped from the analysis. The indicators are ranked and weighted before any addition is performed, using direct, pair-wise comparison and rank ordering weighting methods. Then rankings of nominal types for each of the selected factors and weights are combined to obtain the final flooding risk index map. In principle, at least, the weights can be considered to transform the scores associated with the nominal types for each factor into a common unit of measure.

Although the individual indicators normally consist of quantitative data (e.g., population statistics), the resulting vulnerability, hazard and risk results are scaled between 0 and 1. These relative data allow us to compare the indicators for the various units. These methods can be carried out at different levels, ranging from local communities, cities to countries.

Step 8: Planning of appropriate remediation strategies and actions: A holistic framework for ecological restoration by people and business for next generations.

Flooding and its associated damages have been with us throughout history. Deciding which flood control alternative to be used in any specific watershed is a very important decision within the larger framework of floodplain management. Any general flood control approach might include structural methods, nonstructural methods, or a combination of both. Examples of structural methods include channel modifications (channelization), reservoirs, diversions, and levees or dikes. Those methods that seek to modify the damage susceptibility of developed regions within a floodplain are classified as nonstructural. Nonstructural measures include floodplain land acquisition (also commonly referred to as voluntary relocation of or property buyouts), land use restrictions (zoning), floodproofing, flood warning systems, drainage maintenance programs, and public awareness or information programs (Bedient et al. 2013).

Step 9: Monitoring with ICT applications and periodic evaluation.

The list of sensors and actuators for monitoring stations is shown in Table 1.6. The calculated (or predicted modeling) risks can be compared with observed (or measured) impacts through monitoring over time. The results are subject to periodic evaluation.

- Virtual connection mapping: Assembly of the physical and digital world using flooding risk management with Big Data

Physical and digital world described in the previous section can be assembled as shown in Fig. 1.29.

The technology of IoT helps to monitor the safety and security of the people and environment so that the local government can give responsive protection from climate disasters such as floods. The use of AI programming and algorithm should further be developed for natural hazard and disaster risk management.

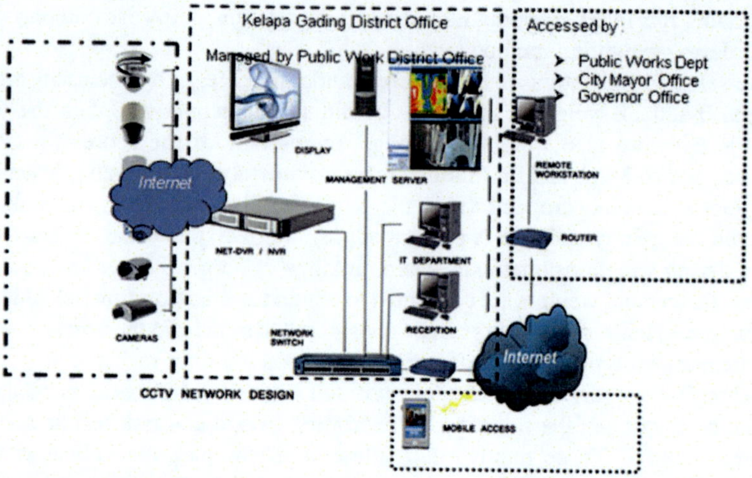

Fig. 1.29 Conceptual IoT Network for cameras and sensors at the district level for virtual digital connection. *Source* Kim (2018)

- Disaster preparedness and recovery measure mapping

Societies have been disaster-stricken, with many people losing their properties and lives. Flood disaster recovery systems are under pressure, the global economy is in decline and vulnerable groups are seriously affected. To help the world navigate toward a substantial recovery, flood risk experts have to monitor the situation and share policy recommendations and recovery measures.

The information on the virtual connection of the flooding risk management together with disaster recovery measures can be shown in the plan layout. Figure 1.30 contains physical and digital disaster recovery measures in a graphic form in terms of structural and nonstructural measures.

3. **The smart platform model for COVID-19 pandemic**

The ravage of COVID-19 is extensive. COVID-19 pandemic presents the opportunity to accelerate climate plans.

Why it matters

Connecting climate change with human health by using the Green New Deal and digital technologies is important to keep COVID-19 under control and implement the COVID-19 green recovery plan.

Cities are bearing the brunt of all the overlapping global crises. Major population centers are disproportionately vulnerable to climate shocks, such as floods, storm surges, and other extreme weather events. They're home to more COVID-19 cases than anywhere else. Businesses are shuttering and urban livelihoods are at risk.

And yet, it would be wrong to think of cities as helpless victims of these cases. In fact, they are a big part of how we solve them. Over half of the globe lives in cities

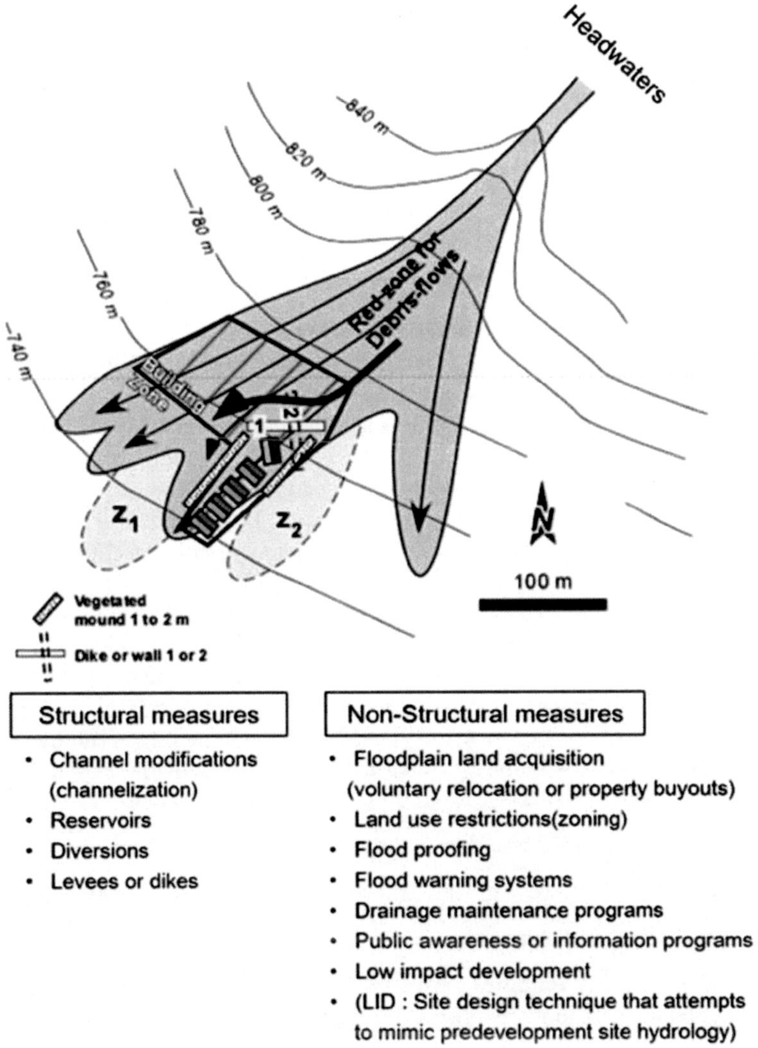

Fig. 1.30 Flood disaster recovery measures

generating 80% of its total economic output. With an outsized share of the population and economy, cities are where our sustainable future can and must be won.

To understand how we can deal with the impacts of COVID-19 with the basic truth, we can consider the strategies used to combat the corona virus pandemic. In early 2020, the term "flattening the curve" exploded across newspapers and websites across the globe. Now that we have learned how to flatten the curve of the infection, we could apply the same principle to our cities. Beyond healthcare, all infrastructure of cities-from roads to energy, water, buildings, and green–blue-white grids- could help the curve-flattening in cities.

It is expected that an integrated smart platform suggested in this section will help shift the trajectory toward climate-smart-resilient-resource wise-healthy development and lay the groundwork for a sustainable recovery from COVID-19.

The author also hopes the platform will demonstrate that a coordinated, holistic approach is the way forward for cities and the planet: sharing living space vs separating living space.

Components

We require a profound transformation of all parts of society to secure a sustainable future. In a rapidly changing and digitalizing world, we must mitigate the social consequences of the urban transformation through green and resource transition, climate transition, COVID-19 transition and digital transition. This platform should describe how to integrate COVID-19 recovery plan in planning urban pattern and density, buildings, workplace, mobility, outdoor and indoor public spaces, public facilities, all infrastructure and technologies of cities. Things before COVID-19 (pre-pandemic) should be analyzed to deal with pandemic responses. Cities that try to return to the way things were before COVID-19 will struggle. They need to reinvent because volatility is inevitable, and urban pattern and density, workplace, mobility, outdoor and indoor public spaces, public facilities and all infrastructure including road, energy, water and green–blue-white grids, and cutting-edge smart disruptive technologies need to be ready to rapidly respond to the next disruption. In particular, the role of the workplace in a post-COVID-19 world is more important than ever.

(https://www.steelcase.com/research/articles/topics/postcovidworkplace/designing-the–post-covid-workplace/).

Criteria for phased-in design ideas for the post-COVID-19 recovery

The post-COVID-19 recovery process can be divided into three waves as follows:

Wave 1: responding in the now

The immediate priority for cities, businesses, universities and health organizations is to bring people back to work as quickly and safely as possible.

- Design ideas:

  A. Urban pattern and density control

  - Shift in urban pattern and density (for example, densely populated, built-up areas, non-urban surroundings, etc.)
  - Dispersed layouts of space could help physical distancing.

  B. Workplace

  - Implementation of sensible strategies including practicing physical distancing, like moving desks and removing chairs, adding barriers,

enhancing cleaning and safety measures, as well as supporting those who are working from home longer
- Office automation
- Connection of untact workplaces and other communal spaces.

C. Mobility

- Enforcement of safe traffic management with Big Data and mobility behavior index
- Expansion of bicycle lanes to keep social distance
- Subsidy for companies acquiring electric and hydrogen fuel cell vehicles
- Increase in bicycle lanes
- Separation of pedestrian ways
- Widening of sidewalk of streets
- Provision of more greenways
- Use of drone delivery: Converting parking lots into drone delivery spots
- Use of driverless autonomous cars.

D. Park and green space

- Distancing of benches, playgrounds, sports facilities, etc.
- Decentralization of hierarchical park systems for untact activities
- Remeandering and ecological restoration of small and medium-sized streams for connecting people with nature
- Creation of green–blue (water)-white (wind) corridor network to enhance airflow, exchange air, recycle air, and reduce urban heat island (UHI) effects in densely populated, built-up areas as the prime driver of risks.

E. Buildings

- Building retrofit to mitigate the risk
- Convertible, adaptable and flexible indoor space
- Promotion of air passage network or corridor between buildings.

F. Public facilities

- Localization of consumption
- Dispersion of public facilities to keep distance
- Remote access to contents for library users and other public facility users, and how to balance life and working remotely.

G. Smart tech for the good

- Digitalization of untact public spaces
- Increase in internet connectivity under lockdown for the city as a whole
- Use of cutting-edge smart technologies for all urban grids'(for example, energy grid, water grid, green–blue–white grid, etc.)
- Development of an integrated Big Data for compound disaster including disease disaster and natural disaster

– Use of digital innovation and connectivity for monitoring COVID-19 cases at different levels of space (sensors and actuators, IoT, AI, etc.)

Some examples of design ideas for wave 1 are shown in Fig. 1.31 at the strategic level.

Wave 2: planning near-term

As the first wave adjust, cities can start to reconfigure the urban elements that citizens will feel more natural and allow more people to return to normal. Scientists will learn more about how this particular corona virus behaves which can help guide cities' plans for the near term.

- Design ideas:

  A. Urban pattern and density control

     – New layouts in the space could help physical distancing feel more comfortable.

  B. Mobility

     – Travel is likely to be limited for some time which means the workplace needs to provide more and better options for videoconferencing.

  C. Workplace

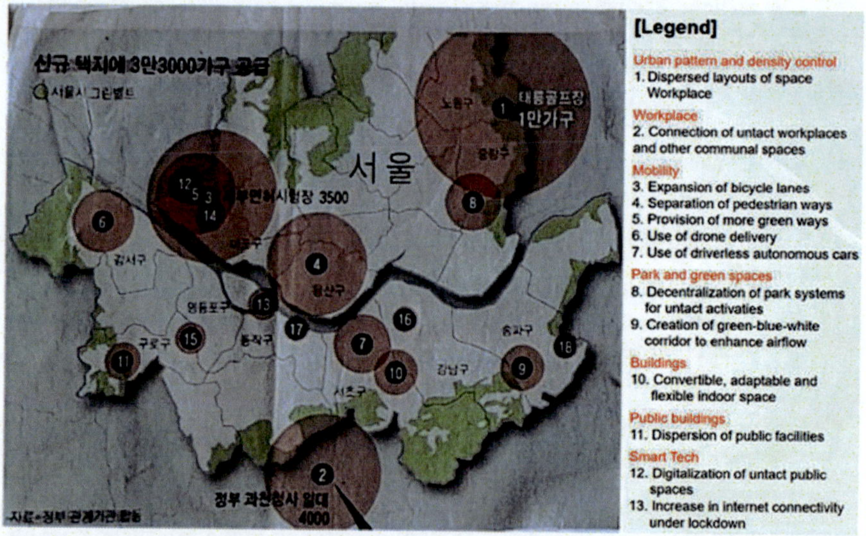

Fig. 1.31 Urban Space-Urban elements-health organization for the post-Covid-19 recovery: managing COVID-19 with particular emphasis on linkages between COVID-19, climate change, Green New Deal and urban elements (illustrations only). *Source* Kim (2020) Untact climate smart, resilient and resource wise city for the post-Covid-19

- Some organizations will continue to support working from home as a strategy to reduce density in the workplace, where others may redeploy real estate in new ways.

Wave 3: looking toward the far

As we reinvent the city, science-based solutions will integrate with emerging technology to deliver citizen safety and an even higher level of citizen wellbeing.

- Design ideas:

In wave 3, people will want to feel a renewed sense of community and have a whole new appreciation for being together and for moving back to communal spaces. Design of layers of the virtual and physical world using Big Data will bring people together in new ways to create a sense of belonging.

A corona infection platform model: a step-by-step toolkit

With the gradual easing of lockdown restrictions in many countries, people will slowly begin to move back into communal spaces. Everyone will need to look at their environment and make critical judgments. We need to develop a tool that extends the capabilities of urban planners to help city administrators and business owners critically assess the risks based on factors described in this section so that they can design the safest and best strategy for their needs.

The followings are steps for a developing corona infection platform model:

- Step 1. Describe a city profile;

Existing layout of the urban pattern (circular, grid and radial), density (densely populated built-up areas, non-urban surroundings), demographic features, income level, seating areas, circulation routes, air-conditioning zones, etc.

- Step 2. Identify high risk areas or clusters of COVID-19 cases/100 K (100,000);
  - High risk areas
  - Medium risk areas
  - Low risk areas

- Step 3. Select corona virus hotspots with the ability to simulate demography, mobility, airflow and circulation;
  - Location of road, people, workplace, buildings, public space and facilities, etc. using GIS and GPS
  - Assessment of domino effects of COVID-19 on people and the urban pattern, etc.

Highlighting the role of demography and appreciation of how population studies can shape efforts to address and understand this public health crisis, starting with the trinity of fertility, mortality and migration and their implications.

- Step 4. Predict severe causes of COVID-19;
  - Use of the decision tree and quantitative models for predicting severe causes of COVID-19 with help from Big Data, AI and machine learning engineers/experts.
- Step 5. Prepare phased-in recovery plan;
  - Intelligent spatial layouts for social distancing generated based on different scenarios
  - New layouts of existing urban elements using AI
  - 3D layout models and visuals: Assembly of the physical and digital world.
- Step 6. Develop implementation tools;
  - Adoption of green deal projects, digital deal projects and COVID-19 stimulus measure package
  - Market analysis (coronomics analysis, twin-demic analysis, carbon market analysis, etc.)
  - Use of new funds for a new urban future (for example, City Climate Finance Gap Fund for finance ready projects in cities)
  - Enforcement tools.
- Step 7.
  - Monitoring cases with digital innovation and developing Big Data
  - Actual or observed cases through monitoring can be compared with predicted ones.
  - Use of the fifth-generation (5G) network, robots, IoT and AI to develop the quality Big Data which impact decision making made by city governments, businesses and householders.

Green project for corona virus stimulus measures

The green projects in cities should navigate the post-COVID-19 stimulus package. Then a question is how best to move fast on the right projects for the COVID-19 in terms of pilot projects.

Green projects for corona virus stimulus measures are identified from different content perspectives in Table 1.7.

The details of the Corona virus stimulus package are shown in Fig. 1.32 in the context of the neighborhood.

Untact climate smart and wise city for the post-COVID-19 recovery needs a systemic change in its approach to urban development and health, in order to improve the wellbeing of the citizens. The implementation of SDGs and the Green New Deal policy would be our best option to recover better and to overcome similar crises in the future.

Table 1.7 Green projects for corona virus stimulus measure package

Category	Strategy	Project Content
1. Wise resource Deal: Spatial planning practices	1.1. Sustainable land and zoning	• Low-carbon urban smart zone • Smart zoning for climate change adaptation • Public space zoning for post-COVID-19 recovery
	1.2. Growth management and urban form	• Energy-saving urban structure • Compact city
	1.3. Eco-cycle and urban metabolism	• Connectivity of green infrastructure: Ecological corridor • Protection and restoration of natural ecosystems
2. Green deal: climate, resource, environment, green space, energy and transport nexus	2.1. Creation of smart water system	• LID • Smart water grid
	2.2. Development of smart transport system	• Smart, green and integrated transport measures • Intelligent transport system(ITS) • Driverless autonomous cars
	2.3. Creation of smart energy system	• Smart energy grid platform including ESS • Clean energy: Urban 100% renewable energy grid system
	2.4. Enhancement of climate resiliency	• Hazard mitigation practices to reduce the physical impact of disasters • Disaster information platform and early warning system • Mitigation of heat island effects
	2.5. Food security with local food production	• AI Urban farming: Ag Tec with AI • Urban food systems • Untact food market
	2.6. Green job	• Development of green industry
3. Digital Deal: Digital urban development innovation	3.1. Digital platform development	
	(1) Sensoring	• Development of network • WIFI Installment of WIFI • Crack gauge to monitor movement across cracks on different rock surfaces

(continued)

Table 1.7 (continued)

Category	Strategy	Project Content
	(2) Communication	• 5G projects: 5G communication equipment and content projects in 5G cities • IoT • Remote working
	(3) Storing	• DLT/Blockchain platform • City data store
	(4) Integrating	• Big data • 3D digital maps
	(5) Prediction	• Development of AI program and machine learning platform for predictions causes of COVID-19
	(6) Intelligent decision making	• Smart controller • Optimization
	3.2. Digital SOC/Infrastructure	• Intelligent street lighting • Hydrogen networks • Smart parking and locker • Automated early warning system • Smart Home IoT • Combined heat and power plant(CHP)
	3.3. Digital industry	• Civic crowdfunding
	3.4. Digital job	• Vocation training centers in the fields of sensors, AI, IoT, Cloud computing, Big data, 5G, DLT, Block technology, etc • Online business (online shopping, working from home space, remote medical examination and treatment, etc.) • Digital urban research and education in the wake of COVID-19
	3.5. Digital event	• Virtual community cultural and physical activities – Sports – Concert • Virtual conference and forums

(continued)

Table 1.7 (continued)

Category	Strategy	Project Content
4. Post-COVID-19 deal: Untact contents in the Coronomics Era	4.1. Health care network	• Digitally connected health care centers • Infections disease disaster platform
	4.2. Empowering communication	• Virtual and digital methods for stakeholder meetings • Provision of public space (parks and bike lanes, etc.) and public facilities (library and community centers, etc.) for social distancing
	4.3. Predicting causes of COVID-19	• Use of big data, AI and machine learning • Digitalized early warning system
	4.4. Legal and basis	• New health standards and guidelines • Smart healthy city ordinances and mandates

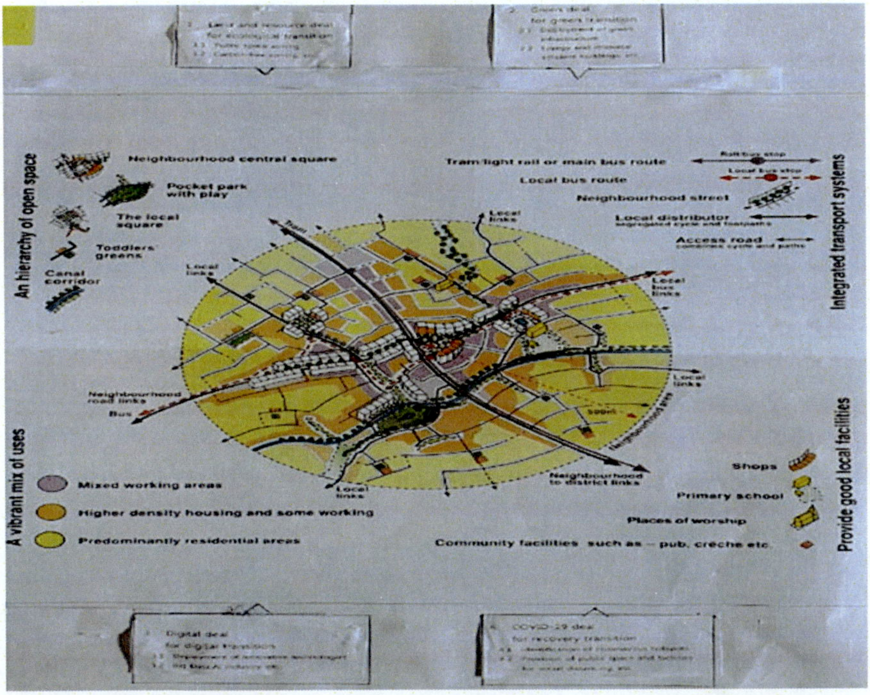

Fig. 1.32 The key components of climate smart and wise cities at the neighborhood level: Innovating tomorrow in the coronomics era

Local leaders across the globe must be engaged in the local response to the COVID-19 pandemic and the Green New Deal movement toward climate smart and wise cities in the COVID-19 pandemic era and beyond.

1.4 Planning Cases: Cases of Climate Smart and Wise City Planning Models for SDGs, COVID-19 Recovery and Green New Deal

The next generation of modern urban professionals will face unprecedented challenges in dealing with the evaluation in existing urban planning models as well as the development of new and innovative ones.

This section explores some world-renowned cases that demonstrate climate smart and wise city planning model applications for land and resource use transition, ecological transition, digital transition and post-corona-19 recovery transition, which are described in this chapter. The cases include best practices and case reports in the US, Europe, Australia, Northeast and Africa, and the Pacific Island countries at a different level of scale and achievement.

The COVID-19 pandemic and economic crisis are likely to push tens of millions of people into poverty and yet this is only a preview of how the climate crisis will threaten human wellbeing. We need to share insights from recent global cases on how the world can respond to the pandemic while advancing the objectives of the Paris Agreement and Sustainable Development Goals.

Evolution of new smart methods and hyper-modern cutting-edge technologies enables to conduct numerous well-known projects that demonstrate various types of integration platform models for climate smart and wise cities.

1.4.1 Smart Platform Project Models for Climate Disaster and Recovery: The Cases of the Thames River in London, England and New York City, USA

The Thames River in England is a good example of a coastal river system that faces both flooding problems and water quality issues.

The Thames River, one of the most famous in all of Europe, flows 205 miles from the Thames Head, in the Cotswold Hills, through southern England before emptying into the North Sea near London. It is the longest river in England and drains approximately 5,500 square miles. The last 65 miles of the river are an estuary where the saltwater and freshwater barrier fluctuate daily. In London, the incoming tide can raise the river up to 7 m during the spring, reversing its flow through the estuary (Encyclopedia Britannica 2011).

The flood risk assessment in the Thames River is part of "London Regional Flood Risk Appraisal 2018" (Environment Agency 2018). The London Flood Risk Appraisal (RFRA) provides an overview of all sources of flooding in London and addresses its probability and consequences. Currently, 6% of London is at high risk (1 in a 30 year event) of tidal, river, or surface water flooding and 11% at medium risk (1 in 100 year event) (Fig. 1.33).

This appraisal report covers introduction, an overview of flood risk to London, spatial implications of flood risk, conclusions and looking ahead, including appendices. The appendices include a list of monitoring recommendations, abbreviations and flood risk maps.

As sea levels rise, coastal flooding associated with storms will very likely increase in intensity, frequency, and duration. Any increase in the frequency or intensity of storms themselves would result in even more frequent flood occurrences relative to 1-in-10 and 1-in-100 year coastal flood events. Storms of a given magnitude will result in higher coastal flood elevations that inundate a larger geographic area. The heavy precipitation associated with these storm systems can cause increased surface runoff, particularly, in urban areas dominated by impervious surfaces, and river surges that can intensify the effects of storm surges coming in from the sea.

New York City Case Study on Climate Change Adaptation is a world-renowned project that demonstrates a coastal flooding and storm surge model application at the highest level of achievement (Fig. 1.34).

Fig. 1.33 Flood risk map from multiple sources in Greater London 2017. *Source* Environment agency, London regional flood risk appraisal 2018, appendix 3—flood risk maps, Map1

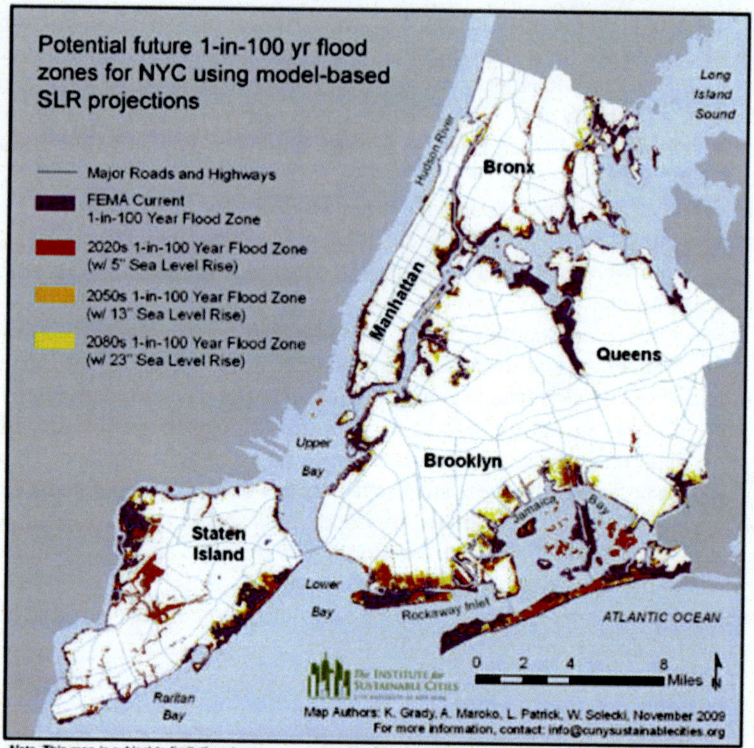

Note. This map is subject to limitations in accuracy as a result of the quantitative models, datasets, and methodology used in its development. The map and data should not be used to assess actual coastal hazards, insurance requirements or property values or be used in lieu of Flood Insurance Rate Maps issued by FEMA.

Interpretation. The floodplains delineated above in no way represent precise flood boundaries but rather illustrate three distinct areas of interest: 1) areas currently subject to the 1-in-100 year flood that will continue to be subject to flooding in the future, 2) areas that do not currently flood but are expected to potentially experience the 1-in-100 year flood in the future, and 3) areas that do not currently flood and are unlikely to do so in the timeline of the climate projection scenarios used in this research (end of the current century).

Fig. 1.34 Areas potentially at risk to the 1-in-100 year flood in NYC due to sea level rise projections derived from global climate models. *Source*: (2010) New York City panel on climate change 2010 report, climate change adaptation in New York City: building a risk management response, annals of the New York Academy of sciences, p. 316

1.4.2 Microgrid and Digital Technology Model: The Cases of Sub-Saharan Africa and Monash, Australia

Microgrid and digital technology models were published by the Institute for Advanced Sustainability Studies (IASS) in 2019, and Monash University and Indra in 2019.

In 2019, the Institute for Advanced Sustainability Studies (IASS) conducted a study on the nexus of mini-grids and digital technologies with particular emphasis on potentials, challenges and options for sustainable energy access in Sub-Saharan Africa. This study aims to shed light on two issues:

- The different use cases of digital technologies across the value chain of mini-grids, and
- How they may assist mini-grids in meeting the requirements of sustainable development.

One of the main features of the study is that the implementation of mini-grids on renewable energies in Sub-Saharan Africa provides a potential application area for digital technologies. This study distinguishes two levels of applications which include different sub-categories:

Technical functionalities and system balancing

- Generation and storage
  - Forecasting algorithms
  - Intelligent battery management
  - Optimal hybrid operation
  - ESS.
- Distribution and control:
  - Remote monitoring
  - Supervisory control and data acquisition (SCADA)/IoT.
- Demand-side management
  - Smart meters
  - Demand limiting devices
  - Smart appliances
  - Demand forecast.

Application on the level of the mini-grid value chain such as

- Finance
  - Crowdfunding
  - Blockchain-based climate action tracing and accounting
  - Solar coins.
- Planning and design
  - Geospatial portfolio planning
  - Drone imaging
  - Demand estimation with AI and GIS
  - Design software.
- Operation and maintenance
  - Smart maintenance

- Cloud-based management platforms.

- Customer management
  - Mobile payments
  - Smart contracts
  - P2P electricity sharing
  - SMS/smartphone.

- Productive use of energy from mini-grids
  - Internet cafes
  - Telecom equipment
  - Telephone charging
  - E-learning services.

Figure 1.35 shows application areas of digital technologies in mini-grids described above.

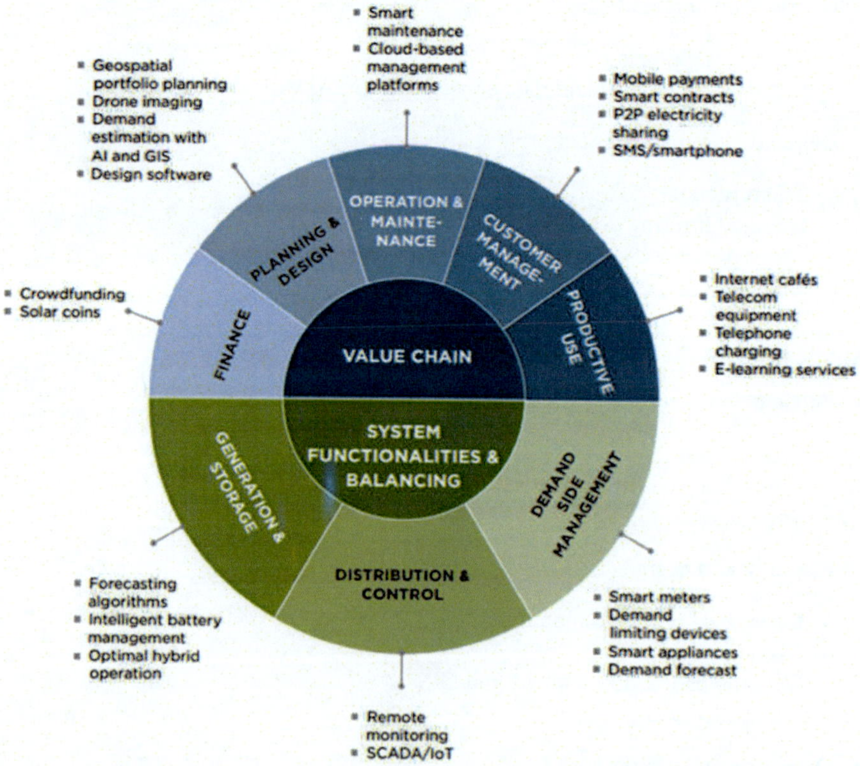

Fig. 1.35 Application areas of digital technologies in mini-grids. *Source* Fritzsche et al. (2019) IASS study

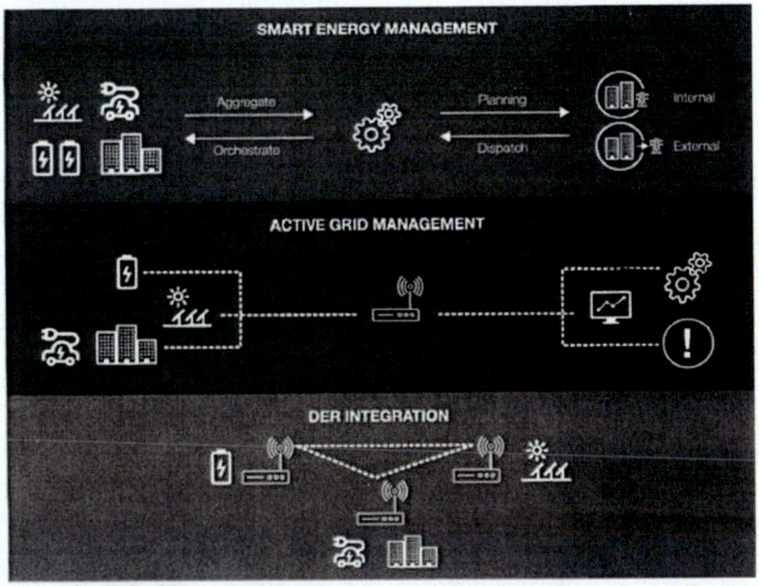

Fig. 1.36 Smart energy framework layers

The smart energy city model developed by Monash University and Indira involves the use of microgrid with numerous digital technologies. A Smart Energy Framework has been developed as a design and implementation method for the introduction of Smart Energy Management to precinct scale microgrids. The three framework layers (distributed energy resources (DER) Integration, Active Grid Management and Smart Energy Management) (Fig. 1.36) have been applied in the Smart Energy City project to enable Monash to manage and orchestrate the energy generation, storage and two-way power flow in the microgrid, including the interface with the broader electricity network, both from an energy and power quality perspective.

The Smart Energy City project aims to demonstrate how a grid-tied 100% renewable-powered net zero city could operate reliably, and outline the value it could provide to both customers and the broader electricity network. Figure 1.37 illustrates core elements of the Monash microgrid model. For the Smart Energy City project, each DER asset is connected to the Active Grid Management layer via a smart gateway node to collect, monitor, analyze and manage the data. Node #1, Indra' Edge Platform and iSPEED, and Indra's Real-Time Databus make the DERs visible and controllable by the Active Grid Management and Smart Energy Management layers. Figure 1.38 shows Indra' onesait platform for DER integration and active grid management with the use of IoT real-time data and setpoints; operation instruction (Fig. 1.37).

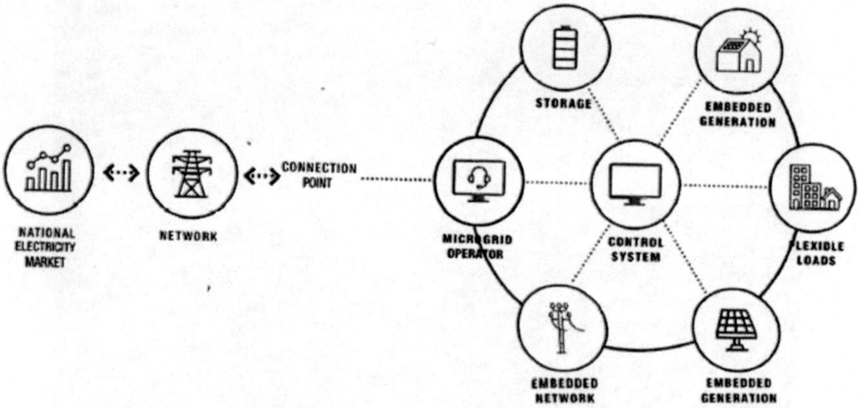

Fig. 1.37 Core elements of a microgrid

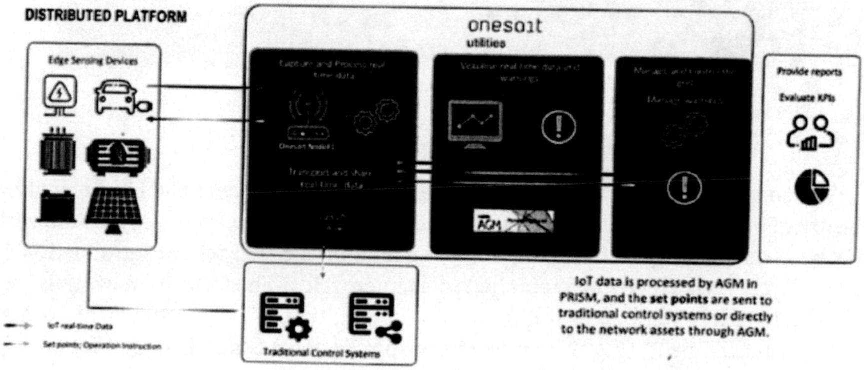

Fig. 1.38 Indra's onesait platform for DER integration and active grid management

1.4.3 Cases of Big Data in Practice

Cities around the world are facing a triple challenge of a growing population, climate change and flooding of information and knowledge. In tackling these challenges, big data and analytics are being put to use.

It can be said that we live in the age of big data. Big data refers to collections of data that are so enormous in size, so varied in content, and so fast to accumulate that they are difficult to store and analyze using traditional approaches. The three "Vs" are the defining features for big data:

- Volume. Data collections can take up petabytes of storage and are continually growing.
- Velocity. Many data sources change and grow at very fast speeds.

- Variety. Relational databases are very efficient for structured information stored in tables, but businesses can benefit from analyzing semi-structured and unstructured data as well (Wallace 2015).

Climate scientists and climate planners have been gathering a great deal of data for a long time, but analytics technology's catching up is comparatively recent.

Big data and the Internet of Things (IoT) are going to work with other software and hardware to lead the vision of climate smart and wise city to fruition (http://www.kdnuggests.com/2015/10/big-data-smart-cities.html).

The idea of "smart cities (http://www.forbes.com/sites/bernardmarr/2015/05/19/how-big-data-and-the-internet-of-things-create-smarter-cities/)" is central to the concept of the Internet of Things – the idea that everyday objects and tools are becoming increasingly connected, interactive, and intelligent, and capable of communicating with each other independently of humans. Many of the ideas put forward by smart-city pioneers are grounded in climate awareness, such as reducing carbon dioxide emissions and energy waste across urban areas. Smart metering (http://data-informed.com/utilities-executives-cite-need-for-analytics-skills-smart-meter-data-integration/) allows utility companies to increase or restrict the flow of electricity, gas, or water to reduce waste and ensure adequate supply at peak periods. Public transport can be efficiently planned to avoid wasted journeys and provide a reliable service that will encourage citizens to leave their cars at home (http://data-informed.com/how-big-data-is-helping-to-tackle-climate-change/).

In terms of big data technologies, databases without boundaries are part of emerging disaster relief. Online databases can help victims find missing family members, organize volunteers, or link people who can provide shelter to those who need it (Wallace 2015).

(1) **The use of big data analytics for smart grid: The cases of China and the United States**

Big Data analytics is critical to the smart grid. Smart grid systems with the use of big data and the Web (the cloud) are published by Qiu and Antonik (2017). They have compared Chinese cases with those of the United States in the systems.

The systems involve networking vast numbers of sensors in transmission and distribution facilities, smart meters, SCADA systems, back-office systems, and devices in the home, which interact with the grid. Large amount of data (hence big data) is generated by meters, sensors, and synchrophasors.

The system is basically the use of the Web as the computing and information management platform. This allows the integration of data and capabilities from multiple, diverse sources to deliver powerful composite applications over the Web. Figure 1.39 provides a conceptual illustration of this model.

In addition to advanced metering and utility-wide communications infrastructure enabling demand response, and distributed resource management, this methodology represents the impacts associated with the smart grid which include operational and enterprise information systems, supervisory control and data acquisition (SCADA),

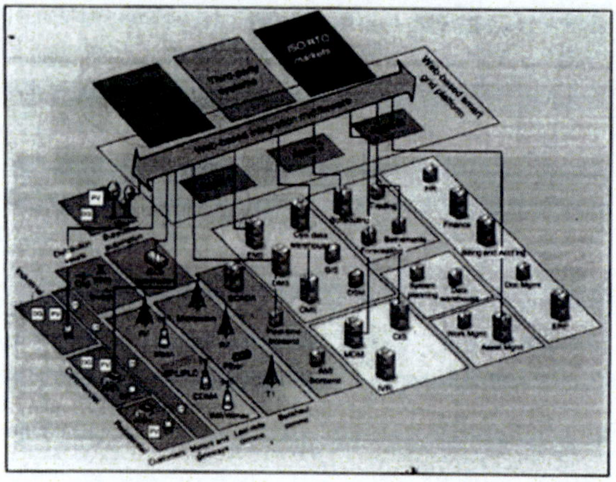

Fig. 1.39 Using the cloud for smart-grid applications

feeder and substation automation, customer-service systems, planning, engineering and field operations, grid operations, scheduling, and power marketing. The smart grid also impacts corporate enterprise systems for asset management, billing and accounting, and business management. As illustrated in Fig. 1.39, many information technology (IT) systems are impacted, including those for distribution management and automation, operations planning, scheduling and dispatch, market operations, and billing and settlements.

China prefers to use a "supply-side policy", with a focus on "public enterprise, scientific and technical development and legal regulatory" policies. The United States, on the other hand, prefers to use an "environmental-side policy", with a focus on "scientific and technical development, financial, political and public enterprise" policies (Fig. 1.40).

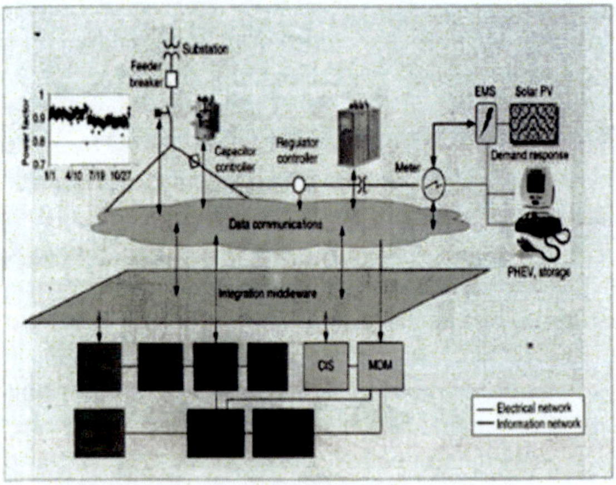

Fig. 1.40 Systems required to support the high penetration of distributed resources

(2) The use of big data to create smarter cities: The case of Milton Keynes, England

Data is becoming an increasingly important input into the decision-making process, and improving decision making is probably the most widespread way cities are using data today. Milton Keynes, England has started one key initiative, developed with the help of the Open University and BT, called MK: Smart. It has acted as a data hub for all of the other projects in the city where their effectiveness and impact can be assessed. Internet of Things (IoT) type, connected solutions have been proposed for transport, energy efficiency, water supply planning, enterprise growth and provision of education. Sensors will be installed at waste disposal facilities, meaning the process of emptying them by truck can be made more efficient. Traffic and footfall through public spaces will also be monitored in order to plan public transport routes as well as a footpath and cycle path infrastructure.

Many of the projects are underway, with a number of homes taking part in trials of energy-saving home appliances and smart meters, in association with energy provider E.ON. Other families have been provided with electric cars in order to carry out a year-long viability study on the technology. And in the near future, the city will see the first UK trials of driverless cars on its grid-like, roundabout-heavy road network.

The latest initiative, CAPE, uses satellite imagery and data on the thermal leakage of houses; the aim is to help citizens manage their own community energy schemes and ultimately reduce their carbon footprint. It will supplement satellite imagery with building and energy data to locate neighborhoods that could benefit from an energy makeover. It's a first in the UK and the latest addition to the MK: Smart program (Marr 2016).

1.5 Evaluation of Sustainability, Smartness and Resource Wisdom of Climate Smart and Wise Cities: An Indicator Approach

It appears that politicians have created a storm by picking up the words "smartness" and "resource wisdom" which are intended to be the driving forces for the climate smart and resource wise development effort. There should be a great need for fuller debates/considerations on user-friendly climate smart and wise city indicators and dataset to suggest ways forward with climate smartness and resource wisdom of cities. We have to strive to develop the indicators to get cities onto a zero-emission, carbon neutral pathway and are open to debate them with the experts and the public for the following reasons:

- There is no consensus about what really makes a city smarter and wiser in terms of climate emergency.
- The increasing diffusion of models, standards and definitions of smart city (SC) is creating ambiguity and makes it difficult to estimate whether SCs are aligned with the original expectations, the ideas and principles claimed by their promoters.

- However, several recent studies have been raising concerns about ongoing SC initiatives (Taamallah et al. 2017): SC programs have been criticized for being too techno-centric, mainly driven by big Information and Communication Technologies (ICT) companies which own their agendas and in turn of lacking enough attention to citizens' needs and real environmental sustainability (Huovila et al. 2019).

In this chapter, we have tried to define the climate smart and wise city notion, principles, solutions, models and platforms, and their implementation. This section proposes a dataset based on empirical analysis to help understand the main drivers that originate the launch of climate smart and resource wise initiatives worldwide.

1.5.1 Review of Existing Dataset

Different types of cities such as smart city, sustainable city and resource wise city use different sets of domains, sub-domains and variables.

(1) **Smart city**

Table 1.8 shows the individual sub-domains organized into 6 domains, and 23 sub-domains.

(2) **Sustainable City**

Table 1.9 shows the individual core indicators organized into 8 themes, 24 core indicators, and 28 supporting indicators.

(3) **Resource wise City**

Table 1.10 Summary of main resource wisdom indicators. The indicators have been selected to correspond to the principles of resource wisdom. The units and accounting instructions of the indicators are listed in the table.

(4) **Comparison of indicators**

A comparative study of 3 different types of indicator sets selected is conducted according to questions that represent the criteria associated with domain or theme and indicator.

Table 1.11 provides a summary of the evaluation of the 3 sets of indicators. This evaluation result could be utilized for the development of an indicator set of climate smart and wise cities in the next section.

The criteria and questions are described as follows:

- Comprehensiveness—Does the method address a full range of indicators?
- Aggregation—Does the method provide a way for aggregation of indices to provide an evaluation of overall total value?
- Key issues—Does the method suggest a way of highlighting key issues?
- Specificity—Are specific city parameters identified?
- Adaptability—Can the method be applied to different basic urban settings?

Table 1.8 Classification of SC application domains and associated sub-domains

Domain	Sub-domain	Indicator	Definition	References	Variables
Natural resource and energy	Smart grids		System aimed at providing power supply and usage more efficiently by adding intelligence to power generation and distribution, through the integration of communications and control technologies	Avancini et al. (2019)	Smart Energy Grids
	Public lighting		New Technologies in public street lighting such as the adoption of LED can reduce energy consumption and at the same time improve the visibility of pedestrians, cyclists and drivers	De Marco et al. (2016)	
	Green/renewable energies		The use of energy from renewable sources—wind, biomass, hydro and solar power—is crucial for achieving a general reduction in energy consumption	Corsini et al. (2019)	
	Waste management		It is a concern for modern cities due to both the operations related to waste collection and the environmental issues of landfills. Innovative technologies seem promising solutions for handling waste collection and recovery operation in SCs(Smart Cities)	Esmaelian et al. (2018)	
	Water management		In this field the integration with innovative technologies can be exploited for monitoring, finding anomalies and the data can be useful in waste distribution network efficiency and effectiveness	Shahanas and Sivakumar (2016)	Both digital and physical

(continued)

Table 1.8 (continued)

Domain	Sub-domain	Indicator	Definition	References	Variables
Transportation and mobility	City logistics		It can be defined as a comprehensive approach to make the last mile logistics services more organized and to mitigate the negative impacts of freight distribution	Cagliano et al. (2017b)	
	Infomobility		It represents a network system mainly characterized by connections, both digital and physical to reduce congestion and carbon emissions	Battarra et al. (2018)	
	People mobility		The urban mobility of citizens is more sustainable by using alternative energy sources, soft transport systems and by optimizing public transports services	Benevolo et al. (2016)	
Buildings	Building services		The efficiency of the building's systems(electricity, water, air) play a crucial role in enhancing the building performance	Leibowiez et al. (2018)	
	Facility management		It refers real estate security, operations and maintenance of buildings. The advancement of technology and automation allows to more proper usage of resources	Roper (2017)	
	Housing quality		It represents the capability of providing high levels of quality for occupants, by meeting their different needs over time	Malakouti et al. (2019)	
Living	Pollution control		Smart sensing unit and any data-driven technique is able to perform more correct pollution control activities in the nowadays cities	Migaita er al. (2019)	

(continued)

Table 1.8 (continued)

Domain	Sub-domain	Indicator	Definition	References	Variables
	Public safety		System for surveillance, detection and identification, improved by using new technologies such as Big Data	Lacinak and Ristvej (2017)	
	Healthcare		The application of technologies such as IoT supports the integration of medical devices and improve the way physician can deliver care. Also, it boosts patients' satisfaction by facilitating their interaction with doctors	Ghani (2019)	
	Welfare services		It is intended a greater state involvement for promoting and supporting social policy goals	Docherty et al. (2018)	
	Public spaces		This aspect is related to the importance of creating spaces for social interaction. In several cases, these areas refer to the urban conversion of industrial sites	Leorke et al. (2018)	
Government	E-government		It is based on the engagement of the citizenry in government in a used-centered manner. Operations and services are delivered so that to be tailored for the citizen's needs	Twizeymana and Andersson (2019)	
	Procurement		The exploitation of ICT may be used to improve procurement processes in terms of more precise forecasting of demand for public utilities, controlling irregularities, and risks	Adjei-Banfo et al. (2019)	

(continued)

Table 1.8 (continued)

Domain	Sub-domain	Indicator	Definition	References	Variables
	Transparency		Initiatives aimed at making decision-making processes more transparent improve the quality of the services that are delivered and reduce the related costs++	Osei-Kojo (2017)	
Economy and people	Innovation and entrepreneurship		It represents the impact of innovative technologies on the way entrepreneurial initiatives are developed, in terms of integration among the assets, selling and evolution of new ventures	Srinivasan and Venkatraman (2018)	GDP
	Entertainment and culture		These aspects, related to the issue of the knowledge economy are very important because they stimulate flexibility, collaboration and tolerance among citizens	Albino et al. (2015)	
	Human capital		It fosters human capital development opportunities as a way to support the knowledge	Camboim et al. (2019)	
	Education and school		Higher shares of citizens with higher education are likely to better support the growth of smart urban areas. In this sense, programs of education and school become relevant	Nilssen (2019)	

- Ambiguity—Is the method ambiguous?
- Explicit indicators—Does the method suggest specific measurable indicators for indicator quantification?
- Technologies—Are special technologies required?
- Digital connection—Does the method provide a way for valuating the digital connectivity of urban elements?
- Data requirements—Does the method use current data or are special studies required?
- Costs—What are the costs of using the method?
- Analysis bias—To what degree will different results occur depending on the analyst?

Table 1.9 ISO37120 Sustainable cities and communities-Indicators for city services and quality of life

Theme	Core indicator	Supporting indicator
Economy	• City unemployment rate • Assessed value of commercial and industrial properties as a % of the total assessed value of all properties • % of city population living in poverty	• % of persons in full-time employment • Youth unemployment rate • Number of businesses per 100,000 population • Number of new patents per 100,000 population per year
Education	• % of female school-aged population enrolled in schools • % of students completing primary education: survival rate • % of students completing secondary education: survival rate • Primary education student/teacher ratio	• % of male school-aged population enrolled in schools • % of school-aged population enrolled in schools • Number of higher education degrees per 100,000 population
Energy	• Total residential electrical energy use per capita(kWh/year) • % of city population with authorized electrical service • Energy consumption of public buildings per year(kWh/m^2) • % of total energy derived from renewable sources, as a share of the city's total energy consumption	• Total electrical energy use per capita(kWh/year) • Average number of electrical interruptions per customer per year • Average length of electrical interruptions
Enviroment	• Fine particulate matter (PM2.5) concentration • Particulate matter (PM10) concentration • Greenhouse gas emissions measured in tonnes per capita	• Nitrogen dioxide(NO_2) concentration • Sulphur dioxide(SO_2) concentration • Ozone(O_3) concentration • Noise population • % change in number of native species
Finance	• Debt Service ratio(debt service expenditure as a % of a municipality's own = source revenue)	• Capital spending as a percentage of total expenditures • Own-source revenue as a percentage of total revenues • Tax collected as a percentage of tax billed

(continued)

Table 1.9 (continued)

Theme	Core indicator	Supporting indicator
Fire and emergency response	• Number of firefighters per 100,000 population • Numbers of fire-related deaths per 100,000 population • Number of disaster-related deaths per 100,000 population	• Number of volunteer and part-time firefighters per 100,000 population • Response time for emergency response service from initial call • Response time for fire department from initial calls
Governance	• Voter participation in last municipal election (as % of eligible voters) • Women as a % of total elected to city-level office	• % of women employed in the city government workforce • Number of convictions for corruption and/or bribery by city officials per 100,000 population • Citizen's representation: number of local officials elected to office per 100,000 population
Health	• Average life expectancy • Number of in-patient hospital beds per 100,000 population • Number of physicians per 100,000 population • Under age five mortality per 1,000 live births	• Number of nursing and midwifery personnel per 100,000 population • Number of mental health practitioners per 100,000 population • Suicide rate per 100,000 population

Table 1.10 Summary of main resource wisdom indicators

Principle	Indicator	Unit	Accounting instructions
Zero climate emissions	Consumption-based greenhouse gas emissions	tCO_2 eqv/res	KASVENER or CO_2 report
Zero waste	Material losses	tonne	Flows into landfills + incineration of non-renewables + export of waste for final disposal
Zero overconsumption	Ecological footprint	gha/res	Global Footprint Network's instructions and presumptions
Sustainable wellbeing	Share of residents who perceive their quality of life to be at least good	% residents	The World Health Organization's quality of life indicator, WHOQOL-8 (National Institute for Health and Welfare(THL) survey)

Table 1.11 A comparative study of 3 different types of indicator sets

	Methods		
	Smart city	Sustainable city	Resource wise city
1. Comprehensiveness	S	S	S
2. Aggregation	S	L	L
3. Key issues	S	L	L
4. Specificity	N	S	L
5. Adaptability	N	S	L
6. Ambiguity	N	S	S
7. Explicit indicators	L	S	S
8. Technologies	L	S	L
9. Digital connection	L	S	S
10. Data requirements	L	S	S
11. Costs	L	S	L
12. Analyst bias	S	L	N

Note L, substantial; S, partial compliance; N, no or minimal compliance

1.5.2 Strategy for Developing Climate Smart and Wise City Indicators

1. **Indicator development process**

The major steps in developing indicators are as follows:

- Defining the purpose of indicators
- Developing the criteria for indicators
- Collecting methodological literature on existing indicators (and measurement methodologies)
- Comparative evaluation of the degree of compliance with the developed criteria of each of the existing indicators (and measurement methodologies)
- Selecting indicators for climate smart and resource wise cities based on the results of comparative methodology evaluation
  - Quantitative indicators and measurements
  - Qualitative indicators and measurements.
- Utilizing feedback to the selected indicators and measurements from experts, the public including residents, and planning and policy documents to select the most desirable alternative indicators for meeting project objectives through workshops, forums, questionnaire surveys, or other public participation techniques
  - Professional judgment
  - Public perception

- Review of planning and policy documents.

- Finalizing a set of indicators and measurements corresponding to the three parameters of sustainability, smartness, environmental integrity and carbon neutrality
- Applying the selected indicators and measurements to testbed areas for assessing the practicality and validity of the indicators and measurements
- Evaluating the effectiveness of various governmental and community climate smart and wise city projects by a means of the tested indicators and measurement.

  - Literature review (e.g., planning and policy documents)
  - Statistical data
  - Sample survey (a combination of a stratified survey and a targeted sample survey)
  - Monitoring data (measuring data)
  - Subjective perceptions
  - Qualitative research
  - Interview
  - Questionnaire survey, etc.

(2) **The Purpose of climate smart and wise city indicators**

Generally, the purpose of climate smart and wise city indicators is a practical one: to monitor the status of the city in terms of the level of smartness, sustainability, resource wisdom (or environmental integrity) and carbon neutrality, and evaluate the effectiveness of governmental and community climate smart and wise city projects.

(3) **Criteria for developing indicators**

Criteria for evaluation of selected indicator methodologies and selection of indicators included in the dataset used should be developed to achieve the purpose of indicators. In addition to the criteria and questions associated with indicators described in 4) Comparison of indicators' above, the followings are two examples of criteria.

- Be comprehensive- with regard to coverage, indicators should be comprehensive but limited. The city contains intricate systems of living and non-living elements bound together by complex interrelationships. They should be comprehensive in the sense of measuring all of the dimensions and determinants of a climate smart and wise city. Especially important in this regard is the development of indicators that measure not only climate and resource status but also climate and resource-related characteristics of the city itself. It also incorporates those conditions of a city which are pertinent to the climate and resources of the city. The number of indicators should be limited to prevent information overload and to facilitate data management and the comparison of research results. Those more interested in monitoring the sustainability, smartness and environmental integrity of one city than in measuring and comparing the common situations of different cities should not feel too constrained by the need to limit severely the number of indicators.

- Be relatively inexpensive and easy- with regard to data source and resource requirements, indicators should be relatively inexpensive and easy to collect and thus be available to serve as a monitoring system within a short period of time. Again, as with the concern for comprehensiveness and limitation, this criterion should be seen as a flexible guide, not a hard rule. Sufficient flexibility must be contained in the indicators since cities of different sizes and functions need different types of indicators.

1.5.3 Steps for the Application of Domains, Associated Sub-Domains and Variables for the Project Evaluation

This application process is suggested as follows.

First step is to decide the taxonomy of climate smart and wise city domains, sub-domains and their variables as provided in Tables 1.8, 1.9 and 1.10. Each domain is subdivided into several sub-domains and variables (or indicators). Different variables are suggested for measuring the level of sub-domains.

Second step is to select sample projects.

Third step is to collect a dataset from the sample projects.

Fourth step is to calculate, based on the dataset, the Coverage Index (CI) for each city of the sample project. Here the CI is defined as a proxy for assessing the level of sustainability, smartness and resource wisdom and carbon neutrality of a city. It accounts for the variables and sub-domains of a given domain that are covered by projects carried out in a city. In other words, a CI equal to one in a given domain means that the city has launched at least one climate smart and wise city project in each of its variables and sub-domains. The CI is computed as the sum of the indices obtained at the level of the individual domains. This index represents the number of domains covered by climate smart and wise city projects for each city of the sample.

Fifth step is to collect data for all context, structural, technological, economic, and environmental profiles of the cities that are expected to influence the CI.

Sixth and final step is to analyze the dataset via descriptive statistics, factorial analysis, cross-tabulation analysis, analysis of variances and correlation analysis. The statistical analysis has several goals. First, the objective is to identify the climate smart and wise city domains with the largest diffusion of climate smart and wise city initiatives throughout the world. Then to understand commonalities and differences in the pattern across continents all around the world. The second phase of the empirical analysis aims at evaluating how the collected city profile might affect the different domains, sub-domains and variables of the CI.

1.6 Conclusions

This chapter has sought to find out how climate smart and wise cities may support and/or require a conceptual framework and key components of the integrated planning system, planning procedure and planning methodology including platform project models, based on the authors' "Integration Platform Theory and Practice in Planning Sustainable Development".

In relation to the planning instruments, the answer seems to be positive. However, the identification of issues, solutions for climate emergency and function of principles of climate smart and wise cities are to require adequate validation for further application. Therefore, the proposed spatial planning system and models for the climate smart and wise cities would seem to be used as integrated tools for testing the practicality, adequacy and legality for implementing them in particular cases and hence require the capacity to set certain limits to the choices of developers.

In relation to the international standardization of climate smart and wise cities, the level of integration for spatial, green and digital deal approach to climate smart and wise cities proposed may not be totally enough in the light of those objectives of this book. It is hoped that other chapters of this book help turn this contribution into more elaboration.

References

AECOM, IBM, KPMG (2017) Integrated smart city framework plan: smart city master planning and sector-specific smart city infrastructure pojects for Visakhapatnam, India. p I-5

Bedient PB, Huber WC, Vieux BE (2013) Hydrology and floodplain analysis. p 707

Cullingworth B, Nadin V (2002) Town and country planning in the UK, 13th edn. Routledge, London and New York, p 76

Emilia R, Anna L, Eeva A, Aulikki H, Toivo M (2017) New resource-wise planning strategies for smart urban-rural development in Finland. Systems 5(1):10. https://doi.org/10.3390/systems5010010

Energypedia (2019) Mini grids, definition and overview, https://energypedia.info/wiki/Mini-Grids//cite-note-o. Accessed 8 June 2019

Environment Agency (2018) London regional flood risk appraisal. p 64

European Commission (2019) The European green deal, COM (2019) 64o final, p 3

Hiroaki S, Arish D, Sebastian M, Nanae Y, Hinako M (2010) Eco2 cities: ecological cities as economic cities. The World Bank, Washington DC

https://www.forbes.com/sites/davidrvetter/2020/4/16/south-korea

https://www.steelcase.com/research/articles/topics/postcovidworkplace/designing-the-post-covid-workplace/

http://data-informed.com/utilities-executives-cite-need-for-analytics-skills-smart-meter-data-integration/

Huovila A, Bosch P, Airaksinen M (2019) Comparative analysis of standardized indicators for smart sustainable cities: what indicators and standards to use and when? Cities 89:141–153

IASS (2019) Exploring the nexus of mini-grids and digital technologies: potentials, challenges and options for sustainable energy access in Sub-Saharan Africa, Potsdam, p 15

Jager J (2020) European union coronavirus proposal, WRI

Kim KG (2009) Cities in the age of low carbon, green growth, The Korea Herald, p 10
Kim KG (2017) Development of an integrated smart water grid model as a portfolio of climate smart and wise cities. J smart Cities 3(1):23–33
Kim KG (2018) Low-carbon smart cities: tools for climate resilience planning. Springer, Heidelberg, p 98, 105, 146
Marr B (2016) Big data in practice. Wiley, Chichester, UK, pp 151–152
Meeting of the Minds (2020) 3 Ways communities can bond with resident in the age of COVID & beyond
Plunz R, Sutto MP (eds) (2010) Urban climate change crossroads, the trustees of columbia university, City of New York, p Vii
Rogers R (1996) How cities may look in the 21st century, p 16–17
Siebel TM (2019) Digital transformation: survive and thrive in an era of mass extinction. Rosetta books, New York, p 11, 27–28
Smart City Research Larissa Romualdo (2013) What is a digital city? 2013, Facebook
Scharffernberg WA, Jeff H (2008) Hydrologic modeling system: interior flood modelling. Hydrologic Engineering Center
Taamallah A, Khemaja M, Faiz S (2017) Strategy ontology construction and learning: insights from smart city strategies. Int J Knowl—Based Develop 8(3):206–228
Wallace P (2015) Introduction to information systems, 2nd edn. Pearson, Harlow, p 149
World Bank Blogs (2020a) Hurricanes don't know borders, Neither does a pandemic. https://blogs.worldbank.org/sustainablecities/hurricanes-dont-know-borders-neither-does-pandemic. Accessed 19 Aug 2020
World Bank Blogs (2020b) Six reasons why demography matters during the pandemic. https://blogs.worldbank.org/sustainablecities/six-reasons-why-demography-matters-during-pandemic. Accessd 14 July 2020
World Bank Blogs (2020c) Cities are on the frontlines of COVID-19
www.beechamresearch.com/files/Br Smart City EXEC Summ....2015

Chapter 2
The Role of Blockchain, Big Data, Internet of Things and Artificial Intelligence for Transformative Climate Action

Harald Rauter

Abstract In light of the climate emergency, mankind is facing the challenge of having to make deliberate choices around how our lives will have to structurally change in order to avert the climate breakdown at an unprecedented urgency and scale. At the same time humanity, through technology, has unprecedented and excellent tools at hand to achieve the just, inclusive, far-reaching systemic change that is needed to avert climate breakdown. While it will be ultimately us, the humans, who will have to take the decisions, technology empowers smarter decision making when decarbonizing our economy and, at the same time, build the desirable and regenerative society of the future. Exponential Technologies will be the major driving force for future socio-economic paradigms and are already shaping our future beyond anything else. It requires a collective retooling of the global cooperation and its institutions to put exponential technologies at the heart of an effective response to future climate risks on a whole system level. This chapter discusses the current misjudgments and categorization errors of exponential technologies and offers disruptive leadership- as well as innovation frameworks aiming to unlock the transformational potential of exponential technologies for far-reaching climate action.

Keywords Exponential technologies · Whole system transformation · Disruptive leadership · Abundant futures

2.1 Introduction

This chapter aims to discuss the role of emerging technologies in the context and in service of whole socio-economic transformation. Existing literature, willingly or not, dominantly gravitates towards discussing the role of emerging technology in optimization efforts and the disruption of incumbent technologies—however, all along with existing path dependencies and value chains. Understandably, less attention is given to the transformational potential of emerging technologies, which, in the

H. Rauter (✉)
EIT Climate-KIC, Limmatstrasse 265, 8005 Zurich, Switzerland
e-mail: harald.rauter@climate-kic.org

K.-G. Kim and M. Thioye (eds.), *Planning Climate Smart and Wise Cities*,
The Urban Book Series, https://doi.org/10.1007/978-3-030-80165-6_2

context of far-reaching climate action is more needed than ever before. The ontological understanding of the collective human "now", including the forces that shape our macro-environment is a critical puzzle-piece to adequately discuss the role of emerging technologies in the context of the necessary ambition for climate action as it is laid out in the IPCC Special Report on Climate Change and Land (Shukla et al. 2019).

Our understanding of complexity, digital transformation and the emerging opportunity space arising from it reshapes the possibility for socio-economic transformation and also redefines the role of agency and leadership beyond the existing notions of leadership. The relevance of the above mentioned cannot be overestimated. It will help to answer and finally address and overcome the question.

Why has the energy and resources allocated to technological progress not translated into the necessary societal transformation that is necessary and sufficient to avert the climate catastrophe? This is particularly striking as current empirical research (Bhowmik et al. 2018) suggests that even, if we were to halt technological innovation immediately, currently existing technological solutions, facilities and resources would be powerful enough to decarbonize our lives, value chains and economies to a sufficient extent so that a sustainable future for humankind would be ensured respecting the remaining carbon budget available.

What we need, more than ever beyond the development and scaling of technological use-case a new integrated approach for technology as a driver and catalyst for transformational change—on the level of the individual as well as human systems, communities, cities, value chains, countries or even humankind as a whole.

This chapter aims to provide a framework on how we can start to think of far-reaching climate innovation from a first-principle perspective (https://en.wikipedia.org/wiki/First_principle) building on the principles of complexity. It will further discuss how emerging technologies enable us for- and can be in the service of far-reaching climate action by amplifying our intrinsic human capacity for societal coordination, deliberation, cooperation and collaboration. It will further discuss how emerging technologies can help us to redefine our relationship with ourselves in the process (agency), our relationship to others (leadership), our relationship to the physical world (Oracle) and our relationship to time (future contracts). The chapter will make the argument that the probably its biggest impact technology can make at this critical point in time is nurturing and building the necessary societal resilience to shift our focus from thinking and innovating in sectors and value chains to thinking in challenges (Mazzucato 2018a) and systems.

Climate change is different and it is, by first principle, a challenge of complexity.

We feel this hidden complexity as the global emergency that it is, now, in a full world (UKEssays 2018), one where externalization is not possible anymore without immediately affecting us again through system feedback loops.

This is a different world that requires a different approach to innovation. When the problem at hand cannot be broken down into less complex sub-problems that can be solved in isolation then we require an innovation system that embraces complexity and doesn't try to ignore it. The big question that this chapter aims to, at least, provide some framework for is: How can we now collectively take decisions that put technological achievements in service of transformational climate action.

This chapter will therefore put a spot-light on the following:

- The need for coherent Systems Innovation Model is on par with the complexity of the challenge at hand, in which technology can be in service of transformational change.
- Establishing an ontological argument for systems innovation and why system innovation must be mainstreamed as a framework for impactful climate action
- Discussing the roles technology and data can and should play to break out of path dependencies and catalyze the structural socio-economic change needed to avert climate catastrophe.
- Discussing the role of disruptive leadership in enabling and catalyzing necessary collective action for climate impact

This chapter is aiming to build an intellectual bridge between the context of ambition and the innovation happening on the ground—it aims to promote a new innovation lens—one where technology is in service of a whole system transformation effort.

2.2 Climate Change and the Complexity Challenges

Climate innovation often targets its energy to solutions that mostly fix the problem or in other words: fight the symptoms. In light of the climate emergency, it is understandable to push for immediate solutions—and to a certain extent, it is also necessary. However, let us pause for a second here and do something different. Let's dig into better understanding the problem first and developing a framework that helps us to design and implement more impactful solutions down the road. It starts with the language we use. Our vocabulary is rich with attributes and descriptions to describe the consequences of climate change. Our vocabulary is less rich when we want to refer to the nature and identity of climate change itself. Therefore, let's spend some time to characterize and understand the problem first. Let's look at the planet and climate change from the perspective of a complex adaptive system with all of its associated properties and physics that allows us to describe this complexity—at least by approximation. Throughout most of human history, human activity could ignore the planetary interdependencies without consequences and could externalize those human actions without disrupting the planetary resilience. Industrialization has changed exactly that—in a dramatic fashion. Suddenly, human action has started to greatly affect the environment and to a point where it started to compromise the overall resilience of the planet as a whole. We can therefore think of Climate change as the externalized energy of human action. In the language of complex adaptive systems theory, climate change is an emerging property of the complex system dynamics of the planet.

This might sound like semantics at worst—or no more than a subtle twist at best. However, characterization of climate change through the lens of complex system dynamics allows us to approach climate change from a first-principles perspective and allows us to better understand the cascade of down-stream events, the role of humans and more importantly—the cause and effect of the societal and economic design principles which we have collectively agreed on. The dominant socio-economic thesis that we have collectively agreed on has advanced human civilization and has served us well. However, now it needs to change dramatically. In order words, solving for climate change is about deliberately shifting the societal and economic paradigms and not about fixing the symptoms. Technology, in service of far-reaching climate action, helps us collectively to evolve the societal and economic contract that is honoring the reality of a finite planet and the imperative for an equitable and climate-just future for all. The alternative, solely innovating along known path dependencies that have their origin deeply rooted in eighteenth-century thinking, is not an option. Instead, what is required is a collective effort for societal progress, which diverges from our known path dependencies and establishes new societal and economic paradigms which allow for an inclusive and climate-just future for all.

At this point of path-dependent innovation versus transformative innovation humans individually as well as the innovation community collectively face a cognitive dilemma which is best expressed by what is widely called "the streetlight effect" (Fig. 2.1).

The way the human brain is wired tricks us into building solutions from a place we know and understand—where there is light. It is tempting to build on the existing, improve it, harvest immediate successes but in the bigger picture we fall short in accomplishing the bigger ambition—finding the quarter. We need to develop a new intellectual muscle and practical toolkit to search for emerging opportunities where the light isn't shining. Let's not kid ourselves—this is hard, very hard and while throughout human history there have always been individuals that have mastered this path we have never found ourselves in a position where we had to undertake

Fig. 2.1 The streetlight effect refers to one of the many human biases of the category of believe-biases; the effectiveness of human action is assessed based on our belief sets and not on their real merit in reality. The streetlight effect is therefore one of the reasons why specifically transformational change is so difficult as it requires us to enter unknown territory

such an endeavor collectively, at an unprecedented magnitude, speed and scale. Technology and digital infrastructure, however, are offering an unprecedented platform to coordinate and enable such collective action at scale.

2.3 Complexity Can Be Navigated by First-Principle Thinking

Climate action is mostly a challenge of managing and preventing externalization of human action in a complex system environment. Understanding the complexity theorems has direct effects on the innovation models and innovation approaches that we apply. Understanding and embracing complexity will help us to better put technology to use in service of far-reaching climate action under this framework. It will further allow us to better understand why supply-driven innovation models are insufficient to deliver climate impact at the speed and the scale needed. Furthermore, it will also equip us with a mental framework for disruptive leadership and the role that each and every one of us can play to help eliminating the innovation barriers to achieve the desired impact.

A key enabler for this discussion is the exponential rate of digitization that has happened over the past 30 years which has massively reshaped the world—not only from the way we live but also from a perspective of complexity. The internet has created a global super network with endless and almost infinite, connections between actors and network participants. In complexity science, we describe such phenomenon through NK-complexity (Valente 2008) measures, where (n) equals the participants/actors of a network and (k) equals the existing connections between those actors.

While the overall increased complexity poses a couple of initial challenges that we have to acknowledge and skilfully embrace first it is ultimately this very opportunity space that gives humankind the most powerful tool into our hands to solve the most pressing challenges of humankind—collectively, coordinated, permission-less and inclusive (Boisot et al. 2011).

Most of our current and dominant thinking and tooling on how innovation works and potentially succeeds originated from a world of lower-order complexity—on path dependency that has its origin in the eighteenth century during the industrial revolution. Collectively, we have intuitively pivoted our innovation principles and models through experimentation and validation for what works and what doesn't. The increased digitization has, however, leap-frogged the overall complexity state of society. It has connected people at scale defying boundaries, national borders and other obvious constraints. To harness the power of this connected super organism calls for more sophisticated system innovation principles that evolve the legacy linear innovation models. Climate change calls for systems—or non-linear—innovation models that are on par with the non-linear environment our realities operate in. This is formalized in what complexity science calls Ashby's law of requisite variety.

Ashby's law of requisite variety (Ashby 1958) was originally formulated by W. Ross Ashby, who is considered one of the pioneering thinkers of Cybernetics, the science of self-regulating systems.

Ashby's law of requisite variety postulates that only variety can absorb variety. In simpler terms, it means that only solutions can change the state of a system if the solution complexity meets the complexity of the problem. Solutions, which do not live up to the complexity of the system will not be able to affect the system and thus its system properties. Why is this relevant for climate action? Practically, it allows us to formulate an intervention thesis and practical approaches that allow us to explore how simultaneous, connected and coordinated action must look like so that change can happen and we can break out of given path dependency or, to overcome the inertia and resistance of any given system to change.

Climate change, as a global, supernational challenge of high complexity must be met with the same global and supernational response. The only regime that humans have ever established that reflects such a global, supernational and highly connected organism is the cyberspace of virtually connected humans. Granted, that this might sound counter-intuitive to solve for physical, biological and ecological challenges in the virtual space—but this is exactly where the power of emerging technologies comes to play. It is this very virtual space where we have the opportunity to design and execute the socio-economic paradigms shifts that are in line with the finite planet and meet the imperative of a climate-just, socially just and equitable future. Looking at climate change from a first-principle perspective allows the re-imagination of a global response and the transformational role the emerging technologies can play in the process.

2.3.1 *Complexity Theorem 1: The Complexity Trap or the Unpredictability of Outcomes Within Higher Order Complexity Systems*

To put it in the simplest possible terms: increased complexity through interdependencies and connections within any given system decreases the degree to which outcomes can be predicted. Simple causality thinking is the enemy of effective solutions in a complex environment. We like the idea that increasing computational power will ultimately allow us to outsmart and therefore predict systems outcomes. Unfortunately, this will not be the case as outlined by Gödel's second incompleteness theorem which shows that a system of sufficient complexity cannot demonstrate its own consistency (Gödel 1931). In other words: we always lived in the illusion of a fully deterministic world. Increased system complexity through increased interconnectedness has just surfaced the illusion as what it is—an illusion. In order to affect system behavior effectively, we need to leave the path of description and statistics and embrace and organize for a world of stochastics and emergence.

Take the weather as an example: with all the available computational power and modeling capabilities we are currently able to predict the weather reliably for a few days. If we were to extend and push the prediction horizon a few more days into the future predictive precision declines already exponentially (Toth 2019). Instead of trying to run a race that cannot be won it is important to shift the perspective and approach of how a solution discovery process can be successful in the face of highly complex and interdependent challenges. As a consequence, we collectively find ourselves in a volatile, uncertain, complex and ambiguous world (VUCA) (Millar et al. 2018). There is an obvious tension arising: Most of our dominant innovation models have their origin in the pre-digital and are therefore lagging behind the realities of our time. Additionally, the generation of leaders and decision makers currently in power grew up in a strong causality mindset with a strong "command and control" management ethos. Consequently, the toolkits from this era, e.g., strategic foresight, planning and management by objectives—to name a few—are now rendered obsolete in a high complexity environment. Navigating VUCA environments requires a different approach, which will be discussed later. Charles Handy puts it bluntly: "*Whereas the heroic manager of the past knew it all, could do all and could solve every problem, the post-heroic manager asks how every problem can be solved in a way that develops other people's capacity to handle it.*"

Innovation in a VUCA environment needs to apply a different lens. Experimentation needs to substitute planning, agile learning and sensemaking need to substitute controlling. Technology, in service of transformational change, is less focused on product and service innovation built rather focused on enabling coordination, deliberation, cooperation and collaboration. It enables human systems to be more responsive in a coordinated manner, take decisions faster and more robust at scale, enables complementary (collaborative) and synergistic (cooperative) action (Fig. 2.2).

2.3.2 Complexity Theorem 2: The Powerbase Shift in Higher Order Complexity Systems

Peter Kruse has described this phenomenon the following way: "*The complexity trap has forced organizations into increasing the interconnectedness (remark: Ashby's law of requisite variety). Subsequently, the increased interconnectedness has forced the process of democratization of power with highly connected systems*" (Failed 2013).

It is one of the most profound consequences of the silent digitization of the planet and the emergence of highly complex artificial human systems powered by technology. The more networked a system is the more it becomes difficult to impossible to describe the system's state (Strogatz 2001). It can exist in multiple states at the same time. With the consequence that it becomes increasingly hard to understand, describe and manage a complex system from a centralized power base. Research from Network Science and Game Theory proposes a new view of how effective network

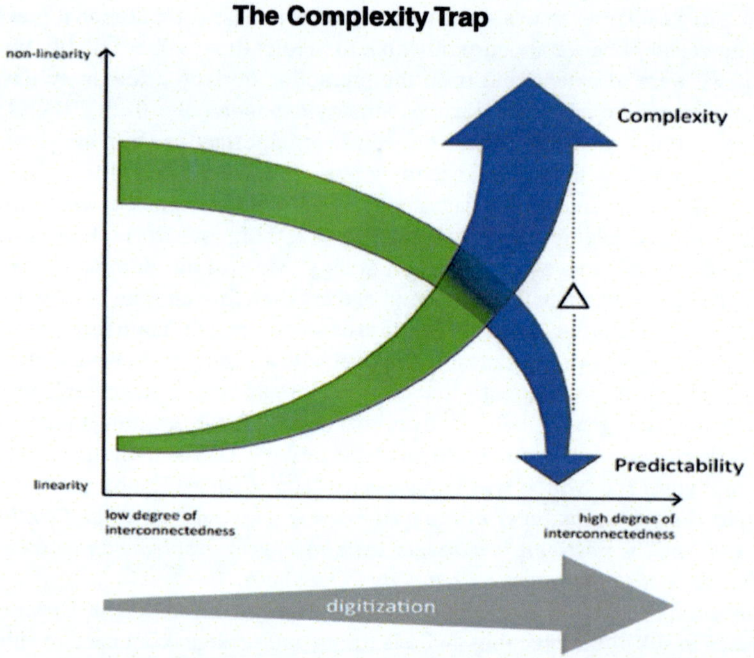

Fig. 2.2 The complexity trap theorem: the degree to which systems outcomes can be predicted is directly linked to the complex state of the system itself. Navigating complexity requires new toolkits based on systems logic—both for innovation initiatives as well as for managerial and organizational purposes. Digitization has changed the playfield, the powerbase and the face of action

state-description and therefore decision making in networks happens (Iranzo et al. 2016). This research has shown that selecting adequate connections leads to cooperative solutions which trumps the importance of single nodes and single actors participating in the network and allows weak networks to overcome the strongest opponents. Adaptation to new network states is therefore determined by the network actor's ability to dynamically establish connections within a given network. This ability of a network to constantly re-examine its own Nash-equilibrium (Holt 2004), the description of the network state where each actors choice is his/her best response to the choices of (n-1) other actors in a network of (n) actors increases the collective network's ability to respond to changes in an agile manner, deep code collective learning and thus increase resilience. Science confirms what we observe on a regular basis. Decentralized decision making is more responsive, more agile and yields superior outcomes over longer periods of time. It is more effective and efficient in problem solving by building new connections (increase decentralization) instead of preventing them (increase centralization).

In real terms, it means that the higher the complexity state of a network the more democratized the agency and the power within the network becomes. The more democratized agency becomes the more it offers the possibility for distributed

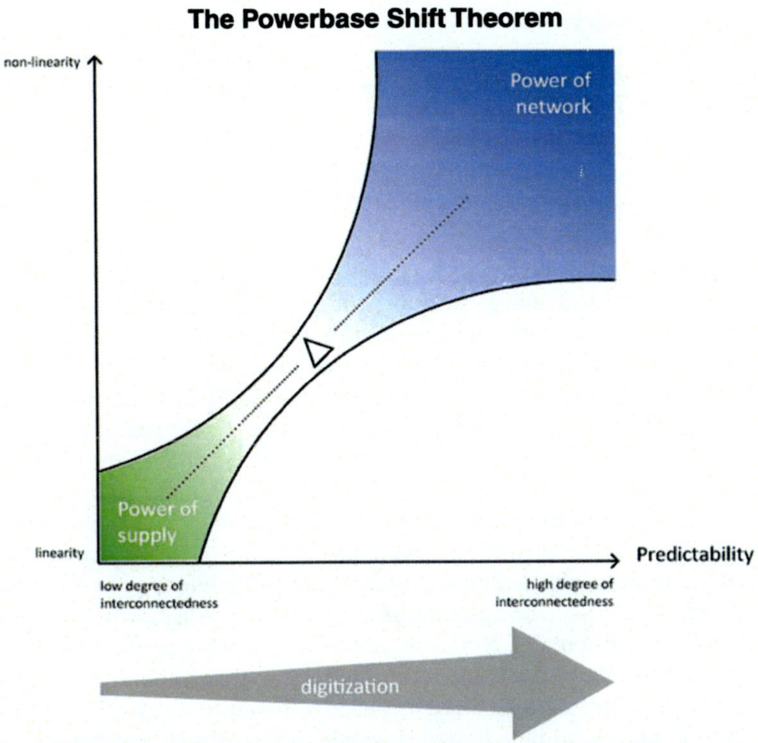

Fig. 2.3 The power base shift theorem: in highly connected networks selecting adequate connections leads to cooperative solutions which trump the importance of single nodes and single actors participating in the network. It allows weak networks to overcome the strongest opponents. Problem solving effectiveness and efficiency increase by constantly examining the optimal Nash equilibrium by building new connections (increase decentralization) instead of preventing them (increase centralization). The collective network's ability to respond to changes in an agile manner increases the overall resilience

leadership in a change process. The more the conditions for distributed leadership are given the more the focus is supposed to shift towards coordination of agency to achieve change. Decentralized networks are the most promising opportunity space to solve humanity's trickiest challenges (Fig. 2.3).

2.3.3 *Complexity Theorem 3: The Identity Crisis*

Complex networks favor distributed leadership over the concentration of agency and power to singular actors. As such networks have the high ground in their ability to appropriately represent and reflect the network state as such. However, highly complex systems cannot be described by one single and permanent identity. Complex adaptive systems can exist in multiple states at the same time, some may only exist very transiently and are in constant flux. Trying to describe the network state by a single, immutable and permanent identity fails to reflect the network's identity and reality. This raises challenges and opportunities at the same time. For actors within a network, it is difficult to affiliate to more than one identity (a structure, a company, a state, etc.) at the same time and to accommodate for the reality that actors within a network hold multiple identities at the same time (Ramarajan 2014). The human identity crisis in face of increased complexity is a direct result of the loss of centralization, a centralized entity that traditionally holds interpretative sovereignty. At the same time, the emerging "identity network" which accommodates the notion of multiple co-existing identities requires the ability to develop an increased degree of ambiguity tolerance. We know from literature (Holt 2004) that the ability to shape transformational outcomes is directly linked to the ability of the network actors to accommodate multiple identities. The collective ability of a network to accommodate for multiple identities is the prerequisite for effective and efficient problem solving through establishing new and non-obvious connections within the network. The more non-obvious connections the network is theoretically able to establish the higher the overall resilience of the network and its requisite variety. In other words: the more actors are collectively able to accommodate for multiple identities at the same time within a network the more complex problem solving capability the network has. For the most wicked and complex problems we know—e.g., climate change—it is therefore important that we re-think and re-image from where transformative innovation will come from in the future. Societal inclusion is therefore not an affirmative action in the process but has to be a design principle in the solution process for climate change (Fig. 2.4).

Emerging technologies are playing a critical role to constructively address and account for the three complexity theorems outlined in the context of socio-economic transformation and the climate emergency. Beyond the role of technology in climate tech (State and of Climate Tech 2020), we argue, that it is vital to examine and cultivate the role of emerging technology as an enabling technology to address the realities of our collective, global human system, the consequences that emerge from it and to use technology to empower collective action that is in line with the Paris Agreement.

2.4 Systems Thinking and Green Growth

What is now the significance of these complexity theorems and how do can they help us in the effort to transform entire complex adaptive systems—specifically, in light of the climate emergency and the deployment of technology in service of it?

The majority of existing impact models (https://hbr.org/2018/04/why-some-of-the-most-groundbreaking-technologies-are-a-bad-fit-for-the-silicon-valley-funding-model)—including social entrepreneurship and classical venture-funded

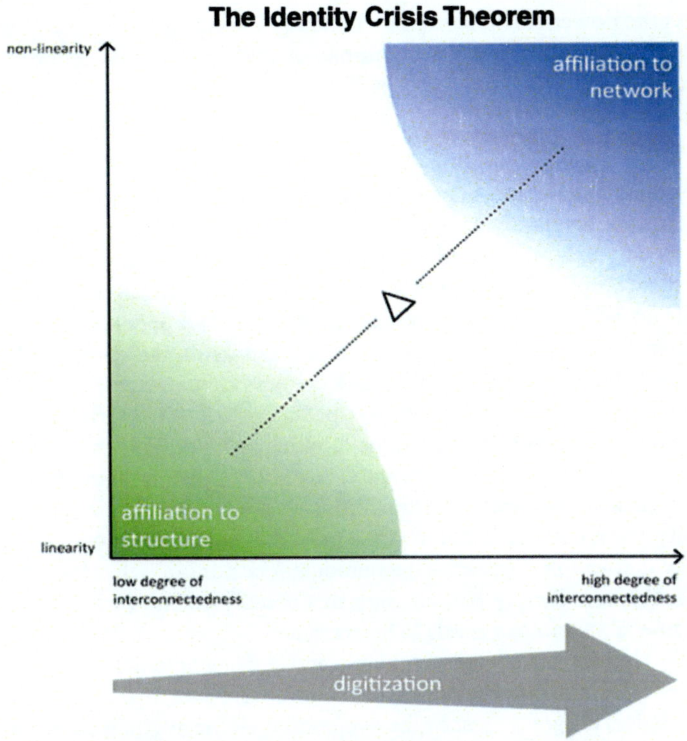

Fig. 2.4 Complex adaptive systems can exist in multiple states at the same time that may exist only very transiently and are in constant flux. The collective ability of a network to accommodate for multiple identities is the prerequisite for effective and efficient problem solving through establishing new and non-obvious connections within the network. The more actors are collectively able to accommodate for multiple identities at the same time within a network the more complex problem solving capability the network has

entrepreneurship models—work through a supply-driven model, single-point intervention portfolios and linear scaling models which pay maximum attention to their individual merit but little attention to the relationships between different approaches and solutions and the system they aim to change. Legacy innovation models also largely ignore any type of externalities and system effects. Single-point interventions and climate solutions - no matter how impactful in themselves - will not be able to overcome the inherent inertia of complex adaptive systems at the speed and scale needed to prevent the climate catastrophe. In order for innovation to be impactful, sustainable and profitable a lens shift is necessary from single-point intervention portfolios to a whole systems approach. Systems innovation approaches:

- Address the complexity trap by pivoting connected experiments, learning and sensemaking instead of a planning and execution approach

- Address the power-shift theorem by building, nurturing and developing bottom-up movements that focus on coordination and deliberation, cooperation and collaboration.
- Address the identity crisis theorem by cultivating fluid identities around challenges and missions as opposed to fixed structures.

The systems innovation approach aims to identify through experimentation, through multiple levers of change simultaneously the tipping points which allow us to collectively enter the decarbonization play which is in the order of magnitude required to accomplish staying within 1.5 °C warming compared to pre-industrial times while ensuring inclusive economic prosperity for all. It continues to pay attention to the position itself (n) but pays specific attention to the relative relationships between the different portfolio positions (k) to provide the necessary and sufficient NK-fitness of the portfolio (https://en.wikipedia.org/wiki/First_principle).

What should we target these systems thinking to? Nothing less than rethinking what green growth means. To put it bluntly: Our collective current model of growth is not actually growth, the unpriced externalities we are generating are self-terminating us. In reality, we are living in degrowth, a great stagnation and a world of zero-sum games. Any innovation approach—technological or nor not—that fails to acknowledge this reality—is simply perpetuating this paradigm. The real challenge, under which the role of technology needs to be examined, is how to chart a new growth that is materially circular and socially equitable and shifting activism from policy to the technologies of human development. The ambition and goal of any innovation model and technological progress must be the overcoming of the ongoing crisis we face that is the structuring of our socio-economic systems as a zero-sum game and a place of rivalry. Discovering a new growth model requires us to recognize that predominantly our governance model has failed and that our human development thesis and the way we have agreed to interact with each other are systematically outdated. Our means of organizing ourselves in a human system is predicated on an outdated industrial society and that our investment in innovation is fundamentally inadequate. It means recognizing that carbon is not the problem of the climate crisis but the underlying structural problem. Addressing this reality has to be the core of any renewal and the mantra under which the meaningful use of technology has to be seen. In this light technology will help us collectively to build a new future. It will require us to invest into and deploy technology in service of (re)-building our collective capacity to not fear the future, build our collective capacity for long-termism, build our collective psychological wealth and last but not least, build trusted deep democracies for and through the people. These intangible societal assets are vital to any socio-economic transition towards a climate-just, equitable and inclusive future. It will require us to cultivate and constantly examine our sense of self, our relationship to others on the one hand as well as scrutinize and review how we use technology to hardwire the rules we agree upon.

We have now established that a reason for the insufficiency of climate innovation lies in the tension between the status-quo of societal organization which has proven to be sufficient to ensure progress in the legacy world and the requirement for new types of societal organization that meets the requirements of the moment. We have established that we must collectively revisit the design principles and paradigms that build the fabric of our current societal and economic models. In doing so we can now open the space to explore and develop a thesis on why and how technology can and maybe should play a critical role in this process of societal transformation and what domains will be massively impacted by the progress of technology.

The solution space to navigate a VUCA environment and provide the infrastructure to execute the basic functions for a bottom-up, self-governed and inclusive society is therefore the digital space. The essential functions to tap into the world described above are coordination, deliberation for decision making and cooperation and collaboration for execution. The space that provides the enabling environment to perform all of these actions, powered by technology is the Decentralized Autonomous Organization (DAO).

2.5 Reshaping Socio-economic Future Through Technology Enabled Bottom-Up Emergence

Reshaping a socio-economic future means exploring what green growth can look like? It means to explore what equitable human progress means? It means to find, cultivate and develop a global societal regime that allows humankind to navigate a VUCA environment collectively, coordinated, self-governed and inclusive. The space that provides the enabling environment to accommodate for the complexity of the challenge and provide the platform to perform all necessary human actions is the virtual space. Let's make the case for what is called a Decentralized Autonomous Organization (DAO) as the ultimate enabler for coordinated climate action and green growth; as the space that allows humankind to revisit the design principles and paradigms that build the fabric of our current societal and economic models; as space where humankind can collectively experiment with superior societal paradigms for new types of societal organization that meet the requirements and the magnitude of the moment.

2.5.1 The Decentralized Autonomous Organization

Decentralized autonomous organizations are powered by blockchain technology (Priyadarshini 2019). It is assumed that the reader has a fundamental understanding of blockchain technology, its properties, functionalities, limitations and potentials. The internet provides a richer source of educational material—including MOOCs from prestigious institutions (https://www.coursera.org/learn/introduction-blockchain-technologies). In short, blockchain protocols allow for a decentralized, permission-less, inclusive, trustless, immutable, self-governed virtual societal organization. These virtual societies can organize around common purpose, goals, rules and incentives. Before we drill deeper into the mechanisms of DAOs and why DAOs matter for far-reaching climate action let's break the most important terminologies for blockchains down—at least on a very high level:

- Decentralized: blockchain protocols have no centralized entity that controls what's going on the blockchain, which rules shall be applied and what the purpose of the participating community is.
- Permission-less: there is a basic difference between a public blockchain and a private blockchain while participation in a private blockchain can be restricted to specific users, participation on a public blockchain is open to everyone who is willing to consent to the governance rules of the blockchain protocol.
- Inclusive: in contrast to legacy organizations where the natural form of organization is the one of the hierarchy, block chains are peer-to-peer and are fully horizontally organized and therefore equitable participation is ensured by programmed code.
- Trustless: the rules of peer-to-peer interaction on a blockchain are hard-coded into the blockchain protocol which is open-source and therefore accessible to everyone. It cannot be changed or tampered with by one single actor. Each interaction can be verified electronically by every network participant which builds on the foundations for horizontal societal self-organization - a base layer for interaction based on radical transparency. Interactions between peers in the network do not require any (trusted) third party but will be validated and verified by a consensus mechanism that aims to make block chains byzantine fault tolerant (Stifter et al. 2019) meaning that the consensus mechanisms of blockchains are solving for the Byzantine General's problem (Dolev et al. 1982), which ensures information symmetry between two counterparties, irrespective of the substance of their interaction. Solving for the Byzantine's Generals Problem eliminated the need for trust between two counterparties in a 1:1 interaction as well as between all participants in a (n:k) network situation. The importance of this cannot be overstated as the blockchain, as the first technology, can now deep code governance rules and incentive regimes that allow for nudging human behavior towards a desired outcome.
- Self-governed: The rulebook of governance for block chains entails multiple dimensions, e.g., how decisions are taken? How exchange between participants is regulated and verified as true or false? How conflicts are resolved? How certain rules can be changed or amended and under which circumstances? Due to the decentralized nature of the blockchain, no single entity or actor is capable of changing the rules of the protocol. While governance rules and principles can be changed it always requires a majority vote and/or consensus.
- Immutable: The blockchain follows the principle of "what is decided, is decided". The blockchain is called the blockchain for a reason. Once, a block, which contains the log of all transactions in a given period, has been validated and agreed upon by the consensus mechanism of the blockchain it is immutably added to the sequence of old blocks and written as the actual ledger onto all participating nodes in the

blockchain network. This final step of copying the latest version of the blockchain to all the participating computers of the network makes the files immutable as no single entity can now tamper with the record that has been created upon consensus.

How do all of these principles now aggregate into an enabling environment that allows for coordination, deliberation as well as cooperation and collaboration and as such allow for coordinated effort to drive inclusive change?

In traditional structures of any kind like companies, cities and even nation states, all participating actors have some sort of contract with the affiliation that regulates their relationship mostly with the central authority who holds the power. These contracts usually also regulate the regimes of peer-to-peer interaction. Usually, those contracts and the legal environment they are embedded in treats humans in the tradition of governance models based on the age of industrialization as "bad robots". This means that legal and governance systems assume the inherent malicious motivation and thus legal and governance systems are based on punishment of undesired behavior. As irrational as this may sound it is important to mention here that we collectively agree to those terms and conditions out of a lack of alternatives that would speak to a different aspect of human nature—the good-will, the well-intended, the helpful, supportive, empathic actor in a human system.

If anything goes wrong, or someone does not stick to their end of the bargain, the legal contract will define who can be sued for what in a court of law.

Decentralized autonomous organizations, however, allow for a different set of rules which incentivize a different set of behaviors based on a self-enforcing open-source protocol.

DAOs flip the game theoretical dilemma of prioritizing the self over the common by incentivizing behavior which helps the common goal by rewarding such behavior usually with aggregating the native network tokens. Such actions may for example be: keeping the network safe by allocating computational power to the network activity, performing other network tasks, working on improvements of the source protocol, etc.

Most importantly, participants and actors within a DAO network are not bound together through legal contracts, nor are any of the actors legally bound to any centralized power authority. What ensures cohesion around the common is the transparent rulebook which represents the governance rules of the network and is deep coded into the source code of the protocol, enforced by machine consensus, immutable to any efforts to tamper with or change the code by a single individual and only changeable by majority vote. The lack of any form of bilateral agreement and full dis-intermediation of any central authority or a third party makes DAOs a true peer-to-peer network of equals, non-discriminatory, inclusive and open to everyone based on a governance rulebook that aims to overcome the limitations of the societal zero-sum architecture we have locked ourselves collectively in.

Now, how would a structural and societal change experiment look like if we were to explore the human potential, human creativity and ability to cooperate and empathize in such an enabling environment?

2.6 The Decentralized WEB3.0 or How to Make Decentralization Useful for Transformative Socio-economic Change

We now have the unprecedented opportunity to re-imagine a societal model and movement that does not require central authorities to function and is purely based on decentralized coordination, deliberation as well as cooperation and collaboration—all geared towards green growth. This opportunity re-defines the role of technology and how emerging technology can and maybe should be in service of transformational socio-economic action. This discussion is deliberately positioning itself in contrast to climate-technological innovation for problem solving which is extensively discussed elsewhere.

This type of technology-driven societal transformation is already happening and is gaining momentum through popular use cases that reflect the width of human expression and human–human interaction. These, so-called decentralized applications (dApps), span from decentralized marketplaces, decentralized use of resources, entertainment, gaming, Machine-to-Machine economy, etc. Most popular, the native cryptocurrency Bitcoin (Dolev 1982) as a decentralized alternative monetary policy and a new form of digital currency to exchange value between peers. Decentralized Finance (Nakamoto 2008), in particular, offer decentralized applications that mimic legacy financial products and services of legacy financial markets including derivative markets, decentralized exchangers and others.

All of these applications and services embody the values of openness, permissionlessness, inclusivity and decentralization leveraging the power of a fully connected peer-to-peer world. Why is this so important for the question of societal change?

As a reminder from above: Complex adaptive systems are constantly evolving and maybe even exist in multiple states at the same time - thus, they are ambiguous and predictive measurement can become statistically meaningless in the analysis. We cannot plan and administer our path into the future. As tempting as this may sound, planning and managing is simply at odds with the increasingly complex world we live in and the complexity we face in light of the climate emergency. We live in a post-managerial world where predictability is at minimum, agency and power is distributed across the network and institutional identities are starting to blur.

If we were to change the socio-economic path into the future beyond our collective path dependency, that we have locked ourselves into, a connected world of determined change agents, connected and empowered by an open network and ready to accept that we collectively start from a point of not knowing what the outcome of this transformational effort is, is proving to be our most promising starting point and deserves all our attention.

These dApps (Schär 2020), layer 2 applications on base-layer blockchain protocols make all the necessary assets and societal regimes instantaneously accessible to everybody in the network with no latency or any sort of access barriers and/or any sort of asymmetry which prevent anybody from participating in this collective effort to achieving change ensuring true inclusivity and agency in the process (Pólvora et al. 2020; Upadhyay 2019).

Orchestration of all of these tools and deploying them through portfolios for strategic synergies and correlations, selecting individual assets not based on their individual merit but based on how they reinforce the portfolio's potential to unlock transformative dynamics is the critical design principle to get right.

Truth be told, our collective starting point and job at hand is the re-imagination and transformation of systems, institutions and infrastructure that arises from a mindset of humans as separate and solely self-interested—leading to systems built on zero-sum game logic and competition; both as competition amongst each other in the present as well as between us and future generations in a big question on what is a fair intergenerational distribution of wealth and resources.

The strategic lens to approach societal transformation through a DAO approach and the different tools (dApps) built on it allows us now to re-think, test, build and refine the fabric of our societal design as we want it to be and not as we know it to be. This process of collective innovation has live experimentation and learning at its core, with a preference for emergent learning over accountability to a predetermined set of specific goals. The instantaneous nature of a connected network drives radical transparency through instant feedbacks which mitigates the obfuscation of flaws we see in our current system. Challenging our current short-termism oriented systems and -institutions means interrogating the fabric—the elements that ensure that short-term views and actions are hardwired into our institutional frameworks from accounting, valuation and rating methodologies, regulatory standards, processes and many more.

Structuring twenty-first century flows of value that has climate justice and social equality built into their core means to radically democratizing the innovation process that brings the world onto this path; it means investing in infrastructure and institutions that help overcoming a highly centralized societal system that has increasingly structurally excluded most of the people and has stripped them of their chances to participate in shaping their own future.

The Web 3.0 allows us to truly democratize this societal innovation process but it only offers the enabling infrastructure which levels the playfield for us. Building further on this technological infrastructure as an enabling environment, emerging technologies will help us overcome the real, dominant and fundamental societal deficit in future-oriented thinking which has permeated our psychological, cultural, technological, legal, financial and political infrastructure. None of the mentioned areas can be viewed in isolation and disconnected; this would inadequately simplify the complexity of the task. All of them are highly interdependent and in sum are amplifying our collective bias towards the present which results in short-sighted and vulnerable subjects, short-term decision making, waste economies and a growing political and spiritual fracture between generations.

How can we now approach this process and what are some of the dimensions which seem to be key—to be at least considered as design principles if we aim to actively curate this socio-economic transformation with a success probability beyond chance.

2.6.1 *Resetting the Rules of the Game—Rethinking Governance and the Role of Regulation*

It may be fair to say that rules and regulations are key pillars to influencing behavior in society and provide the framework for coordination and cooperation efforts of actors within. We take the design principles that underly our rules and regulations mostly for granted but it is worthwhile to examine these design principles in more detail. This exercise is important if we were to interrogate our dominant governance models (of any given system of analysis) and skillfully build new opportunity spaces for rules, regulations and governance models that are designed to generate the maximum amount of value out of a given system aligned with the purpose around which the community has been formed in the first place.

Our current dominant socio-economic governance models, regimes, believes are deeply rooted in the eighteenth century and are dominantly driven by structural thinking, managerial mindsets and fixed (mostly singular) identities. Back then, the dominant societal model was one of the high degrees of decentralization, very loosely connected. Industrialization assigned humans a very specific role as useful executors/robots within industrial processes driving a value model of economies of scale. Along came the incorporation and founding of institutions that further solidified the role of humans within an increasingly industrialized world— e.g., educational systems, insurance- and financial institutions and -regimes, accounting structures, etc. To give an example: educational systems were designed towards specialization in isolation in artificially disconnected terms—as opposed to system thinking and interdependence. All of which happened under the watch and with the consent of the power monopoly of the nation state and its institutions re-enforcing the idea of a structural and managerial world, a clearly defined and accepted power monopoly and a clear, singular identity to affiliate with. Again, the internet and the globally super organism of connected human beings have challenged and are challenging this historical model, yet again.

Defying the traditional rules and regulations—and thus, power structures, unrecognized, the internet has decentralized the ability to learn, innovate, be creative and has thus opened a new window of opportunity to organize and therefore challenge traditional norms, rules and regulations which provides an opportunity space for governance innovation fit for the twenty-first century and its challenges.

This platform of connected humans has given rise to a new type of rules, regulations and governance models that are fundamentally built around shared consciousness of interconnectedness and the role of "me" in a shared world. Decentralized Autonomous Organizations allow for what is widely called "the boring revolution" (Giorgino 2018). The boring revolution on governance innovation is a widely overlooked field of activity that drives socio-economic change. For far-reaching climate action on the structural systems level to succeed we need to experiment, test and find consensus around a new rulebook on how we interact with each other and with the planet around us, based on emergent learning beyond the structural identities we know.

Governance is the secret ingredient that holds human relationships and gives them context. We have collectively become used to the existing paradigms and have adjusted our behaviors and belief systems according to this socio-economic framing. At the same time, we have forgotten that governance can be so much more than just limiting and punishing. Governance innovation can build the condition for progressive realities that embrace emerging realities beyond our multi-generational path dependencies. The infrastructure of the DAO allows us to dare and dream what the societal rulebook can be or should become. It is those very areas where we can collectively use the DAO platform to explore what it means to build governance:

- For entire systems beyond single purpose, a single product, or single service as the putative silver bullets.
- For driving change in a complex world beyond the governance of administration and managing the status-quo.
- Adequately accounting for the realities of a full world where externalities can no longer be forgotten beyond governance for a world of the perceived illusion that externalization has no effect.
- That democratizes agency for socio-economic change.
- That empowers as opposed to controls.
- That allows for and embraces emergence as a reality of complex system behavior beyond governance that fixes legacy regimes and -institutions..
- That nurtures and cultivates consciousness for interdependency within a shared world and reality.
- That accelerates agency for innovation, creativity and the capacity to innovate.

Obviously, this section is aiming to describe and give depth how such a vision of an empowering governance innovation process can look like. The realities of the complex, adaptive world we live in prevent us from postulating how such governance models could look like. The challenge for change agents and decision makers in the now is very much around examining the role and responsibility that each and every one of us can have in this process.

2.6.2 Blockchain, Self-organization, Coordinated Action and Emerging Outcomes

Governance innovation helps us to collectively establish the rulebook based on shared consciousness and interconnectedness. The next big question is now: how to generate growth models that are driving value creation and are capturing value in a way that is respecting the physical boundaries of the planet and protects the legitimate rights of future generations?

This question is fundamentally about the reconciliation of the dichotomy between every person's raw self-interest and the greater human freedom—in other words, selfishness versus altruism. Amongst the myriad of socio-economic pivots that are currently explored only those will emerge as being superior and viable that align every person's raw self-interests and the greater human freedom. The key area of innovation where technology will propel humanity forward in this digital value chain and society is the area of incentive design and the exploration of new notions of value through crypto-economics and digital cryptocurrencies based on blockchain technology. Incentive design and governance rules and regimes are essentially the two sides of the same medal. The following basic concepts are necessary knowledge to understand and capitalize on the potential of crypto-economic incentive design.

Blockchain technology (Strogatz 2001): a distributed data structure that is collectively maintained by a connected network through a consensus mechanism. Blockchain technology is an umbrella term that refers to the totality of blockchain protocols in existence. Different blockchain protocols may vary in their degree of decentralization or their consensus mechanisms.

Cryptocurrency: Cryptocurrency refers to the ability of each blockchain to represent and transfer value on its native network. Cryptocurrencies (mostly referred to as plural) is an umbrella term for all the different network tokens in existence (https://provocations.darkmatterlabs.org/the-necessity-of-a-boring-revolution-a71b1ae6f956). The value which is associated with and captured in a cryptocurrency depends on the individual context of the blockchain protocol and cannot be generalized.

Token: Token refers to the specific network role and functionality of a native blockchain protocol. Native network tokens of a blockchain are associated with a certain value and grant the holder certain predefined privileges within the network for network specific actions. Each blockchain issues and runs its own network token.

Smart contracts: Smart Contracts are consensually agreed pieces of executable code which are executed if preprogrammed conditions are met and produce specific outcomes.

The above mentioned terminologies, practices and principles can now be applied to and put in the context of information- and network theory (https://www.coingecko.com/en). Networks, irrespective of whether they are physical or virtual networks develop and can be maintained if three key features of healthy networks are (Bueger 2013):

- Transactionality: The network viability is dependent on the user's perception of value gained through interacting with the network. Blockchain is decentralized network infrastructures that are peer-to-peer meaning that no intermediary party is acting as a censor or bottleneck in any regard. This means that the value perceived by a network participant is one that is derived from direct peer-to-peer interaction.
- Accountability: Networks function when participants follow the rules of the network. Blockchains operate on a predefined rulebook which is coded into the blockchain protocol—often referred to as "code-is-law". Network rules cannot be tempered with, changed and/or ignored by individual malicious actors. The rulebook addresses a very fundamental flawed principle of human–human interaction which often represents a single point of failure—the requirement for trust. Coded rulebooks that define the level playfield combined with the power of blockchain to

be able to verify individual interactions of network participants obliviates the need for trust. Therefore, blockchain-based networks are often referred to as "trustless" which will become critically important when we speak about incentive design a little later.

- Relative frequency: Networks function when participants are interacting with each other on a regular basis. This is where Incentives come in. There are three critical junctions where incentives play a role with respect to relative frequency. (a) Incentives are critical for potential network participants to opt into the network in the first place, (b) incentives are critical to maintain and even intensify the interaction with a network and its network participants, (c) incentives can attribute weight to emerging outcomes and can therefore provide direction; e.g., incentives can reward favorable outcomes and disincentivize unfavorable outcomes. Incentive alignment is therefore critical to achieving a high degree of actor's alignment around a common purpose.

In summary, blockchain technology, coded rulebooks and trustless peer-to-peer interaction allow for incentive design to unlock the full potential of the four principles that drive human actions and progress: coordination and deliberation, cooperation and collaboration.

Given the complexity challenge we face with regards to the climate emergency as discussed blockchains now unlock the potential of full peer-to-peer coordination and deliberation to overcome the bounded rationality of- and information asymmetry between individuals. Coded rulebooks are unlocking the full potential of inclusive (permission-less), participatory, self-organizing systems which are guided by common principles and solve for complex socio-economic questions not based on top-down planning approaches but on bottom-up emergence approaches—at scale. Those decentralized, self-organizing systems are guided by an individual as well as collective incentive regimes that give directionality and a measure of the usefulness of agreed actions and experiments.

Our times offer a unique constellation and opportunity. On the one hand, we have never in history, seen more knowledgeable, determined and capable connected individuals that understand the climate emergency and are able to collectively assess complex problems from a different and more flexible perspective. On the other hand, we hold, for the first time in history, the technological toolkits in our hands that allow us to coordinate and bundle this human energy and focus it on solving for a climate-just, inclusive and prosperous future.

2.6.3 Actor Alignment Through Incentivization for Noble Behavior and Emerging Outcomes

Before we discuss incentive design, incentive mechanism and value regimes in more detail it is important to take a step back and first discuss how incentivization and value attribution is handled and processed in our current dominant socio-economic regime.

If we look a bit closer we should come to the simple conclusion that money—fiat money in our economic paradigm—is the single most important incentive that drives behavior and action. We have collectively designed a system that uses money as a mono-dimensional measure to incentivize for action and the willingness to exchange with price being the measure for the monetary value of things, goods and services at which we are able to decide whether we want to exchange or not. If we now expand our perspective we will relatively quickly realize that our current and dominant notions of value, our societal systems and beliefs and our societal institutions are very much built around this mono-dimensional notion of value—which is money.

We know from monetary and market theory that systems are very good at a valuation to price goods, services and other things. The idea of price as a collective, social voting mechanism has been made popular by the early Friedrich A. Hayek in "The Use of Knowledge in Society" (Oh et al. 2016). Hayek writes: "*Fundamentally, in a system in which the knowledge of the relevant facts is dispersed among many people, prices can act to coordinate the separate actions of different people in the same way as subjective values help the individual to coordinate the parts of his plan*". Hayek's thesis postulate that information is always distributed and access to information is always incomplete but that selfish monetary incentives and human interaction will find an optimum equilibrium that is optimal for all the participants in the exchange (Hayek 1945) (and along the value chain). Hayek's thesis is convincing for the pricing of goods and services in the now. When it comes to the climate emergency we have empirical proof that the market is extremely inefficient in accounting for and pricing climate risks. How do we price future risks? How do we price negative externalities? How do we value things that cannot directly be priced? e.g., affection, reputation, empathy. How would a valuation system look like that would be multi-dimensional and where we would invent and assign unique expressions of value (token) to all actions as well as tangible and intangible outcomes? How would such a multi-dimensional value system change human behavior and -action as well as human–human interaction? How would a completely tokenized representation of our collective actions and -expressions change our perception of the responsibility we hold for our actions—in the now as well as with respect to future generations? How would such a multi-dimensional representation affect:

- Our sense of transactional value between peers?
- Our sense of accountability with respect to others, the physical world and time?
- The willingness and the frequency of our interaction with others?

This is where incentive design and token economy comes into play. How would Hayek's thesis of the use of information in society change if any information—without limitation—could be expressed explicitly through tokens? How would his thesis on coordination come to life if coordination would not be based on a single value system like money for the price but on a multi-dimensional value system based on tokens—or a stack of tokens using blockchain technology? In modern terms, Hayek's thesis would be called a crypto-economic approach. Although no generalized definition of crypto-economics exists so far a widely accepted definition by

Vitalik Buterin (Hudik 2020) refers to crypto-economics as the "*combination of cryptographic proofs of past events and economic incentives to encourage future events inside a blockchain system*" (Buterin 2017). Incentive design and token economy is the space at the intersection of cryptographic technology and behavioral economics (Filippi et al. 2015a). While the discipline of crypto-economics is still very young and in its infancy, there are a list of interesting approaches and initiatives that aim to apply the potential of blockchain, token economy and crypto-economics to a wider societal transformation angle, e.g., focusing on participatory economics of horizontal organizations (Kanev 2017), designing for future-fit economic spaces and accounting instruments (Davidson et al. 2016), reconciling the notions of public and private value systems in the commons approach (Virtanen 2018) as well as exploring the potential of crypto-economics for designing a sustainable circular economy (Filippi et al. 2015b).

The societal potential of crypto-economics is to design an information representation- and information flow system that represents and accounts for the complexity of the real world to overcome the contradictions that arise from compromised legacy systems that produce linear deterministic outcomes and are thus, gravely insufficient to solve for real-world problems like climate change. In accordance with Ashby's law of requisite variety, we will only be able to address complex problems like climate change and future green growth if we are able to transform the underlying beliefs and principles of the current socio-economic system design. Emerging technologies are putting us collectively in the position to experiment our way into a more sustainable future. However, it is important to apply the necessary transformational lens to the challenge as well as to the solution. "*The grave problems now facing the global economy have been building up over many decades. In large part, they are the byproduct of well-intentioned but mistaken macroeconomic policies, themselves based on mistaken assumptions about the nature of the global economy. A philosopher would say policymakers have made a fundamental ontological error. Changing those underlying beliefs and thus avoiding the disastrous effects of «still more of the same» macroeconomic policies, is the biggest challenge now faced by policymakers.*" (Dapp 2019).

Incentives, however, are only impactful if the incentivized human action achieves impact and change in the real world.

2.6.4 Linking Virtual Communities to the Physical World—The Role of Data

The final aspect in our discussion on coordinated action beyond governance and incentives is the question of how to link virtual DAO communities, their actions and rules to real impact in the physical world?

Cooperation and Collaboration in a virtual community only matter if the actions affect and have an impact on the physical world. Coordination and deliberation are only meaningful if they are a direct response to signals from the physical world. Virtual communities can only achieve impact if they are deeply connected to the physical world. A DAO, as an example of a virtually organized community, is only an enabling platform whose impact can only be called successful if it manages to leave legacy socio-economic developments and -regimes behind and gives way to new forms of structural reforms and developments that span across all aspects of human life and human expression—from how we interact with each other to how we share and exchange value to how we store value to how we invest into the long-term?

How do we now technologically ensure that any given virtually organized community has the necessary feedback loops that link all its actions to the physical world? Bridging the virtual world and the physical world is what is widely called the Oracle problem (White 2019).

Causal decision making on blockchains usually happens through smart contracts—a deep coded causal logic that follows the *if–then-else* logic. However, smart contracts are not as smart as the name would suggest. In order for them to be able to execute a code in a way that results in meaningful outputs, the smart contract requires data or more generally speaking environmental information, that provides the context in which the code is executed and an output is generated.

Providing trusted and reliable information at the right time is therefore key for any smart contract to operate in a meaningful way (Caldarelli 2020). Consequently, the management of providing reliable data as inputs to the blockchain and all the smart contracts that it entails is therefore key and absolutely essential to produce and drive meaningful outcomes.

Solving for the oracle problem is one of the major barriers to the increased utility of blockchains (Hamilton 2020) for the masses across a wide variety of markets and use cases—including climate action. Let's take a step back and discuss first where data can potentially come from that can serve as context inputs for smart-contract execution of a Decentralized Autonomous Organization.

The Internet of Things is widely referred to as the world of interconnected things that are capable of sensing, actuating and communicating amongst themselves and with the environment (Mammadzada et al. 2020). Additionally, it is the engineering discipline that allows us to perceive and affect the physical world through data and sensory devices. Sensors acquire the data that can serve as digital inputs for a smart contract. These sensor webs are expected to produce zeta-bytes of data that reflect the state of the environment and those massive data loads will be collected by billions of sensory devices (Qusay 2018). Interestingly again, that most of these use cases and IoT applications again predominately focus on existing value chains and problems and very few have a transformational agenda on their mind. The opportunity space where IoT and environmental decision making intersect is gaining traction rapidly (https://www.mckinsey.com/industries/semiconductors/our-insights/whats-new-with-the-internet-of-things). However, at this point, the maturity and integration of environmental data as oracles into the decision making of coordinated virtual communities for transformational change are still in its infancy. The development of this space will require significant space building both from the perspective of deep subject matter expertise as well as from the perspective of disruptive leadership which provides the enabling environment for this space to develop further. Bridging the space of environmental data and integrate data-driven decision making into sustainability initiatives based on virtually organized community to drive coordinated, data-driven transformational action (Hart 2015) will be a key asset in any transformational change effort. Making the immense reservoir of data that produces insights into how the world works accessible to the DAO offers the pathway to data-driven transformational socio-economic design. Bridging the two worlds offers the potential of redefining the way multiple parties, that do not necessarily trust each other or don't even know each other can engage in shared contractual agreements and exchange value with each other. Solving for the Oracle problem means paving the way for a new form of social interaction of coordination, deliberation, cooperation and collaboration. It offers the opportunity to explore alternative data-driven business models that do not require trust between interacting parties and it offers the opportunity to explore alternative valuation models and value-representation regimes with clear environmental and socio-economic impact. Where the subjective, individual perception is equalized by a reliable middleware that provides trusted data the quality that human nature can bring to the mix is far from being obsolete. A trustless environment empowers, frees energy that can be allocated to a renewed form of value creation—one that does not necessarily seek to protect outcomes but one that aims to share outcomes. Cultivating the importance of human qualities in a decision making process immersed in a trustless environment of peer-to-peer interaction by design represents the ideal enabling environment to transform our collective mindset from a view of scarcity to one a view of abundance.

In such an environment the oracle acts as a reliable middleware that facilitates exchange between blockchains and any off-chain system, including data providers, web APIs, enterprise backends, cloud providers, IoT devices, e-signatures, payment systems, other blockchains and more. It eliminates the ambiguity of subjective views and potential information asymmetries and allows actors to shift attention to a different, more sustainable and more long-term oriented form of value creation (Oberhauser 2019). Furthermore, when it comes to transforming entire value chains and systems the decisions that are required to embark on this path are inherently complex, ambiguous, non-linear and reach beyond the pure analytical nature of simpler cause-effect decision making. However, transformation happens in the space-and from a place of not knowing and thus the required decisions begin to delve into the realm of subjectivity and/or require external data that are not accessible to every node in the DAO network. With the maturation of powerful, trustable, flexible and scalable Oracle function as a critical tool for better collective decision making using smart contracting another technology will share the spot light of this success story—Artificial Intelligence. AI will be a key technology both outside of the blockchain environment (off-chain) as well as at the interface to the block chains themselves. Having a zeta-byte of data is only useful if the data are compiled and aggregated in a form that allows for their use as an oracle input. Artificial Intelligence will play a critical role in ensuring effectiveness throughout the Oracle process (see Fig. 2.5). Once a query from a smart contract for off-chain data and data-bundles of the blockchain

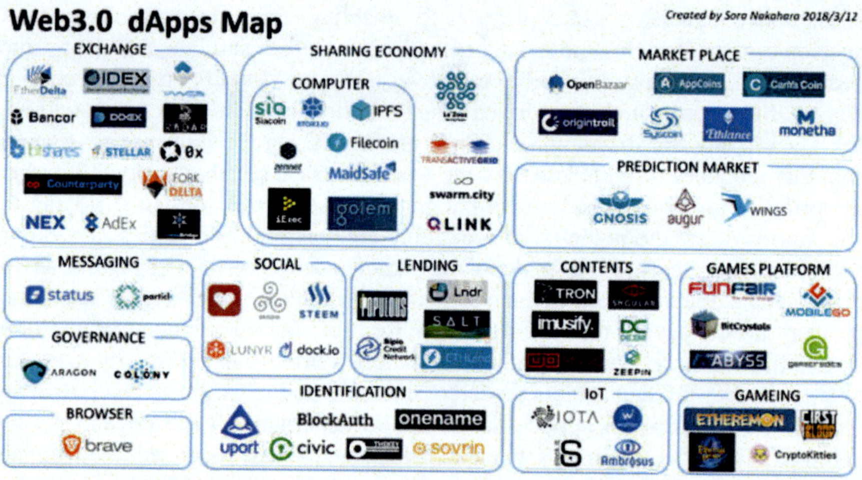

Fig. 2.5 Decentralized applications are built on basic blockchain infrastructure layers. The transition from our currently mostly centralized processes, institutions and regimes to the fully decentralized representation is ongoing and extends into and reflect virtually all aspects of human–human interaction

is detected, AI will play a critical role during the extraction, retrieving and formatting step of the process (Caldarelli 2020). Imagine the smart contract requests very context-specific data which cannot be obtained from one single data source. It will be Artificial Intelligence which ensures that data sources are constantly monitored, clustered, indexed (Teng, et al. 2018). It will be Artificial Intelligence that ensures quality control and robustness of the fetched data sets (Kumar, et al. 2017). It will be specialized AIs that will be able to compile the raw data into robust information bundles which meet the criteria of an input to the Oracle process and can be processed for the smart-contract queries. Furthermore, Artificial Intelligence will also provide housekeeping functions like data forensics (Dai, et al. 2018)—ensuring that the data sets are not tempered with. Specialized AIs will also detect, report and react to malicious actions to ensure the integrity of the process (Joseph 2019) (Fig. 2.6).

It is impossible to predict in which areas the smart use of technology and technology stacks will catalyze societal transformation and in what form. The complexity theorem does not allow us to predict the future with any meaningful statistical confidence. This section aimed at opening up different perspectives of how societal transformation and emerging technologies are intertwined. This fact makes technology always also a political instrument. More than becoming experts in any of those areas described it will be important for change agents going forward to cultivate a responsible relationship to technology—one, that is also a political one. Technology never was and will never be—neutral. Where and how we deploy technology has real consequences for humans and therefore when it comes to just societal transformation we also always have to be aware of the political dimension of technology.

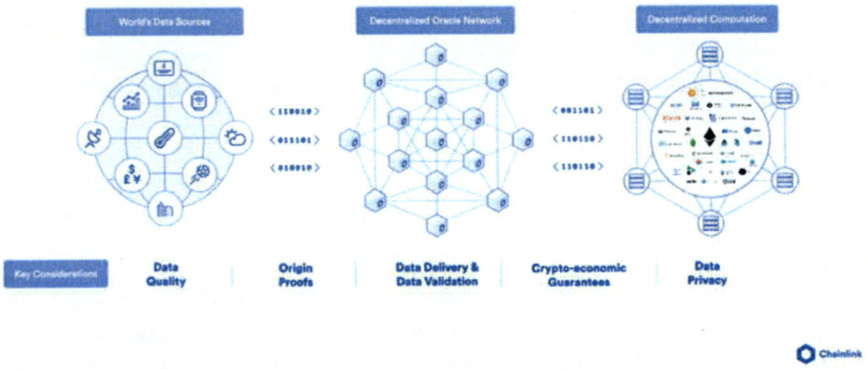

Fig. 2.6 The basic oracle function © by Chainlink: (a) the oracle monitors the blockchain for any potential smart-contract queries that potentially require off-chain data, (b) the oracle then fetches the requested data from any data source or multiple data sources to which it has access through designated APIs, (c) the oracle compiles the fetched data in a way that is readable by the smart contract (d) depending on the nature of the smart contract the oracle may also provide output service to off-chain recipients after the smart contract has been executed

With technological innovation in full swing, it will nonetheless be the human's conscious choice to give those developments a purpose and a direction. It is the irony of the role of leaders that, by committing to and supporting a societal transition that moves from structured to networked societies leaders at the same time disrupt their own future role and future relevance as leaders in the long term. Leadership, as we know it from our structural world, will become a participatory agency in the decentralized future. Transactional authority will become cooperative and collaboration participation. Disruptive leaders have a unique positioning and privilege in the transition process by being able to pioneer a new form of "political" technology use—one, where technology is in service of a just and inclusive society.

2.7 The Leader's Challenge Within Fully Decentralized Society

So what? What makes it so important for disruptive leaders to understand the complexity from a first-principle view in the context of technological progress and societal transformation?

The more the world transitions from a structured to a networked organization the more the traditional forms of structures and leadership will be increasingly marginalized until they will become irrelevant. Both for structures, institutions and leadership the ambition must be clear: either adopt or perish.

The good news is that bold disruptive leadership will be required while the world transitions into the new, networked organization form. It will require leaders to re-invent identity so that their participation in the transition has a catalytic effect.

What does it now mean to be a leader in a VUCA world? It means acquiring new principles, frameworks, toolkits and leading in service of a transition knowing that the world we are transition into will be inherently self-organized and leaderless. It requires bravery to acknowledge that disruptive leadership in the now means working toward a world where leadership as we know it will no longer be needed. Understanding the mechanics and drives of this change is an important start and shall be motivation and inspiration why a decentralized world is desirable as it reflects the real possibility for a more just, inclusive and equitable world. The hard truth is that technology has already initiated an irreversible movement towards true democratization of the world which no institution will be powerful enough to stop. Take the case of Bitcoin (Li 2018): a censorship-resistant, seizure-resistant, border-less, permission-less, pseudonymous, peer-to-peer type of programmable money without any central coordinator or issuer—protected by the most powerful aggregated computer network and safely stored on more than 10,000 distributed computer nodes all over the globe. Bitcoin challenges societies to re-think the relevance of its institutions, it surfaces weaknesses of existing monetary architectures, its adoption, mostly in destabilized countries makes it a tool of freedom for the suppressed. In the Bitcoin network, no single actor is more important than any other. No affiliation, nationality, gender, religion is of relevance. It is a great societal equalizer that levels the playfield and opens new opportunity spaces to re-think notions of value, justice, equality. This leaves us with the hard choice to either opt-out of the change and resist this change or immerse in the process and become part of and own the change from a place of active catalysis.

In a non-linear world of emergence, there cannot be a blueprint for what leaders can and should do. The following section, however, aims to provide action domains, mindsets and toolkits that can help leaders to re-discover what leadership in this emerging environment can be.

2.7.1 Leading the Change

To say it clearly: Leading the change sounds very much like a buzzword. However, on the contrary, it's the most integral, humbling and daunting thing to do when it comes to disruptive change. Leading the change means acknowledging that things are, in fact, changing; that we are tapping into unknown territory when it comes to the necessity of transformational change. How do we now lead the change when stepping collectively into unknown territory where no blueprint and playbook for success is available? It starts from a place of acknowledging that we do not know what is to come and be, if not comfortable, at least at certain ease with it (Complexity Theorem 1). When it comes to transformational change not knowing is a good thing. It means that we are truly opening up new avenues and that we are about to truly unlock new potential. Not knowing is scary. The natural human reaction is to retract into saver land and known territory. But if change is inevitable this safe land is potentially a liability of the future and therefore not so safe as our mind wants to make us believe. Where the past is a liability and the future not yet written then it is usually our imagination that paves the way in which direction to walk. Disruptive leadership means in this context to open, develop and provide the spaces and tools to deal with this uncertainty constructively and skillfully.

2.7.2 *Embracing Challenge-Led Innovation*

The difficulty with operationalizing a systems innovation approach is that it doesn't happen in confined structural boundaries (Theorem 2) and within known identities (Theorem 3) like value chains, industries, public or private actors. The required systems transformation will disrupt existing value chains and will give rise to new ones—some of which we don't even know what they could look like. Because we are moving into the unknown territory it is essential to start from a challenge-based approach (Nakamoto 2008) to face the "grand challenges" of our times. In a challenge-based approach, the direction of travel for the innovation pathway is defined by pivoting and approximation towards and description of desired outcomes or a desired future (normative futures). In a challenge-based approach this normative future functions as the north-star to work against. It provides the necessary strategic guardrails that ensure that any innovation effort is in service of the desired future. Having a direction of travel enables an innovation activity to constantly revisit emerging outcomes, also known as sensemaking (Mazzucato 2020) and ask the so what question along the way (Theorem 1). The challenge-based approach is particularly suitable for the "grand challenge" of climate change as it represents a global challenge bigger than any country, value chain, or industry. The challenge-based approach, instead, argues for an active role of all parties, irrespective of their identity in a legacy system's perspective. It aims to re-think identities and roles (Theorem 3) and provide a level playfield for activities aiming for a certain outcome that is favorable for all (Kurtz 2003).

The role of a disruptive leader in the context of challenge-led innovation is to bridge the silos that legacy systems have brought to the fore. Institutional boundaries are often affirmed by the use of language, subtext and context. For challenge-led innovation to succeed it is vital that the disruptive leaders provide spaces where those different cultures can meet, learn with and through each other and find common linguistic ground that enables communication.

This process of cognitive order formation (Mazzucato 2018b) is quint-essentially important in ambiguous situations. In a siloed world, where every stimulus pattern allows for more than one interpretation, norming of language and understanding is critically important to start an inclusive change process. Common language allows for finding common purpose and it helps to build the necessary ambiguity tolerance

(Kruse 1995) which helps to navigate transformational change in complexity where known identities are starting to blur and faint.

Such reconciliation spaces also cultivate bottom-up resilience by establishing new and non-obvious connections that are required for a transformational journey.

2.7.3 *Beyond Innovation Models—Using Futures Skillfully*

"*Progress is the realization of utopia* (Endres 2009) "and utopia is a result of our imagination (Quote attributed to Oscar Wilde). Progress, therefore, depends in many ways also on our ability to individually and collectively imagine different futures which require us to acknowledge that the future is neither deterministic nor singular. Imagination and anticipation are the magic capabilities that allow us to turn complexity from a liability into an asset (Bregman 2020). Using the anticipatory systems and processes that not only surround us but are critical ingredients of any action in the most skillful way possible allows us to translate human agency and ideas into the desired future we want. Being "Future literate" is helping to cultivate the human need of anchoring the future, which is inevitable, in the present thus fueling desire, seeding excitement and helping to prepare people for the inevitably rocky path of transformational change. Imaging different futures delegates agency in a process and fosters participatory and inclusive action. It is entirely technology agnostic and entirely human centric. One approach, framework and tool that can help in this regard is the discipline of Futures Literacy (Miller et al. 2018).

Future's Literacy provides a practical framework for making sense of the anticipatory assumptions that distinguish different kinds of futures. It is also a practical skill that develops as people gain a better understanding of anticipatory assumptions. It provides individuals and groups with a framework that helps understanding how anticipatory assumptions are a product of path dependencies and how letting go of fixed assumptions unlocks the full transformational potential of the human mind. Using Future skillfully allows actors to cultivate a new and constructive relationship with complexity in itself which in turn, cultivates the sense of self in a transformational process. To bring people along on a journey of transformational change requires everybody to be able to imagine alternative futures beyond the obvious path dependencies and based on the realities that we are familiar with.

The role of disruptive leaders in the challenge of bringing about and acting within a process of inclusive change is to democratize the ability to deal with the future in the most skillful way possible. It is the role of a space builder. Making space for an individual as well as collective experiences like Futures Literacy and enabling participation—irrespective of profession, educational background, or any other classification is a key contribution that leaders with resource access and power can provide to accelerate the change they themselves want to see.

2.7.4 Zero-Gravitational Spaces Experimentation Spaces for Transformational Change

It's daunting to us that our current systems are not future-fit while, at the same time, we are only starting to learn and understand how a transformational journey could look like. We are baking a cake, we know the ingredients but we have to pivot the recipe as we go. It is the privilege of leaders to put their authority and their influence in service of this transition and build the spaces that allow change agents to develop new and future-fit socio-economic paradigms. It also means that disruptive leaders hold a particular responsibility when it comes to including everybody in the process through empowerment and upskilling. A few areas may be highlighted here for inspiration where disruptive leadership will be essential and disruptive innovation will be required. We do know that the change we embark on cannot and should not be planned. We create this world through experimentation and emergence. Outcomes, that have a truly transformational potential are not necessarily compatible with nowadays institutions, value chains and systems—this is a feature of a transformational path, not a bug. Therefore, the merit and utility of emerging outcomes of a transformational journey shall not be assessed in current valuation logics and -metrics but exclusively against their merit and utility in the context of desirable futures. The idea of zero-gravitational incubation spaces is to cultivate the ability to develop transformational approaches and solutions defying the gravitational energy of a dominant market paradigm. Zero gravitation experimentation spaces shall open the path towards building the desirable future new instead of investing the energy in fighting the old legacy systems. It requires disruptive leaders to embrace a much wider angle to societal innovation, which such a zero-gravitational space needs to embrace. A few examples where rethinking is required:

- *How to develop the institutional frameworks in the context of desirable futures?*
- *What democratic governance capabilities and institutions do we need to manage desirable futures long-term?* Requiring to re-think notions of value
- *How to cultivate the required beliefs, mindsets and cultures that are supportive of a desirable future?*
- *How to construct a portfolio of real-world interventions that drive transformational change?*

Critically, the answers to these questions cannot be taken in isolation. As complex adaptive systems are constantly evolving and maybe even exist in multiple states at the same time - they are ambiguous and predictive measurement can become statistically meaningless in the analysis. Therefore, it matters that we view those different lenses as a portfolio of strategic synergies and correlations, selecting individual assets not based on their individual merit but based on how they reinforce the portfolio's potential to unlock transformative dynamics.

2.7.5 *Deep Demonstration Spaces for Orchestrated Experimentation and Real-World Testing of a Transformative Theses*

Deep Demonstration differs from Zero-gravitational spaces, that they take experimentation and experimenting portfolios into the real world.

In order to show and demonstrate what is possible in terms of societal transformation on a level of whole cities, whole regions or even of whole countries or whole of business models and value chains we need to immerse in those systems and test the theses and develop it further through experimentation, learning and sensemaking. Until we see it, until we feel it, we do not really get to know what's possible at the speed and scale we need. Deep Demonstrations surface resistances, frictions, potentials and opportunities. In short, they give us a reality check of what is possible and what we are missing to get there. Across Europe and the world, we witness cities with massive decarbonization ambitions, ports across Europe with a similar challenge, or entire regions that want to overcome their dependencies on legacy energy systems and carbon-intensive value chains. Deep Demonstrations are intended to build a bridge between the initial experimentation and thought leadership on change and the real world. It builds a proposition to address the challenges of systems on their decarbonization pathways through strategic experimentation and integrated learning. It requires disruptive leadership on both sides to hold this uncertainty and provide the spaces that provide the evidences step-by-step that transformational change is possible and can be accomplished through real-world experimentation.

2.7.6 *Reconciliation Spaces for Learning and Sensemaking*

In an agile innovation approach, based on sensing and hypothesis testing and experimentation across connected portfolios a key process that allows a community of innovations and change agents to reconcile hypothesis and learnings is the process of sensemaking (Miller 2021). Sensemaking, as a capability, provides an extraordinarily influential perspective that produces insights on how individuals or communities appropriate and enact the reality. When it comes to agile innovation based on hypothesis testing it is an extraordinarily important step towards generating impact to make sense of:

- Which insights and learnings can be drawn from experimentation and how they relate to the original hypothesis?
- How different learnings from experimentation across connected portfolios relate to each other? Is there complementarity, synergy, antagonizing effects?
- How do learnings related to the desired outcomes and
- Which insights can be drawn from the sensemaking process that will refine the next iteration of orchestrated experimentation?

Sensemaking is the process of building intelligence as a function of an initial hypothesis. Intelligence, in its original Latin meaning *"inter-legere"* meaning between the lines or to put things into perspective. By design, moving into and within uncertainty surfaces a lot of ambiguities, conflicting signals and aspects that are hard to reconcile. The process of sensemaking is designed to support the reconciliation where reconciliation is needed. Reconciliation could potentially be required between a wide array of conflicting aspects, e.g., between different technological approaches, the relationship between technological implementation and different regulatory frameworks, more generally speaking between bottom-up initiatives and top-down policies to ensure that coherent principles are also vitally considered and accounted for in the local context. Sensemaking between the observed outcomes and the bigger ambition. To illustrate, once we enter the space of technology in the context of transformational change we are immediately also exposed to potential policy/implementation friction that arises from one or more of the following potential interfaces: (a) ensuring interoperability between different private and publicly used blockchains, (b) the acknowledgment and use of different technical standards, (c) ensuring the integrity and privacy of personal data, (d) ensuring that the use of Artificial Intelligence algorithms follows the recommendation for the Ethical Use of AI (in the European Context (Alderman 2005)), (e) the handling of bad data and quality assurance of data records, f) formulate compliant processes for malicious acting and data theft to name a few. Reconciliation and sensemaking are the tools to eliminate barriers for transformational change in a skillful and constructive way and it is a means to ensure that different innovation communities and actors are in constant exchange with each other with an ethos of mutual learning and common purpose (Fig. 2.7).

2.7.7 *Initiate and Lead Upskilling and Re-Tooling Initiatives*

The nexus of emerging technologies and far-reaching climate action holds enormous potential to shape a de-carbonized future that is inclusive, climate-just and prosperous for all. However, experimenting with emerging technologies and harnessing their potential is currently a privilege of the few. Real inclusive change starts with the possibility of everyone participating in the process. The digital divide that exists throughout the world between the privileged few and the underprivileged many is a real barrier for inclusion in the process of societal transformation.

It is the role and responsibility of the disruptive leader to acknowledge this fact, initiate and lead the necessary upskilling initiatives that equip everybody with the tools that are necessary to participate in the process. Disruptive leaders should also keep in mind that the educational environment is dramatically changing as well. In order to make education meaning- and impactful upskilling initiatives must meet the velocity and agility of the technological advancements themselves. Otherwise, they risk lagging behind. Traditional educational structures are running the risk of being marginalized by more effective and efficient peer-to-peer and distributed learning

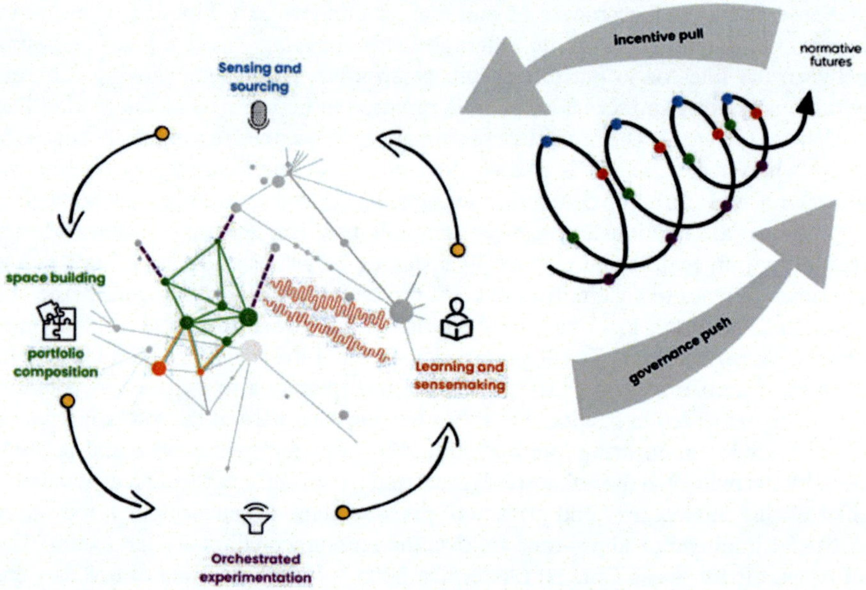

Fig. 2.7 The role of a disruptive leader in the context of transformational change is post managerial and post expert; it is about providing space, perspective and context. The key capabilities that a disruptive leader makes space for a network to perform its essential innovation tasks: (a) sensing and sourcing to ensure that network dynamics are appropriately recognized and acknowledged, (b) strategic portfolio building of potentially connected portfolio experiments (impact hypothesis) is formulated, (c) testing of innovation effectiveness through orchestrated experimentation and (d) learning and sensemaking ensure directionality of innovation activities. The disruptive leader (e) cultivates the prospect of what normative/desired futures can and should look like and focuses and ensures that (f) incentives nudge a system into a desired direction while (g) governance rules hardwire the emerging new rulebook

initiatives—mostly driven through the democratization of- and (mostly) free access to all the relevant information and know-how. In a post-expert and post-managerial world providing the information is insufficient to overcome the access barrier to permission-less participation in a transformative change process. Disruptive leaders should be less worried about what type of expertise and skillsets are required in the future. Peer-to-peer learning is extremely efficient in identifying resources and developing necessary skills to adjust to emerging situations and evolve from there. It is the leader's responsibility to provide the spaces and infrastructure for peer-to-peer learning. In a VUCA world, collective resilience and stamina emerge from complementary skillsets, -expertise and -partnerships are collective in strong partnerships and ecosystems of equals.

2.7.8 Capital, Financing and Democratized Futures

Our current economic paradigm to finance does have a view on the long-term, but it seeks to hardwire and dictate the future—to colonize the future. However, in complex systems, we now know that we cannot predict long-dated future values or risks with meaningful statistical significance. Long-financing requires us to rebalance our relationship with long-term predictions and our capacity to agilely respond. In order to finance for a just socio-economic transition, we need to build 21st-century facilities to finance an evolutionary future, not just a pro-longed, extended present. Long-financing must aim to build financing instruments and -logics that help to decolonize and democratize finance to take Long-funding and long-term orientation as a privilege of a few to making it the privilege and dominant mode of action of the many. An overly centralized finance system has structurally excluded most people from investing in and profiting from valuable assets and participating in financial markets—entrenching inequality by privileging those who have already accumulated capital whether through our inheritance laws or fiat currencies for exchanging value, which perpetuate wealth accumulation. At the same time, we are witnessing an explosion of alternative models of value creation and exchange unlocked by distributed ledger technology which could pave the way for a fundamentally distributed mode of financing which is by design permission-less, trustless and equitable. For example, using smart contracting infrastructures to accelerate the transparent tokenization of long-term value and risk and of civil legal instruments such as Trusts. The whole movement of decentralized finance (https://ec.europa.eu/digital-single-market/en/news/ethics-guidelines-trustworthy-ai) (DeFi) aims to replicate the traditional finance markets instruments, products and services but in a decentralized, inclusive and permission-less fashion. Align the democratization of financing structures, instruments and vehicles with transformational change initiatives will enable us to develop and structure twenty-first-century flows of value to have justice and equality built into the core of socio-economic models.

2.8 Synthesis

We are already witnessing the transformational socio-economic disruption which we have discussed in this chapter. This process, that mostly through digitalization and an increasing degree of interconnectedness across the entire globe, is irreversible. This transformational experience changes the world and changes us, who we participate in it. We are no longer able to plan for future outcomes, the power over process and outcomes shifts from a centralized few to the decentralized many. We no longer have a single identity in this process. We are both leaders as well as actors. We are both experts as well as generalists. We are both enablers as well as executors. Sometimes, not too unlikely, we are more than one identity alone. More than trying to seek refuge in known safe spaces this technology-driven socio-economic transformation demands from us to embrace the uncertainty of the dynamic. In these uncertain times, it is hard to keep track of developments and difficult to keep orientation. In those days it is important to return to the first-principle thinking which aims to establish the foundation and the framework through which any other observation can be described and explained. First-principle thinking is key to make sense of both the climate emergency as well as the dysfunctionality of our current socio-economic systems. As we know, they are essentially the same thing. Climate justice means social justice. While a significant number of climate innovations in all aspects that matter— e.g., technology, regulatory frameworks, accounting, policy, etc.—is happening already there is a need for an increased leadership to initiate, enable and curate the structural transformation towards a circular, green-growth economy. This is where leadership will matter in the future and also where technology can and will play a critical role. This chapter aimed to shed some light on a number of elements where we can witness the fruition of the initiatives that aimed to put technology in service of just socio-economic transformation. Yet, more still needs to be done and more energy, resources and attention is required to accelerate this just transformation. Climate action, therefore often has an enabling identity and our actions have an indirect positive climate effect. It shouldn't matter to us. Climate action is altruistic. When we serve us we serve others and the other way round. The climate emergency is the great equalizer that affects us all and reminds us that we only have this planet. Luckily, the technological tools at hand have never been more powerful than now to address this global threat through technology in a coordinated, deliberate, cooperative and collaborative manner. It will require to let go of many beliefs and convictions that gave us safety and comfort but it will also open our perspective for other convictions that hopefully are in line with our twenty-first-century realities of a globally connected, horizontally, permission-less world that develops and institutionalizes a green economy based on abundance and sharing instead of scarcity and ownership.

References

Alderman N (2005) Sense-making as a process within complex service-led projects. Int J Project Manage 23(5):380–385

Ashby WR (1958) Requisite Variety and its implications for control of complex systems. Cursos Congr Univ Santiago De Compostela 1(2):83–99

Bhowmik A et al (2018) Powers of 10: a cross-scale optimization framework for rapid sustainability transformation

Boisot M et al (2011) Complexity and organization–environment relations: revisiting Ashby's law of requisite variety

Bregman R (2020) Humankind. 978-1-4088-9894-9

Bueger C (2013) Actor-network theory, methodology, and international organization. Int Political Sociol 7(3):338–342

Buterin V (2017) Introduction to cryptoeconomics, ethereum foundation. https://youtu.be/pKqdjaH1dRo

Biffl S, Eckhart M, Lüder A, Weippl E (eds) Security and quality in cyber-physical systems engineering. Springer, Cham
Caldarelli G et al (2020) Overcoming the blockchain oracle problem in the traceability of non-fungible products. Sustainability 12:2391
Dolev D (1982) The Byzantine generals strike again. J Algorithms 3(1):14–30
Dai W et al (2018) Improving data quality through deep learning and statistical models. In: Latifi S (eds) Information technology—new generations. Advances in intelligent systems and computing. vol 558. Springer, Cham
Dapp MM (2019) Toward a sustainable circular economy powered by community-based incentive systems. In: Treiblmaier H, Beck R (eds) Business transformation through blockchain. Springer
Davidson S et al (2016) Economics of blockchain. https://doi.org/10.2139/ssrn.2744751
De Filippi P et al (2015a) Measuring value in the commons-based ecosystem: bridging the gap between the commons and the market. In: Moneylab reader: an intervention in digital economy. Amsterdam: Institute of Network Cultures, pp 74–91
De Filippi P. et al (2015b) Measuring value in the commons-based ecosystem: bridging the gap between the commons and the market. In: Moneylab reader: an intervention in digital economy. Amsterdam: Institute of Network Cultures, pp 74–91. https://ssrn.com/abstract=2725399
Endres ML et al (2009) Ambiguity tolerance and accurate assessment of self-efficacy in a complex decision task. Journal of Management and Organization. Lyndfield, Vol. 15. Iss 1:31–46
Giorgino, V.M.B. (2018): In Search of a New Compass in the *Great Transition*: Toward Co-designing the Urban Space We Care About. In: Giorgino V., Walsh Z. (eds) Co-Designing Economies in Transition. Palgrave Macmillan, Cham.
Gödel K (1931) Über formal unentscheidbare Sätze der Principia Mathematica und verwandter Systeme. I, Monatshefte Für Mathematik Und Physik 38(1):173–198
Hamilton M (2020) Blockchain distributed ledger technology: An introduction and focus on smart contracts. J Corp Acct Fin. 31:7–12
Hart JK et al (2015) Towards an environmental Internet of Things. Earth and Space Science 2:194–200
Hayek, F.A. (1945): The Use of Knowledge in Society, American Economic Review. XXXV, No. 4. pp. 519–30. American Economic Association
Holt CA (2004) The Nash equilibrium: a perspective. Proc Natl Acad Sci 101(12):3999–4002
https://ec.europa.eu/digital-single-market/en/news/ethics-guidelines-trustworthy-ai
https://en.wikipedia.org/wiki/First_principle
https://en.wikipedia.org/wiki/NK_model
https://hbr.org/2018/04/why-some-of-the-most-groundbreaking-technologies-are-a-bad-fit-for-the-silicon-valley-funding-model
https://provocations.darkmatterlabs.org/the-necessity-of-a-boring-revolution-a71b1ae6f956
https://www.coingecko.com/en
https://www.coursera.org/learn/introduction-blockchain-technologies
https://www.mckinsey.com/industries/semiconductors/our-insights/whats-new-with-the-internet-of-things
Hudik M (2020) Equilibrium as compatibility of plans. Theory Decis 89:349–368
Iranzo J et al (2016) Journal of competition among networks highlights the power of the weak. Nat Commun 7:13273
Joseph DP (2019) An analysis of digital forensics in cyber security. In: Bapi R, Rao K, Prasad M (eds) First international conference on artificial intelligence and cognitive computing. Advances in intelligent systems and computing, vol 815. Springer, Singapore
Kruse P et al (1995) The function of meaning in cognitive order formation. Springer Ser Synergetics 64:5–21
Kurtz CF et al (2003) The new dynamics of strategy: sense-making in a complex and complicated world. IBM Syst J 42(3):462–483
Kanev D et al (2017) Behavioral economics: development, conditions and perspectives. Int E-J Adv Soc Sci. 3(8):413–423

Kumar A et al (2017) Data management in machine learning: challenges, techniques, and systems. In: Proceedings of the 2017 ACM international conference on management of data (SIGMOD '17). Association for computing machinery. New York, NY, USA, pp 1717–1722

Kruse P (2013) Zukunft von Führung "Kompetent, kollektiv oder katastrophal?". Messe Zukunft Personal

Li Jh (2018) Cyber security meets artificial intelligence: a survey. Frontiers Inf Technol Electronic Eng 19:1462–1474

Mazzucato M (2018a) Mission-oriented innovation policies: challenges and opportunities. Ind Corp Chang 27(5):803–815

Mazzucato M (2018) A problem-solving approach to fuel innovation-led growth https://ec.europa.eu/info/sites/info/files/mazzucato_report_2018.pdf

Mazzucato M (2020) Challenge-driven innovation policy: towards a new policy toolkit. J Ind Compet Trade 20:421–437

Millar C et al (2018) Management innovation in a VUCA world: challenges and recommendations. California Management Review. p 61

Miller R et al (2018) Transforming the future: anticipation in the 21st century. https://doi.org/10.4324/9781351048002

Mammadzada K et al (2020) Blockchain Oracles: a framework for blockchain-based applications. In: Asatiani A et al (eds) Business process management: blockchain and robotic process automation forum. BPM 2020. Lecture Notes in Business Information Processing, vol 393. Springer, Cham

Miller R (2021) Futures literacy, media literacy and the capability based approach to freedom

Nakamoto S (2008) Bitcoin: a peer-to-peer electronic cash system. https://bitcoin.org/bitcoin.pdf

Oberhauser D (2019) Blockchain for environmental governance: can smart contracts reinforce payments for ecosystem services in Namibia?, Frontiers in blockchain, 2, 21. https://doi.org/10.3389/fbloc.2019.00021

Oh P et al (2016) Network theory and models, Whiley Online Library

Pólvora A et al (2020) Blockchain for industrial transformations: a forward-looking approach with multi-stakeholder engagement for policy advice. Technological forecasting and social change, vol 157, 120091, ISSN 0040-1625

Priyadarshini I (2019) Introduction to blockchain technology. Cyber security in parallel and distributed comupting: concepts, techniques, applications and case studies. https://doi.org/10.1002/9781119488330.ch6.

Quote attributed to Oscar Wilde

Qusay FH (2018) "Introduction to the internet of things," in internet of things A to Z: technologies and applications. IEEE, pp 1–50

Ramarajan L (2014) Past, present and future research on multiple identities: toward an intrapersonal network approach. ANNALS 8:589–659

Schär F (2020) Decentralized finance: on blockchain- and smart contract-based financial markets. https://doi.org/10.13140/RG.2.2.18469.65764

Shukla PR et al (2019) IPCC, climate change and land

Stifter N et al (2019) Revisiting practical byzantine fault tolerance through blockchain technologies. In: Biffl S, Eckhart M, Lüder A, Weippl E (eds) Security and quality in cyber-physical systems engineering. Springer, Cham

Strogatz S (2001) Exploring complex networks. Nature 410:268–276

Teng X et al (2018) Research on application of machine learning in data mining. IOP Conf Ser: Mater Sci Eng 392:062202

The state of climate tech (2020) PWC. https://www.pwc.com/gx/en/services/sustainability/assets/pwc-the-state-of-climate-tech-2020.pdf

Toth Z et al (2019) Chapter 2: weather forecasting: what sets the forecast skill horizon. Elsevier, Sub-Seasonal to Seasonal Prediction, pp 17–45

UKEssays (Nov 2018) An empty world to a full world environmental sciences essay. Retrieved from https://www.ukessays.com/essays/environmental-sciences/an-empty-world-to-a-full-world-environmental-sciences-essay.php?vref=1

Upadhyay N (2019) Challenges and opportunities. Transforming Social Media Business Models Through Blockchain, Emerald Publishing Limited, pp 31–36

Valente M (2008) Pseudo-NK: an enhanced model of complexity

Virtanen A et al (2018) Economics back into cryptoeconomics. https://medium.com/econaut/economics-back-into-cryptoeconomics-20471f5ceeea

White W (2019) New approaches to economic challenges: the financial system. OECD, Paris

Part II
Nature-Based Approach in Cities

Chapter 3
Climate Change and Urban Nature: Impacts and Policies at the Urban Level

Edoardo Croci and Benedetta Lucchitta

Abstract The chapter highlights the potential of nature in cities in mitigating and adapting to climate change. The concept of nature-based solution (NBS) has gained more and more attention in relation to its capacity to provide a wide array of ecosystem services (ES) in different contexts, including urban areas. NBS can address several urban challenges, as ecosystems in healthy condition provide a variety of functions and deliver multiple services, so contributing to climate commitments and more in general to social welfare.

Targeted policies and instruments can be adopted by local governments to foster the implementation of NBS in urban areas to enhance resilience and mitigate emissions. In particular, payments for ecosystem services (PES) can protect and enhance natural assets connecting users and providers of ES through voluntary or regulated transaction schemes. Specific urban PES case studies are analyzed taking into account the stakeholders engaged, the revenues generated, the ES generated and the transaction mechanisms. Results show that PES have proven to be effective in investing and managing urban NBS. The complexity of urban environments and the interactions between many ES requires to adequately value the benefits generated.

Keywords Nature-based solutions · Urban climate policies · Climate change mitigation · Climate change adaptation · Ecosystem services · Payment for ecosystem services

---

E. Croci (✉) · B. Lucchitta
GREEN Research Center, Bocconi University, Milan, Italy
e-mail: edoardo.croci@unibocconi.it

B. Lucchitta
e-mail: benedetta.lucchitta@unibocconi.it

K.-G. Kim and M. Thioye (eds.), *Planning Climate Smart and Wise Cities*,
The Urban Book Series, https://doi.org/10.1007/978-3-030-80165-6_3

3.1 Natural Capital and Climate Change

3.1.1 *The Contribution of Natural Capital and Ecosystem Services to Climate Mitigation and Adaptation*

Natural capital is defined as the world's stocks of natural assets, which include geology, soil, air, water and all living things. It is from natural capital that humans derive a wide range of services, often called ecosystem services (ES), which make human life possible (Natural Capital Coalition 2012). Ecosystem services are defined as "*direct and indirect benefits (goods and services) provided by ecosystems to human well-being*" (MA 2005). Several authors analyzed the benefits generated by ecosystem services and the damages resulting from ecosystems degradation and loss. Starting from Costanza's work on the value of ecosystems worldwide (Costanza 1997), a growing number of studies proposed methods and applications for the assessment of the impacts generated by healthy ecosystems. The common international classification of ecosystem services (CICES)[1] capitalizing on the work made by the millennium ecosystem assessment (MA 2005) defined three[2] main categories of ES:

- Provisioning services: food production, water provision and raw materials;
- Regulating and maintenance services: carbon sequestration, local climate regulation, water purification, air quality regulation, noise reduction, erosion prevention, regulation of water flow, extreme events mitigation;
- Cultural services: nature-based recreation, intellectual and esthetic appreciation, spiritual and symbolic appreciation.

The variety of services described contribute directly (e.g., food production, etc.) or indirectly (e.g., pollination) to human well-being through a cascade model described in Fig. 3.1. Natural capital provides in fact several functions through ES, so benefitting society and generating social value. The consideration of the values generated by ES and the losses connected to their degradation is essential to make policy and production/consumption choices in coherence with sustainable development (Dasgupta 2021).

Services provided by ecosystem generate, among others, benefits that can influence the stability of the climate, mitigating and adapting the impacts generated by climate change (Turner et al. 2009; Epple and Dunning 2014; Locatelli 2016; IPCC 2019). Ecosystems can mitigate climate change since they act as carbon sinks through the process of CO_2 capture and storage. At the same time, they contribute to adaptation, providing services that reduce the impacts of extreme events (e.g., floods,

[1] https://cices.eu/.

[2] The MA (MA 2005) includes the "supporting services" category that in CICES is mostly integrated in "maintenance services." In fact, in literature, supporting services are defined as intermediate ES. Intermediate ES are ES with no direct benefits to human well-being and "consist of the biophysical structures and processes that maintain ecosystems in a favorable state for the provision of final ES" such as regulating ES (Lamonthe and Sutherland 2018).

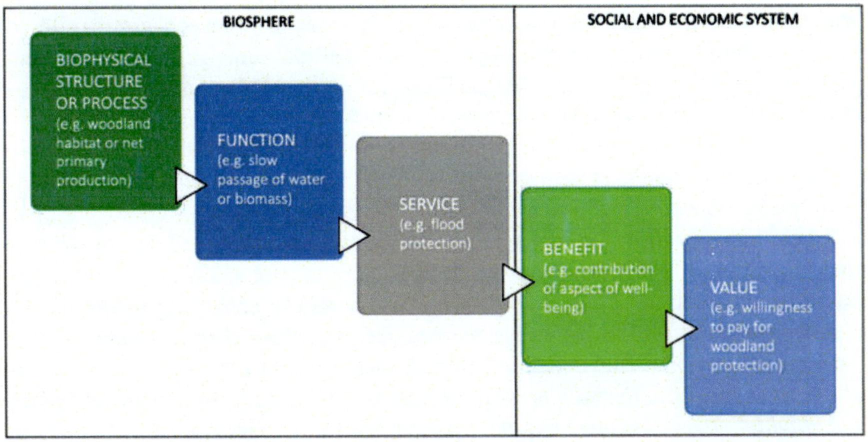

Fig. 3.1 ES cascade model (authors elaboration from Potschin and Haines-Young 2016)

storms, etc.) and contribute to local climate regulation. The idea that natural ecosystems can help us fight both the drivers and impacts of climate change has been gaining traction over the past few years, including in the IPCC Special Report (IPCC Special Report on Global Warming of 1.5 °C (IPCC 2018).

The IPCC climate change and land report (IPCC 2019) emphasizes that the mitigation potential from terrestrial ecosystems comes from restoration and management of forests. The report states a mitigation potential range of 0.4–5.8 Gt CO_2 yr − 1 from avoided deforestation and land degradation, as well as a carbon sequestration potential of 0.5–10.1 Gt CO_2 yr − 1 in vegetation and soils from afforestation/reforestation (IPCC 2019). More recently, Harris et al. (Harris et al. 2021) found that the world's forests sequestered about twice as much carbon dioxide as they emitted between 2001 and 2019. In other words, forests provide a "carbon sink" that absorbs a net 7.6 billion metric tons of CO_2 per year.[3]

At the same time, ecosystems reduce disaster risks and vulnerability of people and infrastructures, and increase territorial resilience. The concept of resilience is not new. Holling et al. (1973) described it as the capacity of complex and dynamic ecosystems' ability to absorb shocks or disturbances, like forest fires or hurricanes, and continue to function (Alexander 2013; Folke et al. 2010; Holling 1973). More recently, the term resilience has been used for natural ecosystems to describe the "ability of a social or ecological system to maintain basic structural and functional characteristics over time despite external pressures. This use of the term acknowledges the fact that all systems undergo continual change, both as a result of internal processes and in response to external factors" (Epple and Dunning 2014). The ability of ecosystems to mitigate shocks and to ensure the functioning of social-economic systems can be

[3] The study estimates that global forest were a net carbon sink of -7.6 ± 49 $GtCO_2e\ yr^{-1}$, reflecting a balance between gross carbon removals (-15.6 ± 49 $GtCO_2e\ yr^{-1}$) and gross emissions from deforestation and other disturbances (8.1 ± 2.5 $GtCO_2e\ yr^{-1}$).

defined as their insurance value (Green et al. 2016), meaning that ecosystems allow to reduce and buffer against negative impacts, such as climate change, or extreme events (Gómez-Baggethun and Barton 2013; Bateman et al. 2014). An ecosystem-based adaptation approach aims to "use activities such as the sustainable management, conservation and restoration of ecosystems as part of an overall strategy to adapt or mitigate climate change effects."

In the last decade, a growing consensus has emerged on the importance of ES and their integration in the management of natural resources and territorial planning (Daily et al. 2009; de Groot et al. 2010; Tratalos et al. 2007). In particular, nature-based solutions (NBS) can play a relevant role in enhancing natural capital at different scales and in providing several ES, including climate change mitigation and adaptation. NBS is defined as "*actions that conserve, manage or restore nature to support biodiversity to help address societal challenges, empower people and provide job and business opportunities; NBS can be powerful tools for combatting biodiversity loss and supporting climate change mitigation and/or adaptation and disaster risk reduction, while delivering further benefits to human well-being (e.g., health). Depending on their context, NBS is also framed as ecosystem-based adaptation (EbA), green infrastructure (GI), ecosystem-based disaster risk reduction (EcoDRR), or natural water retention measures (NWRM)*" (IEEP 2020). Three main types of NBS can be individuated (Eggermont et al. 2015):

- Solutions that involve making better use of existing natural or protected ecosystems;
- Solutions based on developing sustainable management protocols and procedures for managed or restored ecosystems;
- Solutions that involve creating new ecosystems.

The first typology is aimed to maintain or improve the delivery of some ES through small interventions on ecosystems (e.g., protection of mangrove systems, definition of protected natural areas, etc.). The second typology consists in the design and adoption of approaches aimed to "develop sustainable and multi- functional ecosystems and landscapes which improve the delivery of selected ES compared to what would be obtained with a more conventional intervention" (e.g., agricultural diversity to improve multifunctionality, increase tree species genetic diversity, etc.). The third typology includes interventions aimed to reinforce or create new ecosystems and it is linked with the concept of green and blue infrastructures (e.g., green roofs and walls, sustainable urban drainage systems, etc.).

The idea of NBS is being used to repackage policy debates on biodiversity conservation, climate change (adaptation and mitigation) and sustainable use of natural resources, air pollution together with public health, social justice and green economic opportunities. By enhancing and reinforcing ES action, NBS can range from measures to prevent flooding (such as natural water retention measures, dyke relocation, re-naturing rivers, buffering areas, restoration of wetlands, woodlands, floodplain, re-meandering, de-poldering, set back of estuarine defenses, maintaining dunes, beaches and salt marshes) to responding to heat island effects in cities (through multifunctional green public spaces) and droughts and erosion in rural areas (through sustainable agricultural practices and irrigation systems, reforestation, rainfall water management, torrents and river management) (EC 2015).

3.1.2 The Role of NBS in the Paris Agreement

The importance of natural capital and ES has been highlighted by several international initiatives. In 1994, the United Nations Convention to Combat Desertification (UNCCD) has been established with the aim to develop policies and measures to fight desertification and protect ecosystems. The new UNCCD 2018–2030 strategic framework is a comprehensive global commitment to achieve land degradation neutrality in order to restore the productivity of vast expanses of degraded land. In October 2010, the strategic plan for biodiversity for the 2011–2020 has been adopted during the COP10 of the convention on biological diversity. The plan provides an overarching framework on biodiversity for all countries and all stakeholders engaged in biodiversity management and policy development.

In September 2015, the UN approved the Agenda 2030 and the 17 sustainable development goals (SDGs) that aim to achieve sustainability at all levels and sectors. The preservation and enhancement of natural capital are included in several goals (SDG12—Responsible consumption and production, SDG13—Climate change, SDG15—Life on land) also affecting the urban context. Furthermore, SDG 11 is dedicated to the sustainable development of urban areas and it addresses several aspects also related to the potential of ecosystems for the quality of life and well-being improvement in cities including their importance in mitigating and adapting to climate change. In October 2016, UN-Habitat presented the New Urban Agenda that promotes a new model of urban development to integrate environmental protection, sustainable economic growth, and social equity. One of the Urban Agenda objectives is to integrate disaster risk reduction and climate change adaptation and mitigation considerations and measures into resilience-based and climate-effective design of spaces, buildings and construction, services and infrastructure and NBS.

In December 2015, at COP21 in Paris, Parties to the UNFCCC (United Nations Framework Convention on Climate Change) decided to strengthen the global response to the threat of climate change through the adoption of the Paris Agreement (PA). The PA) aims to catalyze action and investment to limit the increase in the global average temperature to 2 °C, or preferably to 1.5 °C, above pre-industrial levels by the end of the century. Countries that signed the PA commit to undertake and communicate their efforts in nationally determined contributions (NDCs), contributing to the achievement of the PA objectives. NDCs embody efforts by each country to reduce national emissions and adapt to the impacts of climate change. All Parties have to communicate their NDCs to the UNFCCC, to be revised every five years and demonstrate increased ambition over time. Additionally, countries have to define and communicate long-term strategies (LTS) that describe phase out pathways to low-carbon transition. So, the NDCs can be defined as "stepping-stones" to achieve mitigation and adaptation targets and the LTS as the instruments that provide a country with direction and vision for increasing ambition in future. In this context, ecosystems protection and enhancement play a relevant role in national strategies and commitments in order to lead to carbon neutrality thanks to their capacity to remove carbon dioxide from the atmosphere and to store it. Different mitigation strategies are coherent with a pathway that limits global warming to 1.5 °C. All pathways use, in different measure, Carbon Dioxide Removal (CDR), and Bioenergy with Carbon Capture and Storage (BECCS) and removals in the Agriculture, Forestry and Other Land Use (AFOLU) sector (IPCC 2019).

The PA explicitly recognizes "the importance of the conservation and enhancement, as appropriate, of sinks and reservoirs of the GHG referred to in the Convention" in light of the significance of ecosystems for climate change mitigation and adaptation, as well as the societal value they provide. Various articles include direct reference to ecosystems. Table below (Table 3.1) summarizes the PA articles that make a clear reference to the importance of ecosystems for mitigation and adaptation to climate change.

Several NDCs include NBS as measures to achieve the targets defined by the PA. More in detail, 66% of the PA signatories commit to restoring or protecting ecosystems in their NDCs (Nature-based solutions Initiative, 2021[4]). Out of 168 NDCs, 104 include commitments to working with natural ecosystems in their adaptation plans, while an additional 27 describe NBS actions in their mitigation plans (76 nations include NBS actions in both their adaptation and mitigation plans). NBS is particularly present in the adaptation plans of low or lower-middle income tropical and subtropical nations. Seddon et al (2019) carried out an analysis of the NDCs presented by the different countries to understand the role of NBS in the achievement of the mitigation and adaptation targets at the national level. NDCs analyzed include actions aimed to improve the management and restoration of terrestrial forests and to fight deforestation. Instead, the percentage of NBS that consider grasslands, drylands, coastal and/or marine ecosystems and other wetlands is quite low in the NDCs. NBS is considered to have an important role in the NDCs, but only around 17% of NDCs with current or planned actions involving NBS for adaptation are associated to evidence-based targets. Actions related to the forest sector are widespread into the NDCs (over 70% of NDCs contain action aimed to enhance and protect world forests), but only the 20% of these NDCs include quantifiable targets. Furthermore, only a small part of the NDCs includes consideration about synergies between mitigation and adaptation actions. More in detail, only 17 countries aim to address adaptation and mitigation together or have sections in the adaptation components of their NDCs that explicitly highlight the mitigation benefits of adaptation action.

Although the benefits generated by NBS for mitigation and adaptation to climate change are now recognized, the results reported above show that there is still a need to improve the integration of these measures within national strategies. The role of national policies appears to be fundamental for the protection and strengthening of ecosystems, but at the same time local authorities can catalyze the changes defied at

[4] https://www.naturebasedsolutionsinitiative.org.

Table 3.1 Ecosystem services in the Paris agreement (authors elaboration)

Preamble	Recognizing the importance of the conservation and enhancement, as appropriate, of sinks and reservoirs of the GHG referred to in the Convention. Noting the importance of ensuring the integrity of all ecosystems, including oceans and the protection of biodiversity
Article 5	5.1 Parties should take action to conserve and enhance, as appropriate, sinks and reservoirs of GHG as referred to in Article 4, paragraph 1(d), of the Convention, including forests 5.2 Parties are encouraged to take action to implement and support, including through results-based payments, the existing framework as set out in related guidance and decisions already agreed under the Convention for: policy approaches and positive incentives for activities relating to reducing emissions from deforestation and forest degradation, and the role of conservation, sustainable management of forests and enhancement of forest carbon stocks in developing countries; and alternative policy approaches, such as joint mitigation and adaptation approaches for the integral and sustainable management of forests, while reaffirming the importance of incentivizing, as appropriate, non-carbon benefits associated with such approaches
Article 7	7.1 Parties recognize that adaptation is a global challenge faced by all with local, subnational, national, regional and international dimensions, and that it is a key component of and makes a contribution to the long-term global response to climate change to protect people, livelihoods and ecosystems, taking into account the urgent and immediate needs of those developing country Parties that are particularly vulnerable to the adverse effects of climate change 7.9 Each Party shall, as appropriate, engage in adaptation planning processes and the implementation of actions, including the development or enhancement of relevant plans, policies and/or contributions, which may include: (c) The assessment of climate change impacts and vulnerability, with a view to formulating nationally determined prioritized actions, taking into account vulnerable people, places and ecosystems. And (e) building the resilience of socio-economic and ecological systems, including through economic diversification and sustainable management of natural resources
Article 8	8.4 Accordingly, areas of cooperation and facilitation to enhance understanding, action and support may include: (h) Resilience of communities, livelihoods and ecosystems

national level. It is increasingly evident that action at the subnational level (regional and municipal) plays a crucial role in following up and implement the strategies and objectives set at the national or international level. The governance of climate change is a complex and multilevel process. Consequently, it can be said that the traditional analytical divisions between international and domestic politics, between local, national and global scale and between state and non-state actors are no longer sufficient (Betsill and Bulkeley 2006). Indeed, global environmental policies are not only a matter of international negotiation and development of national policies but also take place at the local level (Widerberg and Pattberg 2015). Furthermore, the Paris Rulebook reaffirms the key role that a broad range of stakeholders, including, regions, cities, play in ensuring action for climate. It also outlines that specific projects that may contribute to mitigation or adaptation should be included in the NDCs, which may cover, but are not limited to, key sectors, such as energy, resources, water resources, coastal resources, human settlements and urban planning, agriculture and forestry.

3.2 Climate Action at the Urban Level

Recent estimates quantify a relevant potential of cities to contribute to global decarbonization, with a reduction potential of 17.3 $GtCO_2eq$ in 2050 (Coalition for urban transition 2019). Several voluntary initiatives have been launched to involve local governments in mitigation and more recently in adaptation—such as the Covenant of Mayors by the European Commission and 100 Resilience cities by the Rockefeller Foundation. Local governments are increasingly being recognized as having high potential to drive climate change mitigation and adaptation actions thanks to their knowledge of the territory, their responsibilities and powers. Mitigation and adaptation options vary by city characteristics and development level. Major options available for rapidly developing cities include shaping their urbanization and infrastructure development paths. These strategies affect urban GHG emissions by shaping the individual and collective behaviors (Croci et al. 2017). Figure 3.2 shows the plethora of action cities can adopt to achieve climate change and mitigation targets, and NBS gives a strong contribution for both mitigation and adaptation targets.

While the role of natural capital has been investigated by scholars and policy makers with reference to the global, regional and national scales since several decades, only recently its relevance at the urban scale has been analyzed. NBS can address several urban challenges, thanks to their multi-functionality, as, ecosystems in healthy condition provide a variety of functions and deliver multiple services. Main

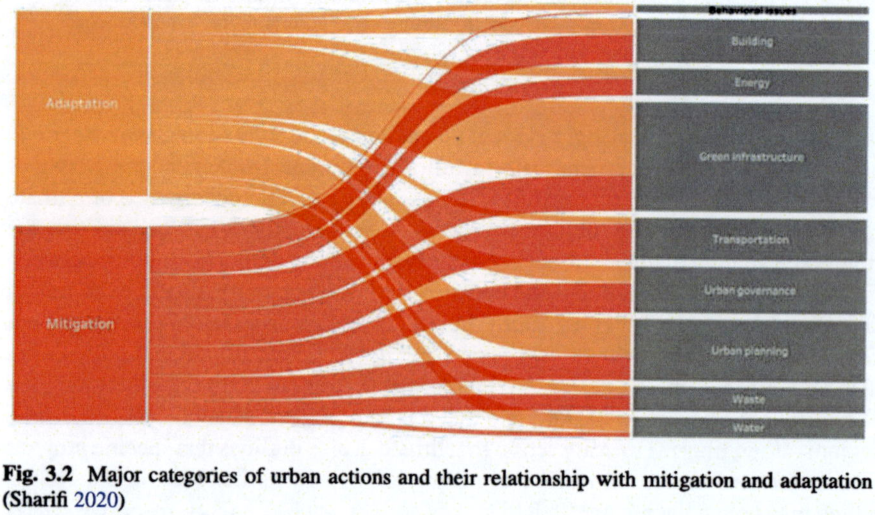

Fig. 3.2 Major categories of urban actions and their relationship with mitigation and adaptation (Sharifi 2020)

functions at the urban scale regard: improving the environment, making cities more attractive, enhancing human well-being, restoring degraded ecosystems, developing climate change adaptation and mitigation, improving disaster risk management and resilience (EC, 2015). In fact, functioning ecosystems allow cities to build adaptive capacities and cope with several urban challenges, providing several ES, including reduction of local air pollution, microclimatic regulation (heat island phenomenon reduction and temperature increase due to climate change), direct health benefits, such as a lower prevalence of asthma in early childhood), mortality reduction, and general health improvements, flood risk reduction, quality of life improvement, social inclusion, safety and cultural benefits (Croci and Lucchitta 2019). Several studies (Escobedo and Nowak 2008; Churkina et al. 2017) confirm that urban vegetation has positive effects on air quality by eliminating polluting determinants, altering the urban microclimate and reducing temperatures through the produced shadow, the evapotranspiration processes, attenuating the winds and decreasing the energy consumption of the buildings, also generating a reduction in CO_2 emissions from power plants. These services are generated by a diverse set of land uses, including parks, waterways, avenues, gardens and courtyards, docks, municipalities, green roofs, and facades, etc. Furthermore, the dependence of cities on the surrounding landscape and its biodiversity is essential to support the production, enhancement, and maintenance of ES and to guarantee the resilience of urban systems as a whole. Actually, NBS can generate positive impacts on environmental, economic and social dimensions, addressing four principal goals at the urban level:

1. Stimulate economic growth as well as improving the environment, making cities more attractive, and enhancing human well-being.
2. Improve the resilience of ecosystems, enabling them to deliver vital ecosystem services and also to meet other societal challenges.
3. Provide more resilient responses and enhance the storage of carbon.
4. Offer synergies in reducing multiple risks.

NBS has the ability to reinforce ES at urban level creating or enhancing the connections between built and natural areas. When considering NBS in relation to ES, Dorst et al. (2019) describe them as able to address resource shortages (e.g., water shortages), floods and heat risks and ecosystem degradation caused by processes of urbanization and climate change. On the other hand, Kabisch et al. (2016) focus on the maintenance, enhancement and restoration of biodiversity. Ecological systems provide a wide range of functions which benefit urban environments. NBS has emerged as a concept, or umbrella term, for ecosystem-based approaches to address the societal challenges of climate change, natural disasters, food and water security, human health and well-being and economic and social development (Cohen-Shacham et al. 2016; EC 2015). The classification of urban ES has been deeply investigated (Bolund and Hunhammar 1999; TEEB 2011; Gómez-Baggethun and Barton 2013; Sieber and Pons 2015). In particular, all the services provided by ecosystems that are linked with mitigation and adaptation to climate change are included in the regulatory category. The following ES are fundamental for mitigation purposes: air quality regulation, local climate regulations and carbon sequestration and storage. Instead, the following ES are relevant for adaptation purposes: waste-water treatment, moderation of extreme events, regulation of water flows, and local climate regulation. Figure 3.3 summarizes the list of the ES provided by NBS at the urban scale that are able to contribute to the achievement of climate goals at the local level.

Several authors analyzed the impacts of ES at the urban scale (Bolund and Hunhammar 1999; Lovasi et al 2008; De Groot et al. 2010; Naumann et al. 2011; Gómez-Baggethun and Barton 2013; Costanza et al. 2014; Kabish et al. 2016; Raymond et al. 2017; EC 2020; Almenar et al. 2021). The mitigation potential of NBS is widely recognized in literature, in particular with reference to the carbon sequestration potential of green areas and trees, the regulation of urban temperature and the decrease in the urban heat island effect (UHI), and the reduction of buildings energy demand. Several authors analyzed the capacity of urban greening to capture and store CO_2 with a specific focus on urban forestry and trees. Table 3.2 summarizes the CO_2 sequestration capacity of urban trees in different cities.

Other urban green spaces are valuable for storing and sequestering carbon (Barò et al. 2014). The measurement of the CO_2 storage capacity of different land uses has been performed in several studies. The InVEST (Integrated Valuation of Environmental Services and Trade-offs) uses maps of land use along with stocks in four carbon pools to estimate the amount of carbon currently stored in a landscape or the amount of carbon sequestered over time. Results show that parks and (semi) natural urban green areas can store 32.6 kg CO_2/m^2, green roofs and walls 5.4 kg CO_2/m^2, allotments and community gardens 27.3 kg CO_2/m^2 and green areas for water management 12.5 kg CO_2/m^2. Another significant impact generated by NBS

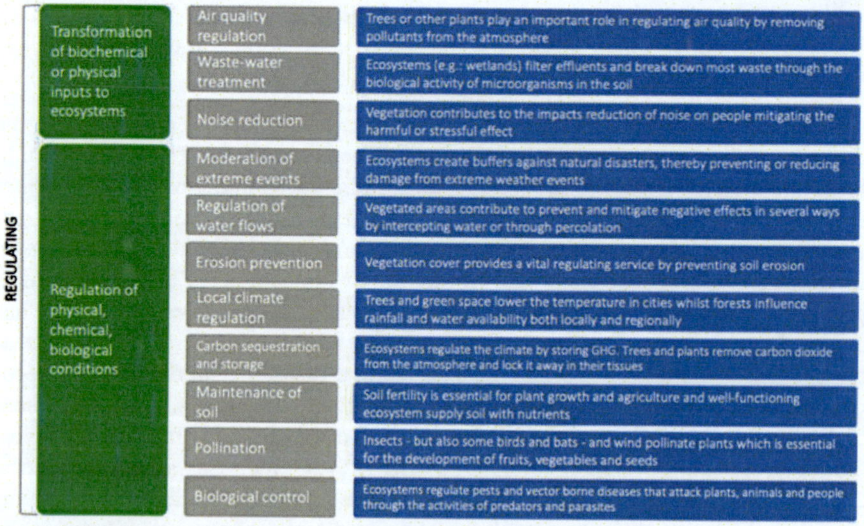

Fig. 3.3 Regulation services at the urban scale (authors elaboration from Croci et al. 2021)

Table 3.2 Urban forest CO_2 sequestration capacity (authors elaboration from EC 2020)

City	CO_2 sequestration
Barcelona, Spain	0.54 t/ha/year (Barò et al. 2014)
Syracuse, USA	0.54 t/ha/year (Nowak and Crane 2002)
Baltimore, USA	0.52 t/ha/year (Nowak and Crane 2002)
Averaged over 28 cities and 6 States in the United States	0.28 kg C/m^2/year (av. gross CO_2 sequestration rate per unit tree cover) (Nowak et al. 2013)
Averaged over 5 European cities	2.43 t/ha/year (Naturvation 2019)

is the regulation of urban climate thanks to their capacity to mitigate urban temperature. Seddon et al (2020) found that the temperature of urban green spaces in daytime is on average 0.94 °C cooler than in other urban spaces. Furthermore, Elmqvist et al. (2016) showed that a 10% increase in tree canopy cover may result in a 3–4.8 °C decrease in ambient temperature. Finally, NBS can be used directly to reduce energy demand in the built environment. In. fact, green walls and roofs can provide an effective means through which to reduce energy demand. Besir and Cuce (2018) found green roofs can reduce the heat penetration by up to 80%, reducing energy consumption by 2–17%., while green walls energy demand reduction potential is estimated in the range 10–30% (Besir and Cuce 2018).

At the same time NBS can improve the adaptation potential in urban areas. This is particularly true for flood mitigation potential (e.g., rain gardens, bioswales, SUDS). Green roofs show large effects on annual stormwater runoff and also on peak flows (Stovin et al. 2013; Keesstra et al. 2018). Bioswales and biofilters are alternative solutions to handle stormwater on ground if space is available. Functionally, these systems also have a potential for infiltration and evapotranspiration (Daly et al. 2012). Chapman and Horner (2010) report that a street-side bioretention facility in Washington can achieve 26 –52% of runoff retention for certain rainfall events. Finally, Holden et al. (2007) highlight the importance of wetlands in the attenuation of flood events, in particular the observations have shown that vegetation cover can affect the velocity of water flowing across wetlands and hence flood generation. It must be said that—as highlighted by Emilsson and Ode Sang (2017)—urban NBS has little impact on large-scale catastrophic rainfalls. Thus, it is necessary working at different scales to define a wider adaptation strategy to reduce extreme events. The literature review results on the impacts generated by different NBS at the urban scale for climate change mitigation and adaptation have been summarized in Table 3.3.

Beyond climate, NBS generates benefits that can contribute to achieve a wider set of goals contributing to environmental, economic and social sustainability, especially at the local and regional levels. Furthermore, "the many advantages of NBS are often not immediately apparent, but instead accumulate over a longer period" (Naumann et al. 2014).

Table 3.3 Regulation services at the urban scale (authors elaboration)

	Air quality regulation	Waste-water treatment	Noise reduction	Moderation of extreme events	Regulation of water flows	Erosion prevention	Local climate regulation	Carbon sequestration and storage
Green roofs	■		■		■		■	■
Green walls	■		■			■	■	■
Urban trees	■		■	■	■	■	■	■
Urban green areas	■		■	■	■	■	■	■
SUDS	■			■	■	■	■	■
Wetlands	■	■		■	■	■	■	■
Bioretentions basin		■		■	■	■	■	■

3.3 Policy Instruments to Enhance Urban NBS

3.3.1 Payment for Ecosystem Services

The concept of value linked to NBS is multi-faceted since it comprises different types of benefits. These benefits are perceived by different stakeholders and they are delivered over different periods of time. Therefore, the assessment and evaluation of the value created by NBS is a complex task.

So it is difficult to establish commonly adopted methods to precisely allocate the share of investment contributions of each stakeholder according to the benefits generated by the measure implemented. The benefits provided by NBS include (i) the reduction of environmental harmful externalities (e.g.: CO_2 capture and storage, reduction of air pollutants), (ii) the reduction of climate related risks (e.g.: flood risk reduction, moderation of extreme events) and (iii) the provision of social benefits (e.g.: improvement of health, job creation, urban regeneration).

The capacity to catch the values generated by NBS through market transactions or policy instruments is defined as "value capture." Based on the economic theory on the classification of goods (Ostrom 1990) four categories of goods can be individuated: private goods, public goods, tool goods and common pool resources.[5] NBS often show characteristics of public goods or common pool resources and therefore can generate market failures, since the price mechanism does not guarantee an optimal level of their production and allocation. In order to correct market failures in the presence of public goods a solution can be the intervention of the state in their

[5] A private good is excludable, i.e., its owners can exercise private property rights, preventing those who have not paid for it from using the good or consuming its benefits; and rivalrous, i.e., consumption by one necessarily prevents that of another. A public good is non-excludable, non-rivalrous, and open to all in its consumption (clean air, soil water storage that yields flood control and beautiful views over a landscape). A toll good is excludable, but non-rivalrous (such as access to private parks). A common pool resource is rival and non-excludable (such as fish stocks in an ocean)

production and management, through fiscal instruments (general or purpose ones). In the case of common pool resources, a solution can be the introduction of governance tools to regulate its use through an institutional framework. In this way, the public sector can introduce policy instruments aimed to internalize positive externalities, catching the value generated by NBS which are not reflected by market prices.

The enhancement of ES depends also on the capacity of local governments to introduce instruments to protect and facilitate the implementation and management of NBS in cities (Naeem et al. 2015). The economics of ecosystems and biodiversity (TEEB 2011) provide a classification of the instruments that can be adopted at the urban level for the integration of NBS and for the enhancement of the ES provided by them into local policy-making and decision-taking. The classification includes three different typologies of instruments: (i) management and regulation; (ii) planning and regulation and (iii) market-based instrument. Market-based instruments influence actors' behavior by changing their economic incentive structure.

A particular focus is dedicated to payments for ecosystem services (PES) which aim to stimulate the production of positive externalities, transforming them into goods that can be bought and sold on a market. In other words, PES are tools that aim to change behaviors having negative impacts on ecosystems by introducing the economic value of ecosystem services in decision-making. A PES can be defined as a voluntary transaction between service users and service providers that are conditional on agreed rules of natural resource management for generating offsite services (Wunder 2015). The concept of PES is quite new, and it is based on the idea that the beneficiary of an ES should reward the owner of the natural resource generating the service to ensure its maintenance or to improve its conditions. Wunder (2005) defines a PES as "a voluntary transaction where a well-defined service (or a land use likely to secure that service) is being "bought" by a (minimum one) ES buyer from a (minimum one) ES provider if and only if the ES provider secures ES provision (conditionality)." Such a definition has been revised by the author himself after receiving a series of criticisms, mainly concerning the market-linked nature of PES (Tacconi 2012; Muradian et al. 2013). Hence, Wunder re-defined PES as "voluntary transactions between service users and service providers that are conditional on agreed rules of natural resource management for generating offsite services" (Wunder 2015). PES schemes allow service buyers to pay for the safeguard of an environmental service on which they are reliant, hence preventing the risk it becomes scarce, and to compensate for the negative externalities caused by production or consumption activities to the society (UNEP 2009). In this way, it is possible to change those behaviors which harm ecosystems (Rojas and Aylward 2003).

Economic, institutional and cultural conditions are necessary for the realization of a PES scheme. The economic condition implies that the users' maximum willingness to pay corresponds to or is higher than the providers' minimum willingness to accept the compensation. If this did not occur, then the service provided would be assessed less than the estimated cost provision landowners are facing for deviating from their land use plan. Therefore, benefits derived from preserving natural resources via PES need to exceed the costs of incremental service provision. The cultural condition implies that the service users are willing to pay for the ES, and the providers must be willing to change their own management habits of the natural resource that produces a specific ES. The institutional condition implies that a third party—such as external intermediaries or facilitators—is able to create trust between actors to ensure the correct PES scheme execution. So a PES scheme requires complex social interactions over time: if all parties were to act independently in perfectly competitive markets, transaction costs would be enormous given the number of different actors to be involved, and no PES would ever take place. Most importantly, however, PES cannot function without tenure clarity and security: if such well-defined stewardship does not pre-exist, then local land users will lack the crucial right to exclude third-party access, making them unreliable service providers with insufficient control over service delivery.

More recently, the voluntary character of a PES has been challenged, considering the option of a stronger role of public entities, which can go beyond the role of intermediary or facilitator and act as a regulator. Salzman et al. (2019) individuated three typologies of PES. The first one is called *User-financed PES*. Users of ES compensate landowners to maintain or enhance the delivery of ES. In this case the users (public or private) are direct beneficiaries of ES protection. The second typology is the *Government-financed PES*, public entities on behalf of users compensate landowners to maintain or enhance ES delivery. The last category of PES is the *Compliance PES* in which parties facing regulatory obligations compensate other parties for activities to maintain or enhance comparable ES or goods in exchange for a standardized credit or offset that satisfies their mitigation requirements (water quality trading, wetlands mitigation banking and emissions trading scheme for GHG).

3.3.2 Case Studies of PES at the Urban Scale

PES diffused in several countries over time and gained attention in climate policy. In 2008, the UN introduced the REDD + program which aims to protect forest in developing countries through the definition of PES schemes in developing countries. However, PES have been implemented mainly on large scales. Through an extensive literature review,[6] five case studied of urban PES schemes have been selected: Victoria Business Improvement District (BID) case study (London), Adopt a green spot case study (Milan), Catskill watershed case study (New York), the Payments for Watershed Services—PSHA case study (Mexico City) and the Wimbledon and Putney Common Levy case study (London). The case studies have been analyzed taking into account the scheme mechanisms, the stakeholders engaged the revenues generated, the ES provided through the scheme and the other ES generated. The case studies have been summarized in Table 3.4 to compare their structures and peculiarities.

[6] The review has been performed by using Scopus. The keywords used for the research are: urban, PES, stewardship and ecosystems voluntary agreements. 63 papers have been analyzed but only few of them are focused on the implementation of PES schemes at the urban scale. The case studies have been selected based on two criteria: (i) involvement of local stakeholders in the PES scheme and (ii) ES sold generate direct impacts at the urban scale.

Table 3.4 PES case studies description (authors elaboration)

Victoria BID (London, UK)	**Description**: The Victoria BID was established in 2010 after a successful ballot of all the businesses liable to pay the business rate levy in the area. A BID is a defined area in which a levy is voluntary charged on all business rate payers in addition to the business rates bill. This levy is used to develop projects which will benefit businesses in the local area. There is no limit on what projects or services can be provided, the only requirement is that it should be something that is in addition to services provided by local authorities. The BID is managed by a Business Improvement District body in partnership with the local authority. In Victoria BID the taxes revenues have been used for the implementation of some NBS, such as trees plantation, green wall realization, new park realization **PES typology**: user-financed PES **ES seller**: municipality **ES buyer**: local business and the BID **ES bought**: cultural **Revenues**: NA **Other ES provided**: provisioning and regulation
Adopt a green spot (Milan, IT)	**Description**: Milan Municipality in 2005 launched the "Adopt a green spot" initiative. This initiative foresees the engagement of various stakeholders in the management of urban green areas. Citizens, NGOs, private companies, universities can, directly or indirectly, provide for the maintenance of green areas within the city through standardized voluntary agreements designed by the Municipality. A special municipal office was established to promote, manage and monitor the initiative. The duration of the contracts varies from 3 to 5 years based on the agreement typology. The areas under the initiative are flowerbeds, urban parks and gardens, recreational areas, roundabouts, tree lines and trees. Through the sponsorship agreement, the private partner (the contractor) provides the Municipality an economic contribution to the project implementation and maintenance. Through a collaboration agreement the direct implementation of the green area project is taken on by the private partner who also takes care of the maintenance costs during the duration of the agreement. In both cases, the municipality gives public recognition to the contractor through a plate placed in the green area **PES typology**: user-financed PES **ES seller**: municipality **ES buyer**: Various (citizens, universities, firms, NGOs, condominiums) **ES bought**: cultural **Revenues**: 2.6 million euros **Other ES provided**: provisioning and regulation

(continued)

Table 3.4 (continued)

Catskill watershed (New York, USA)	**Description**: The agreement was signed, in 1990, between the NYC municipal company for water services provision and forest owners of the catchment basin. Owners are committed to manage forests in accordance with a program that includes forest management practices which have positive effects on the qualitative and quantitative consistency of water runoff. Compensation for ES performed is paid through surtax on water tariff, paid by end users. The City spent 1.5 billion $ in watershed protection projects against 6 billion $ to build the plant plus another 250 million $ per year for maintenance **PES typology**: compliance PES **ES seller**: land-owners **ES buyer**: municipality and citizens **ES bought**: provisioning **Revenues**: 1.3 billion euros **Other ES provided**: regulation
PSAH (Mexico City, MX)	**Description**: Mexico's National Forestry Commission manages the PSAH program. The program allows land owners to propose projects aimed to better manage the forests to enhance and protect the ecosystems provided. PSAH participants' activities should preferably focus on forest cover conservation aimed at recharging aquifers and halting soil erosion. Operational rules select the applications best fitting such purposes by prioritizing lands that fall in officially declared overexploited Forest and Soil National Inventory. Each year the National Forestry Commission select the projects that will be financed, the National PSAH funding comes from an increase in the water tariff. Furthermore, the National Forestry Commission is in charge to check that the projects are carried out following the standards and criteria previously defined **PES typology**: compliance PES **ES seller**: farmers **ES buyer**: National Forestry Commission and citizens **ES bought**: provisioning and regulation **Revenues**: 15.5 million euros **Other ES provided**: cultural

(continued)

Table 3.4 (continued)

Commons Levy (London, UK)	**Description**: Wimbledon and Putney commons contain about one million trees and provide open green space to the local community. The Wimbledon and Putney Commons Act 1871 lays a duty on the Conservators to *preserve the "forest functions and to ensure its public and local use, for the purposes of exercise and recreation"*. To fulfill these duties, the conservators, have access to a budget of around £1.3 million, two thirds of which comes from the Commons Levy, which is paid by council tax payers in the Levy area. The Common Levy tax has been established on order to maintain the forest close to Wimbledon and Putney. The Levy can be considered a PES arrangement as all those who choose to live within 3/4 of a mile of the commons and therefore benefit from their proximity to numerous recreational and cultural services, have to pay a tax for the maintenance of the forest **PES typology**: compliance PES **ES seller**: municipality **ES buyer**: citizens **ES bought**: cultural **Revenues**: 1 million euro **Other ES provided**: provisioning and regulation

To better analyze the different structures of the PES considered, four main elements that characterize the PES schemes have been evidenced: (i) ES buyers; (ii) ES sellers; (iii) the payment defined for the provision of the ES and (iv) the benefits generated by the ES. Figure 3.4 summarizes the analysis of the PES scheme structures.

The case studies presented show specific characteristics regarding the voluntary or mandatory nature of the agreement. The Catskill watershed case study, the PSAH program case study and the commons levy case study are based on a regulatory framework defined at the municipal or national level. In fact, the main buyer of the ES is the municipality that catches the positive externalities generated by the sustainable management of natural resources through an additional tax that falls on citizens as final beneficiaries of the ES provided through the agreement between the municipality and the landowners. Instead, the Victoria BID and Adopt a green spot case studies are voluntary and are based on an agreement that various stakeholders (citizens, local businesses, universities, associations, etc.) decide to sign with the

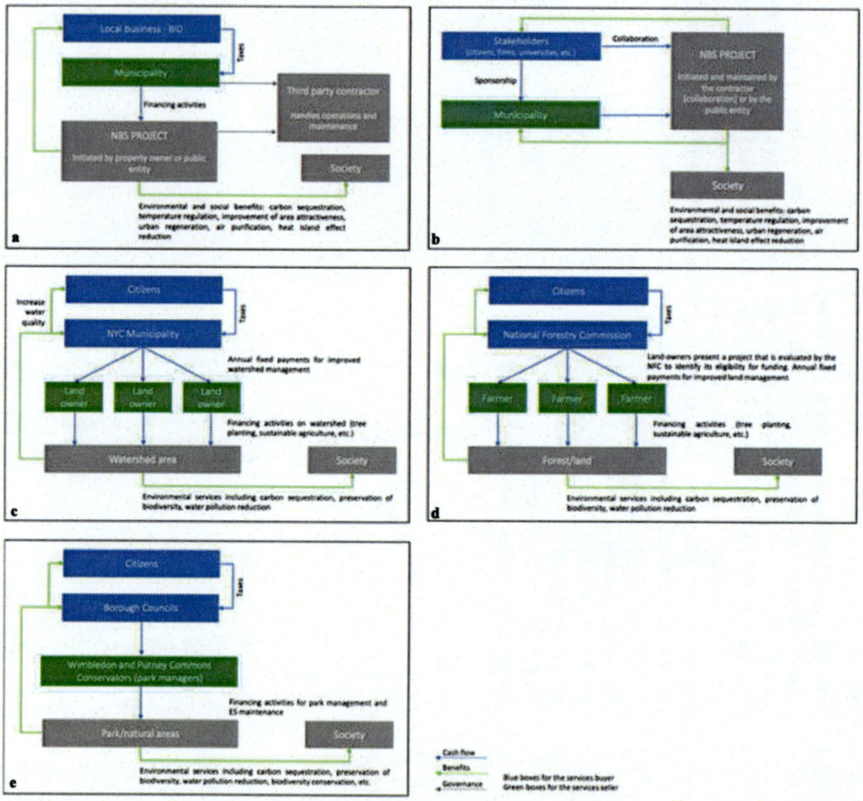

Fig. 3.4 **a** Victoria BID PES scheme; **b** Adopt a green spot PES scheme; **c** Catskill watershed PES scheme; **d** PSAH program PES scheme; **e** Commons Levy scheme (authors elaboration)

municipality to contribute to the implementation and management of NBS. The case adopt a green spot case is particularly interesting as the municipality has defined two types of agreements with different characteristics: sponsorship and collaborations. The first type of contract—sponsorship—implies the direct payment to the municipality of the amount necessary for the creation and management of the NBS. The second type of contract—collaboration—foresees that the contractor takes care of the implementation and management of the NBS following the standards defined by the municipality. The flexibility of the agreements defined allows for the involvement of different types of stakeholders. The Victoria BID case study is also based on a voluntary agreement that is defined by local business for the implementation of different measures aimed to improve the area attractiveness. Through the BID several NBS such as green roofs and green walls have been implemented. The BID contractors decide on a voluntary basis to pay an additional tax for the implementation of the NBS.

Two main categories of urban PES schemes have been identified: (i) user-financed PES, where users of generated ES pay implementers of NBS through voluntary agreements (Adopt a green spot; Victoria BID); (ii) compliance PES, where users of generated ES pay a public authority or a delegated implementer of the NBS through mandatory taxes or charges (Catskill watershed, PSHA, Common levy). The common factor among the case studies analyzed is the role of promoter and guarantor that is carried out by the municipality or by a public entity, even if the PES schemes owns to two different categories. In all cases, the municipality/public entity collects the funds paid by the users of the ES and distributes them to the provider of the ES. In addition, the municipality/public entity deals with defining the standard of service (e.g., "Adopt a green spot" and Victoria BID) or the rules for the sustainable management of ecosystems (e.g., Catskill watershed and PSAH).

These characteristics seem peculiar of the urban PES analyzed. In fact, in literature there are several PES case studies in which a public authority is not included in the scheme. These are self-organized private deals in which beneficiaries of ES undergo contracts directly with service providers. The PES is designed through the definition of an ad hoc voluntary agreement between the two parties (buyer and seller). Including the characteristics of the contract and the standards of service that the contractors must comply with. In the case of urban scale PES, the role of the public body (or of a company partially controlled by the municipality that manages the execution of the scheme) plays a fundamental role. In fact, the presence of a public body can ensure greater trust in the execution of the PES, so providing greater guarantees to stakeholders who decide to or are obliged to join the initiative.

Although PES are instruments that have proven to be suitable for the implementation of NBS and for the engagement of different stakeholders in the sustainable management of ecosystems, the adoption of these instruments still presents some challenges. In the case studies analyzed it emerges that the PES are structured for the payment of a single-ecosystem service, while economic, social and environmental values of the other ES generated are not captured by the PES. In fact, not all ES produced by NBS are accounted in the PES schemes, but only one or a few services are traded, while the values generated by other services are not captured. The correct enhancement and protection of the ES generated by NBS requires to identify and adopt systems for monitoring and measuring the impacts generated considering all the benefits and trade-offs (Salzman et al. 2018). The assessment of NBS in cities should take into account the plethora of impacts generated with attention also to the secondary impacts generated by NBS. For example, the implementation of a green roof can deliver several regulation services, such as climate regulation, air quality improvement and runoff mitigation. At the same time, green roofs can also improve the energy efficiency of a building or increase its attractiveness. Another challenge in the implementation of PES scheme in urban areas can involve social inequalities. At the urban level it must be considered that there will be spatial variation in the provision of the ES (Baró et al. 2016). According to some authors (Barnaud and Antona 2014), PES can improve the provision of services only in certain areas of the city. For example, related initiatives such as land use planning for climate change adaptation have been found to exacerbate socio-spatial inequalities (Wen et al. 2013; Kabish et al. 2015; Anguelovski et al. 2016). Such variation raises challenges for designing PES schemes and impacts the feasibility of PES in urban areas (Richardson and Thompson 2018). Hence, it is necessary to ensure that participation in PES is accessible to different stakeholders and benefits are well-distributed in different parts of the city. Finally, it must be considered that ES demand can vary based on several drivers such as climatic conditions (Mora et al. 2017), cultural background (Akbari et al. 2001), and socio-economic characteristics of urban residents, (Casado-Arzuaga et al. 2013; Zezza and Tasciotti 2010). Applying PES in an urban context is far more challenging than in rural areas, where there is a less complex set of stakeholders and where computing benefits and costs is more straightforward (Davies et al. 2018). Indeed, the complexity of urban environments and the interactions between many ES requires a holistic perspective to ensure that the values generated by NBS implementation can be adequately accounted and allocated.

3.4 Conclusions

The chapter contributes to highlighting the benefits generate by NBS at the urban scale to mitigate and adapt to climate change, such as local climate regulation, extreme event mitigation, CO_2 sequestration and store, mitigation of the heat island effect. Nonetheless, the role of NBS is not limited to this. NBS is multifunctional and generate several benefits that are perceived by different stakeholders. NBS at the urban scale can provide a wide range of benefits so providing relevant private and social values. The enhancement NBS at the urban scale, requires that can capture the generated values. At this purpose, PES aim to stimulate the production of positive externalities, transforming them into goods that can be bought and sold on a market.

Five case studies of urban PES schemes are analyzed, pertaining to two different categories (i) user-financed PES (Adopt a green spot; Victoria BID); (ii) compliance PES (Catskill watershed, PSHA, Common levy). In both schemes, it emerges a key role of municipalities or public entities in the implementation of the PES schemes, in setting rules, managing transactions and ensuring trust between parties.

PES application in an urban context still presents some challenges. Since cities include a complex set of stakeholders, it is necessary to identify flexible schemes that allow to engage them considering their social, economic and cultural diversity and ensuring social equity. Another relevant challenge is the difficulty in assessing the benefits generated by NBS at the urban scale and consequently performing the economic valuation of these benefits.

References

Alexander DE (2013) Resilience and disaster risk reduction: an etymological journey. Nat Hazard Earth System Sci 13(11):2707–2716

Almenar J, Elliot T, Rugani B, Philippe B, Navarrete Gutierrez T, Sonnemann G, Geneletti D (2021) Nexus between nature-based solutions, ecosystem services and urban challenges. Land Use Policy 100:104898

Baró F et al (2014) Contribution of ecosystem services to air quality and climate change mitigation policies: the case of urban forests in Barcelona, Spain. Ambio 43:466–479

Bateman IJ, Harwood AR, Abson DJ, Andrews B, Crowe A, Dugdale S, Fezzi C, Foden J, Hadley D, Haines-Young R, Hulme M, Kontoleon A, Munday P, Pascual U, Paterson J, Perino G, Sen A, Siriwardena G, Termansen M (2013) Economic analysis for the UK national ecosystem assessment: synthesis and scenario valuation of changes in ecosystem services. Environ Resour Econ 57(2):273–297. https://doi.org/10.1007/s10640-013-9662-y

Besir A, Cuce E (2018) Green roofs and facades: a comprehensive review. Renew Sustain Energy Rev 82:915–939

Bolund P, Hunhammar S (1999) Ecosystem services in urban areas. Ecol Econ 29:293–301

Bulkeley H, Kern K (2006) Local government and the governing of climate change in Germany and the UK. Urban Studies 43(12):2237–2259

Costanza R, D'Arge R, de Groot R, Farber S, Grasso M, Hannon B, Limburg K, Naeem S. O'Neill RV, Paruelo J, Raskin RG, Sutton P, van den Belt M (1997) "The value of the world's ecosystem services and natural capital". Nature 387

Chapman C, Horner RR (2010) Performance assessment of a street-drainage bioretention system. Water Environ Res 82(2):109–119

Cohen-Shacham E, Walters G, Janzen C, Maginnis S (eds) (2016) Nature-based solutions to address global societal challenges. IUCN, Gland, Switzerland

Churkina G, Kuik F, Bonn B et al (2017) Effect of VOC emissions from vegetation on air quality in berlin during a heatwave. Environ Sci Technol 51

Coalition for Urban Transitions (CUT) (2019), Climate Emergency, Urban Opportunity. World Resources Institute (WRI) Ross Center for Sustainable Cities and C40 Cities Climate Leadership Group. London and Washington, DC.

Costanza R, de Groot R, Sutton P, van der Ploeg S, Anderson SJ, Kubiszewski I, Farber S, Turner RK (2014) "Changes in the global value of ecosystem services". Global Environ Change 26

Croci E, Lucchitta B (2019) Introduction to special section nature-based solutions (NBS) for urban resilience, economics and policy of energy and environment. ISSN 2018:2280–7659

Croci E, Lucchitta B, Janssens-Maenhout G, Martelli S, Molteni T (2017) Urban CO_2 mitigation strategies under the covenant of mayors: an assessment of 124 European cities. J Clean Prod 169:161–177

Croci E, Lucchitta B, Penati T (2021) Valuing ecosystem services at the urban level: a critical review. Sustainability 13(3):1129

Daily GC, Polasky S, Goldstein J, Kareiva PM, Mooney HA, Pejchar L, Ricketts TH, Salzman J, Shallenberger R (2009) «Ecosystem services in decision making: time to deliver», Frontiers Ecol Environ 7

Daly E, Deletic A, Hatt BE, Fletcher TD (2012) Modelling of stormwater biofilters under random hydrologic variability: a case study of a car park at Monash university, victoria (Australia). Hydrol Process 26(22):3416–3424

De Groot RS, Alkemade R, Braat L, Hein L, Willemen L (2010) "Challenges in integrating the concept of ecosystem services and values in landscape planning, management and decision making". Ecol Complex 7

Dorst H, van der Jagt A, Raven R, Runhaar H (2019) Urban greening through nature-based solutions—key characteristics of an emerging concept. Sustain Cities Soc 49:101620

Droste N, Lima GR, May PH, Ring I (2017) Municipal responses to ecological fiscal transfers—a microeconometric panel data approach. Environ Policy Governance

European Commission (2015) "Towards an EU research and innovation policy agenda for nature-based solutions & re-naturing cities"

European Commission (2020) "Nature-based solutions—state of the art in EU-funded projects"

Eggermont H, Balian E, Azevedo JMN, Beumer V, Brodin T, Claudet J, Fady B, Grube M, Keune H, Lamarque P, Reuter K, Smith M, van Ham C, Weisser WW, Le Roux X (2015) Nature-based solutions: new influence for environmental management and research in Europe. GAIA—Ecol Perspect Sci Soc 24:243–248

Elmqvist T et al (2016) Benefits of restoring ecosystem services in urban areas. Curr Opin Environ Sustain 14:101–108

Escobedo FJ, Wagner JE, Nowak DJ et al (2008) "Analyzing the cost effectiveness of Santiago, chile's policy of using urban forests to improve air quality". J Environ Manage 86

Emilsson T, Ode Sang Å (2017) Impacts of climate change on urban areas and nature-based solutions for adaptation. In: Kabisch N, Korn H, Stadler J, Bonn A (eds) Nature-based solutions to climate change adaptation in urban areas. Theory and practice of urban sustainability transitions. Springer, Cham. https://doi.org/10.1007/978-3-319-56091-5_2

Epple C, Dunning E (2014) Ecosystem resilience to climate change: what is it and how can it be addressed in the context of climate change adaptation? Technical report for the mountain EbA project. UNEP world conservation monitoring centre, Cambridge

Folke C et al (2010) Resilience thinking: integrating resilience, adaptability and transformability. Ecol Soc 15(4):20

EC (2020) Nature-based solutions—state of the art in EU-funded projects

Gomez-Baggethun E, Gren Å, Barton DN, Langemeyer J, McPherson T, O'Farrell P, Andersson E, Hamsted Z et al (2013) "Urban ecosystem services". In: Elmqvist T (ed) Urbanization, biodiversity and ecosystem services: challenges and opportunities. A global assessment

Gómez-Baggethun E, Barton DN (2013) Classifying and valuing ecosystem services for urban planning. Ecol Econ 86:235–245. https://doi.org/10.1016/j.ecolecon.2012.08.019

Green TL, Kronenberg J, Andersson E et al (2016) Insurance value of green infrastructure in and around cities. Ecosystems 19:1051–1063. https://doi.org/10.1007/s10021-016-9986-x

Harris NL, Gibbs DA, Baccini A et al (2021) Global maps of twenty-first century forest carbon fluxes. Nat Clim Chang 11:234–240. https://doi.org/10.1038/s41558-020-00976-6

Holden J, Gascoign M, Bosanko NR (2007) Erosion and natural revegetation associated with surface land drains in upland peatlands. Earth Surf Proc Land 32(10):1547–1557

Holling CS (1973) Resilience and stability of ecological systems. Annu Rev Ecol Syst 4(1):1–23

IEEP (2020) Workshop on mobilizing up-scaling of nature-based solutions for climate change throughout 2020 and beyond

IPCC (2018) Summary for policymakers. In: Masson-Delmotte V, Zhai P, Pörtner H-O, Roberts D, Skea J, Shukla PR, Pirani A, Moufouma-Okia W, Péan C, Pidcock R, Connors S, Matthews JBR, Chen Y, Zhou X, Gomis MI, Lonnoy E, Maycock T, Tignor M, Waterfield T (eds) Global warming of 1.5°C. An IPCC special report on the impacts of global warming of 1.5°C above pre-industrial levels and related global greenhouse gas emission pathways, in the context of strengthening the global response to the threat of climate change, sustainable development, and efforts to eradicate poverty. World Meteorological Organization, Geneva, Switzerland, p 32

IPCC (2019) Climate and land: an IPCC special report on climate change, desertification, land degradation, sustainable land management, food security, and greenhouse gas fluxes in terrestrial ecosystems

Kabisch N, Frantzeskaki N, Pauleit S, Naumann S, Davis M, Artmann M, Haase D, Knapp S, Korn H, Stadler J, Zaunberger K, Bonn A (2016) Nature-based solutions to climate change mitigation and adaptation in urban areas: perspectives on indicators, knowledge gaps, barriers, and opportunities for action. Ecol Soc 21

Keesstra S et al (2018) The superior effect of nature-based solutions in land management for enhancing ecosystem services. Sci Total Environ 610:997–1009. https://doi.org/10.1016/j.scitotenv.2017.08.077

Lamothe KA Sutherland IJ (2018) Intermediate ecosystem services: the origin and meanings behind an unsettled concept. Int J Biodivers Sci Ecosyst Ser Manage

Locatelli B (2016) Ecosystem services and climate change. In: Potschin M, Haines-Young R, Fish R, Turner RK (eds) Routledge handbook of ecosystem services. Routledge, London and New York, pp 481–490. ISBN 978-1-138-02508-0

Lovasi GS, Quinn JW, Neckerman KM, Perzanowski MS, Rundle AJ (2008) "Children living in areas with more street trees have lower prevalence of asthma". Epidemiol Community Health

Muradian R, Arsel M, Pellegrini L, Adaman F, Aguilar B, Agarwal B, Corbera E, Ezzine de Blas D, Farley J, Froger G, Garcia-Frapolli E (2013) Payments for ecosystem services and the fatal attraction of win-win solutions. Conserv Lett 6(4):274–279

Millennium Ecosystem Assessment, (MA) (2005) "Ecosystems and human well-being: the assessment series". Island Press, Washington DC

Natural Capital Coalition (2012) "Natural capital declaration"

Naturvation (2019a). NBS Navigator, technical note: carbon storage indicator. NATURVATION project

Naumann S, Anzaldua G, Berry P, Burch S, Davis M, Frelih-Larsen A, Gerdes H, Sanders M (2011) Assessment of the potential of ecosystem-based approaches to climate change adaptation and mitigation in Europe. Final report to the European commission, DG environment, ecologic institute and environmental change institute, Oxford University Centre for the Environment

Naumann S, Kaphengst T, McFarland K, Stadler J (2014) "The challenge of climate change—partnering with nature nature based approaches for climate change mitigation and adaptation". German Federal Agency for Nature Conservation

Nowak DJ, Crane DE (2002) Carbon storage and sequestration by urban trees in the USA. Environ Pollut 116(3):381–389. https://doi.org/10.1016/s0269-7491(01)00214-7 PMID: 11822716

Nowak DJ, Greenfield EJ, Hoehn RE, Lapoint E (2013) Carbon storage and sequestration by trees in urban and community areas of the United States. Environ Pollut 178:229–236. https://doi.org/10.1016/j.envpol.2013.03.019

Potschin M, Haines-Young R (2016) Defining and measuring ecosystem services. In: Potschin M, Haines-Young R, Fish R, Turner RK (eds) Routledge handbook of ecosystem services. Routledge, London

Raymond CM, Frantzeskaki N, Kabisch N, Berry P, Breil M, Nita MR, Geneletti D, Calfapietra C (2017) A framework for assessing and implementing the co-benefits of nature-based solutions in urban areas. Environ Sci Policy 77:15–24

Rojas M, Aylward B (2003) What are we learning from experiences with markets for environmental services in costa Rica? A review and critique of the literature 2

Seddon N, Sengupta S, García-Espinosa M, Hauler I, Herr D, Rizvi AR (2019) Nature-based solutions in nationally determined contributions: synthesis and recommendations for enhancing climate ambition and action by 2020. IUCN and University of Oxford, Gland, Switzerland and Oxford, UK

Seddon N, Daniels E, Davis R, Chausson A, Harris R, Hou-Jones X, Huq S, Kapos V, Mace GM, Rizvi AR, Reid H, Roe D, Turner B, Wicander S (2020) Global recognition of the importance of

nature-based solutions to the impacts of climate change. Global Sustain 3:E15. https://doi.org/10.1017/sus.2020.8
Sharifi A (2020) Co-benefits and synergies between urban climate change mitigation and adaptation measures: a literature review. Sci Total Environ 141642
Sieber J, Pons M (2015) Assessment of urban ecosystem services using ecosystem services reviews and GIS-based Tools. Procedia Eng 115:53–60
Somarakis G, Stagakis S, Chrysoulakis N (eds) (2019) ThinkNature nature-based solutions handbook. ThinkNature project funded by the EU Horizon 2020 research and innovation programme under grant agreement No. 730338
Stovin V, Poë S, Berretta C (2013) A modelling study of long term green roof retention performance. J Environ Manag 131:206–215
The Economics of Ecosystems and Biodiversity (TEEB) (2011) "TEEB manual for cities: ecosystem services in urban management"
Tacconi L (2012) Redefining payments for environmental services. Ecol Econ 73:29–36
The economics of ecosystems and biodiversity (TEEB) (2010) The economics of ecosystems and biodiversity ecological and economic foundations. Earthscan, London, Washington, DC
Tratalos J, Fuller RA, Warren PH, Davies RG, Gaston KJ (2007) "Urban form, biodiversity potential and ecosystem services". Landscape Urban Plann. 83
Turner WR, Oppenheimer M, Wilcove DS (2009) Aforce to fight global warming. Nature 428:278–279
United Nations Environment Programme (2009) UNEP 2008 annual report
Walker B, Holling CS, Carpenter SR, Kinzig A, (2004) "Resilience, adaptability and transformability in social-ecological systems". Ecol Soc 9
Widerberg O, Pattberg P (2015) 'Nonstate actors in a Paris agreement. Are cities and companies bridging the ambition gap?' Policy brief, IVM (Institute for environmental studies) and FORES
Wunder S (2005) Payments for environmental services: some nuts and bolts
Wunder S (2015) Revisiting the concept of payments for environmental services. Ecol Econ 117:234–243

Edoardo Croci is Professor of Practice at Bocconi University and the Director of the Sustainable Urban Regeneration Lab. He is also member of the Managing Committee and Senior Research Fellow at GREEN research center—Bocconi University, where he coordinates several research projects. His research and teaching activity are focused on environmental and climate policy and evaluation, with a focus on the urban scale. He chairs UERA—Urban European Research Alliance and he is member of the Thematic Group on Goal 11 (Cities) of UN—SDSN. He served as Deputy Mayor of Milan Municipality and Administrator of Lombardy Environmental Protection Agency.

Benedetta Lucchitta is Researcher at GREEN center—Bocconi University where she has been working since 2014 focusing on nature-based solution and climate change adaptation and mitigation policies at the urban level. She is also a researcher at the Sustainable Urban Regeneration Lab of Bocconi University. She holds an MSc in "Planning and policies for cities and environment" and she is a PhD candidate in urbanism at IUAV in Venice.

Chapter 4
Vulnerability and Risk Assessment of the Port Moresby (Papua New Guinea) Coastline—A Case Study

Regina Kiele, Gertrude Miria, and Eddly Joel

Abstract Climate change is becoming a serious environmental and developmental challenge in recent years. The risk of climate change is the result of interaction of climatic hazards, exposure and vulnerability of the communities and the physical systems. Studies have identified that adaptation to future climate change is to reduce vulnerability and exposure to present climate variability. Assessment of vulnerability and risk is important to develop adaptation policies, strategies and practices.

Port Moresby is a coastal city of Papua New Guinea. The coastal environment is heavily built up with increased demand for coastal landfill developments depleting mangroves and reef ecosystems which increases the risks and vulnerability of built structures and communities exposed to sea level rise. Sea level rise in PNG is 7 mm between 1993 and 2010 and is predicted to rise between 40 and 150 mm by 2030 (NCDC National Capital District Commission (2020) Port Moresby towards 2030. National district development plan revive 2020. NCDC). This means that land use developments, mangroves and communities within 100 m buffer would be affected.

This chapter assesses coastal risk and vulnerability of Port Moresby and identifies the likely impact and persisting vulnerability of biophysical and socio-economic systems due to climate change. The qualitative and quantitative assessments of the potential risks and vulnerabilities from literature are also discussed.

Keywords Port Moresby · Papua New Guinea · Climate change · Vulnerability · Hazard and risk assessment · Coastal region · Coastline · Sea level rise urban · Coastal erosion

---

R. Kiele (✉)
School of Natural and Physical Sciences, University of PNG, National Capital District, Port Moresby, Papua New Guinea
e-mail: kielerh@upng.ac.pg

G. Miria · E. Joel
National Capital District Commission, National Capital District, Moresby, Papua New Guinea

K.-G. Kim and M. Thioye (eds.), *Planning Climate Smart and Wise Cities*, The Urban Book Series, https://doi.org/10.1007/978-3-030-80165-6_4

4.1 Introduction

Coastal regions form the transitional zone between land and the marine environment, and it is unique in its natural and socio-economic features that encourage human activities (Tragaki et al. 2018). Coastal and marine areas contain some of the world's most diverse and productive systems and include extensive areas of complex and specialized ecosystems, such as enclosed seas and tidal systems, estuaries, salt marshes, coral activities, impacts and interventions. These coastal environments are rich in biodiversity and contain sensitive ecosystems such as mangroves, seagrass, wetlands, coral reefs, dunes, beaches and terrestrial trees. Pressures on these world's coastal systems are growing more intense as rapid development and population growth continue in coastal areas, increasingly placing heavy demands on the natural resources and remaining natural habitats along the coasts.

Papua New Guinea has one of the world's longest coastlines (5152 km) and largest areas of territorial sea in the Pacific and lies within the most biologically diverse region for coastal marine biodiversity on earth. Unfortunately, the country has very limited capacity for designing and implementing the coastal policy as there is very little awareness about the importance of coastal ecosystems and the need to manage impacts of land use developments and threats imposed by human activities.

As described by Mukhopadhyay et al. (2012), coastal areas performed important functions such as regulate water exchange between land and sea, influence the chemical composition of sediment and water, storage and recycling of nutrients and human wastes, maintenance of biological and genetic diversity and provide space for agriculture, transportation and navigation. The fragile ecosystems also provide means of recreation and tourism, in addition to providing information on aesthetic, historical, cultural and scientific aspects (Mukhopadhyay et al. 2012).

These coastal environments are exposed to the continuous actions of both natural- and man-induced, operating on different time scales (de Serio et al. 2018, Bukvic et al. 2020). With its moderate climates and access to the seas, coastal regions worldwide are being threatened by rapid urbanization and climate change-related hazards (Lang 2003).

Coastal hazards and risks are becoming a growing concern to governments and communities as a result of disaster loss in terms of costs, human lives, infrastructure and land areas. Coastal hazards can be defined as occasional or chronic destructive natural system events that affect coastal areas, including meteorological (strong winds and coastal flooding); geological (tsunamis and landslides); climate change impacts (sea level rise, temperature and rainfall changes) and land use and coastal change (beach replenishment and land reclamation). About 23% of the world's population live within 100 km of the coast, and about 10% of the population live in extremely low-lying areas (<10 m above mean sea level) (Nelson 2018; Mukhopadhyay et al. 2012). Urban expansion occurs when population and developments increases within a confined boundary, thus putting pressure on the use of coastal resources and exposing their vulnerability to the threats posed by impacts of climate change (Handayani et al. 2017; Nguyen et. al 2016). Therefore, as Bukvic et al.

(2020) stated, 'growing populations, accompanied by significant increase in coastal development, is a trend that is likely to continue for the foreseeable future'.

A study by Boori (2010) in Northern Brazil stressed that due to diverse human pressures, many coastal areas are experiencing acute environmental problems, such as coastal erosion, pollution, degradation of dunes and saline intrusion of coastal aquifers and rivers. In addition, the accelerated sea level rise will intensify the stress of these areas, causing flooding of coastal lowlands, erosion of sandy beaches and destruction of coastal wetlands, as climate change is one of the main impacts of climate change (Boori 2010). Gallina et al. (2020) used a multi-hazard assessment of the influencing matrix to analyse the relationships among hazards (sea level rise, coastal erosion and storm surge) and their disjoint probability. The multi-vulnerability considers the susceptibility of the exposed receptors (wetlands. beaches and urban areas) to different hazards based on multiple indicators (dunes, shoreline evolution and urbanization rate in North Adriatic coast, producing a ranking of multi-vulnerability risks by means of GIS maps and statistics. Another study undertaken by Boori (2010) on northern eastern Brazil used GIS and remote sensing techniques to identify and assess the risk on coastal zones impacted by sea level rise and its impact on climate change.

Assessment of the physical sensitivity and exposure of the coastal developments to coastal inundation and erosion, is essential components for proper coastal vulnerability study (Gallina et al. 2020; Tragaki et al. 2018). Vulnerability assessment includes the susceptibility of the coastal zone to physical changes resulting from climate change, anticipated impacts on socio-economic and cultural, and ecological systems and the available adaptation options (Dolan and Walker 2006). Similarly, this study incorporated the techniques of GIS and remote technology and field trip assessments in the spatial and temporal assessment of vulnerable resources and ranks the vulnerable areas of Port Moresby coastal environment most vulnerable to disaster risk. Therefore, this vulnerability and risk assessment would identify, quantify and prioritize the risks and vulnerabilities of coastline and identify threats and its causes and its implications on the environment.

As stated, this research provides a case study in assessing coastal risk and vulnerabilities in Port Moresby as an urban environment. The scope of this coastal vulnerability and risk study extends from Fairfax Habour to the Bootless Bay (Fig. 4.1), encompassing a total length of 88.8 km along the Port Moresby urban environment.

The aim of this study is to use the geographical information system (GIS) and remote sensing techniques with existing literature and field assessments to assess the physical vulnerability of the coastal region of Port Moresby to the impacts of sea level rise and coastal flooding including qualitative and quantitative assessments of socio-economic developments currently affecting the coastal ecosystems.

The following objectives will be pursued in order to achieve the above aim.

- To identify and map out spatial extent of vulnerable land uses and marine resources at risks of sea level rise within a 100 m and 500 m buffer.
- To model and estimate the influence of global climate warming on sea level rise by temporal periods.

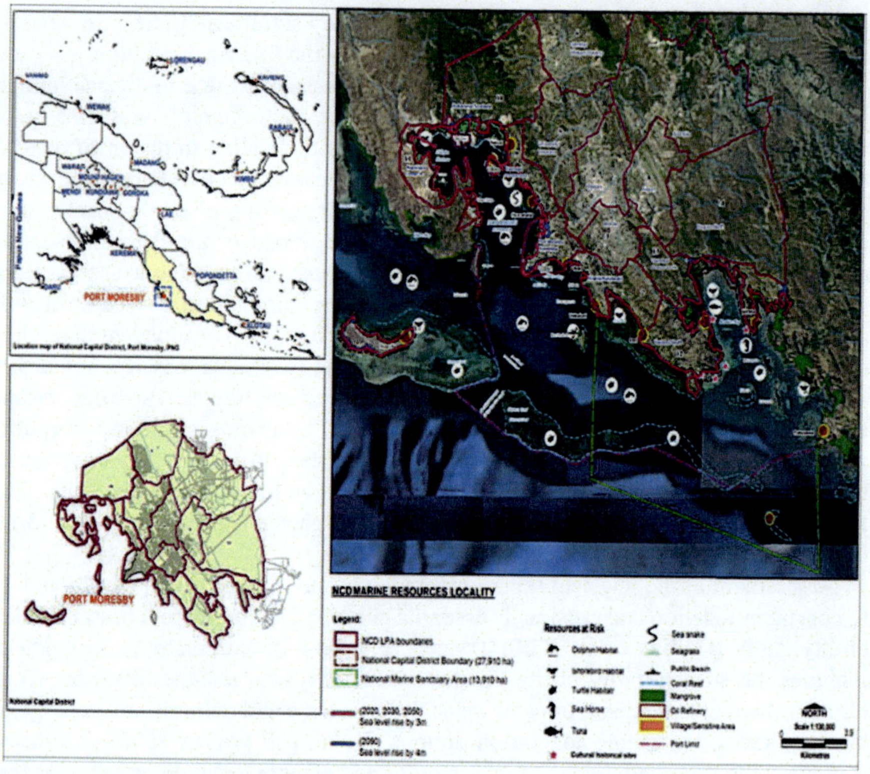

Fig. 4.1 Location of Port Moresby, capital city of PNG

- To evaluate the socio-economic implications of predicted change.
- To determine the trend, nature, rate, location and magnitude of land use land cover change.
- To identify the most reasonable options available to mitigate risks of sea level rise.

4.2 Methodology

There were two methods used for this assessment, and this includes a desktop analysis and field observation. The first step in the desktop analysis involves the use of available spatial databases using GIS techniques to develop base maps and calculate land areas. The MapInfo Professional (Ver. 11.0), geographical information software, was used for the GIS assessments. A buffer of 100 m and 500 m was overlaid over the coastline and the NCD cadastral map to identify spatial extent of settlements and infrastructure (residential, commercial, light industry buildings and

Table 4.1 Calculation of SLR value

Year	Est. SLR under A2 emissions scenario (cm)[a]	Equinox high tide (cm)[b] *Highest tide for NCD occurred on 9 March 2020*	Neo-tectonic (cm)[c]	Final SLR value	
				Cm	M
2020	0.7	291	6.7	298.4	2.984
2030	15	291	6.7	312.7	3.127
2055	29	291	6.7	326.7	3.267
2090	60	291	6.7	357.7	3.577

[a]Data source National Climate Compatible Development Management Policy (2014), GoPNG
[b]Data source Papua New Guinea-Port Moresby (2020) Tide Predictions Calendar (www.bom.gov.au/pacific)
[c]Data source Seismo-tectonic model and probabilistic seismic hazard assessment for Papua New Guinea, Ghasemi (2020). https://doi.org/10.1007/s10518-020-00966-1

transport systems) ecosystems and industries in these coastal zones and calculate land uses within 100 m and 500 m buffers. Information on population was derived from the National Census database (NSO 2011). A Google Earth image (2020) with a resolution of <5 m was used as the base map for mapping and as the backdrop for overlay of existing land uses, road infrastructure and vegetation boundaries and other natural resource assessments. The existing land uses, road infrastructure and vegetation boundaries were assessed from the National Capital District (NCDC) spatial databases. The spatial information on biodiversity was also derived from the NCDC GIS Databases.

The second process was to calculate sea level rise using data employed from literature review. The data sources used in this calculation include

- A2 emissions scenario of SLR data in column 2 of Table 4.1.
- Equinox highest tide data for NCD in 2020 in column 3 of Table 4.1.
- Neo-tectonic[1] data in column 4 of Table 4.1.

Given the base data of 0.7 cm (7 mm) sea level rise for PNG in 2020, this data was calculated by adding 0.7 cm to 291 cm (highest tide data in NCD in 2020) and 6.7 cm (movement of fault line for NCD) which equates to the final SLR value in column 5 of Table 4.1. The same calculation applied to years 2030, 2055 and 2090 and the values were converted to meters to use GIS application to map out areas vulnerable to the projected 3 m sea level rise for periods 2020, 2030 and 2055 and 4 m for 2090.

[1] Neo-tectonic refers to young tectonic movements commonly associated with areas of active seismicity and active faulting, such as plate margins. At microscale, it deals with the movement of individual faults (see Fault) and folds, with dimensions as small as a few kilometers (https://www.researchgate.net/publication).

The second method was the field observation which involved verification assessment of the desktop study results and identification of socio-economic developments and coastal marine resources at risk and vulnerable to the impacts of sea level rise. Photographs were captured during the field assessment on 18 February 2021 as most would be displayed in this report.

4.3 Study Area—Port Moresby

The study area is the coastal region of the capital city of the PNG, Port Moresby, currently accommodating the country's administrative and commercial functions of the National Government. Port Moresby is located on the shores of the Papua coastal lagoon on the south-eastern coast of the New Guinea Island at latitude −9°26′ 35.30″S and Longitude 147°10′ 46.99″E (Fig. 4.1). The islands of Daugo (Fisherman Is), Daunagena, Gemo, Lolorua and Manubada are also within the boundary of the city (UN 2012). The coastal region includes marine habitats enclosed from Gemo Island to the Pyramid Point and past Pari Village (Fig. 4.7). The embayment includes several smaller bays such as the Fairfax (Port Moresby) Harbour, Walter and Joyce Bays. The embayment is separated from the open sea by a discontinuous coral line barrier reef that occurs at the edge of a narrow continental shelf approximately 3 km offshore (NCDC 2007). There are two major inlets, namely the Fairfax Harbour and the Bootless Inlet (Figs. 4.9 and 4.10).

Month	Jan	Feb	March	April	May	June	July	Aug	Sept	Oct	Nov	Dec	Total
Mean	187.0	191.1	224.4	147.8	112.3	22.6	16.2	29.1	16.7	31.7	48.5	88.2	1115.6
Max	420.4	303.0	366.2	325.6	403.6	174.2	90.2	72.8	62.2	109.2	201.4	162.0	1534.2
Min	80.6	73.4	49.8	48.0	5.0	0.0	0.0	0.0	0.0	0.0	0.0	0.0	648.0

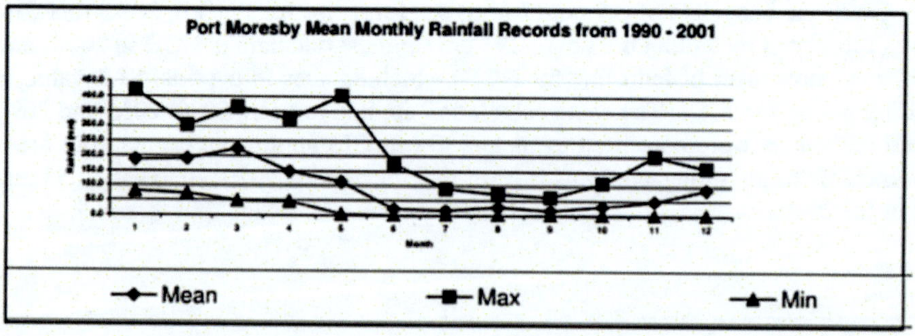

Fig. 4.2 Average annual and monthly rainfall records–Port Moresby 1990–2000 (NCDC 2007)

The total area of Port Moresby is 267.6 km^2 with a total coastline length of 88.14 km that extends from Fairfax Habour to the Bootless Bay and is enclosed by the Central Provinces land boundaries (Fig. 4.1). The coastal resources of Port Moresby (in Fig. 4.7) contain high biodiversity locations including the mangroves, the coral reefs and bays. The coastline is 88.14 km in length including the reclaimed land, bleaches, traditional villages, unplanned settlements, road and seawall infrastructures, industries and commercial developments (Figs. 4.3, 4.4, 4.5 and 4.7).

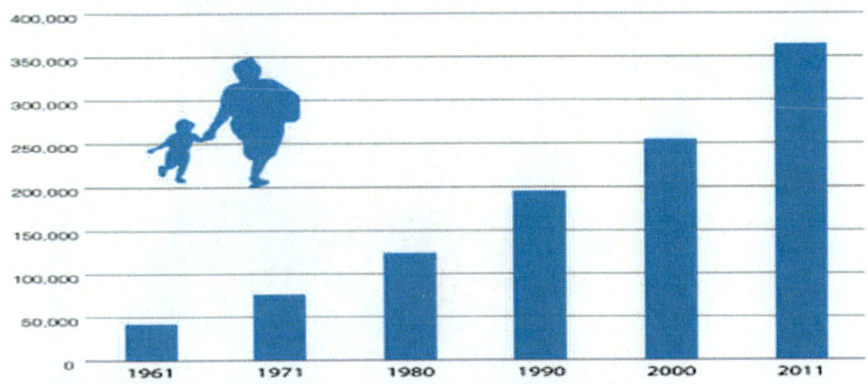

Fig. 4.3 Population of Port Moresby by census dates (NCDC 2020)

Fig. 4.4 Fenced mangrove sites purchased from customary landowners through informal land dealings

Fig. 4.5 High demand of heavy industrial developments at Fairfax Harbour

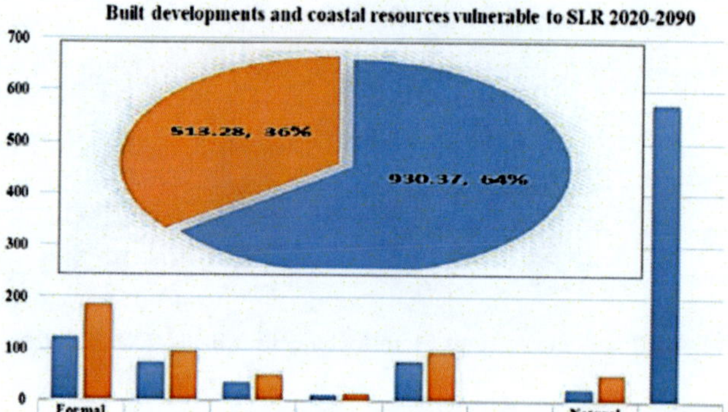

	Formal urban landuses	Informal settlements	Urban villages	Land reclamation	Unused land	Swamp	Natural hillslope areas	Mangroves
Land uses (ha) vulnerable to 3m SLR for years 2020, 2030 and 2050.	123.3	74.7	37.2	12.1	77.2	0.3	23.8	575.07
Land uses (ha) vulnerable to 4m SLR for year 2090.	188.1	96.1	51.4	14.2	97.4	2.91	51.7	0.22

Fig. 4.6 Built and natural environment and resources vulnerable to 3 m and 4 m SLR

4.4 Background Overview of Port Moresby

4.4.1 Physical Environment

4.4.1.1 Climate and Rainfall

Port Moresby is one of the driest regions of the country with a mean annual rainfall range between 1000 and 1500 mm (Shearman and Bryan 2007). Port Moresby experiences two distinct seasons, a wet season from November to April and a dry season from May to October. The average yearly rainfall is 1270 mm of which much

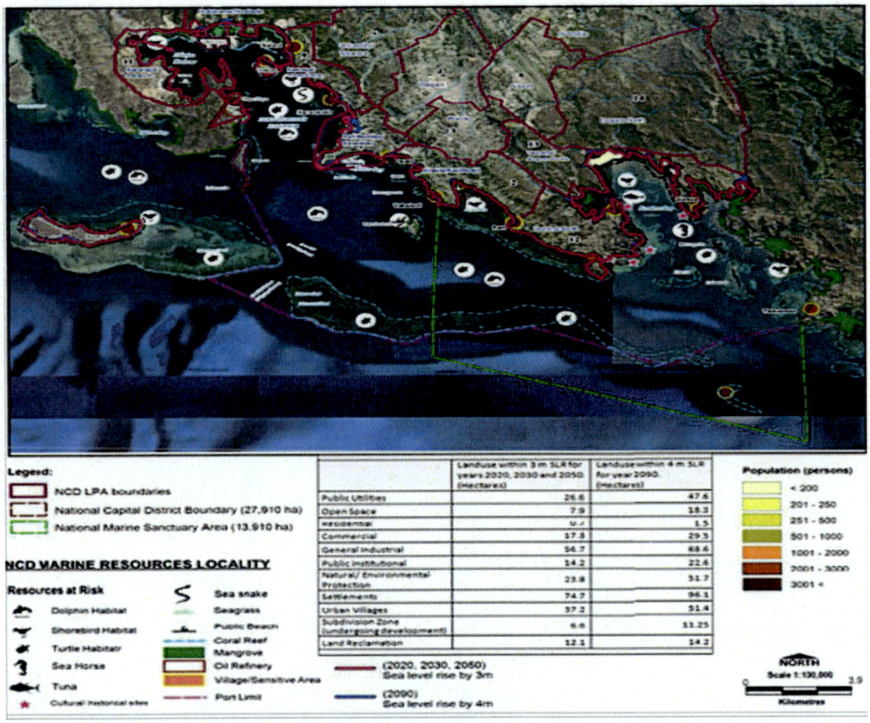

Fig. 4.7 Marine biodiversity resource map of NCD (NCDC datasets 2020)

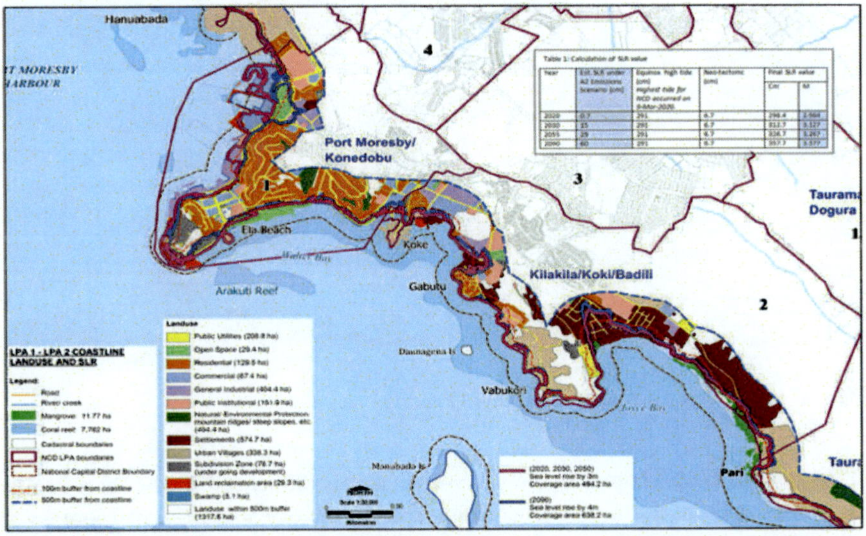

Fig. 4.8 Coastal land use and SLR in LPA 1 and LPA 2

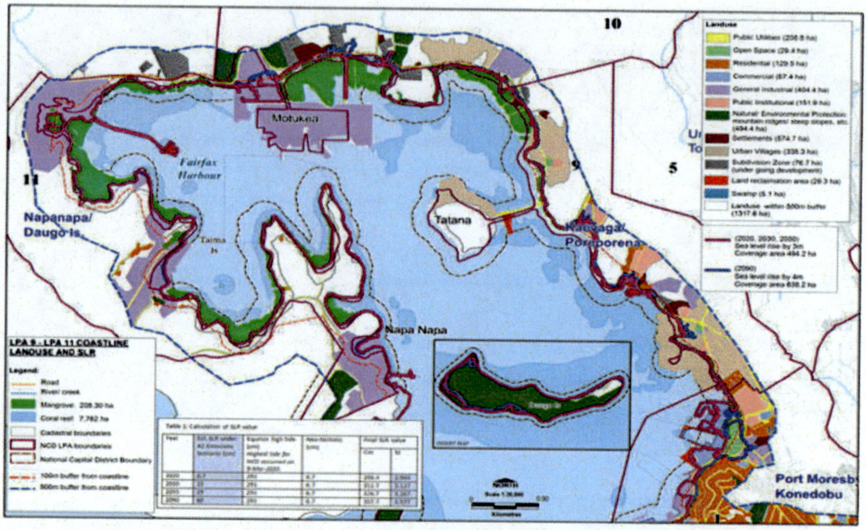

Fig. 4.9 Coastal land use and SLR in LPA 9, 10, 11

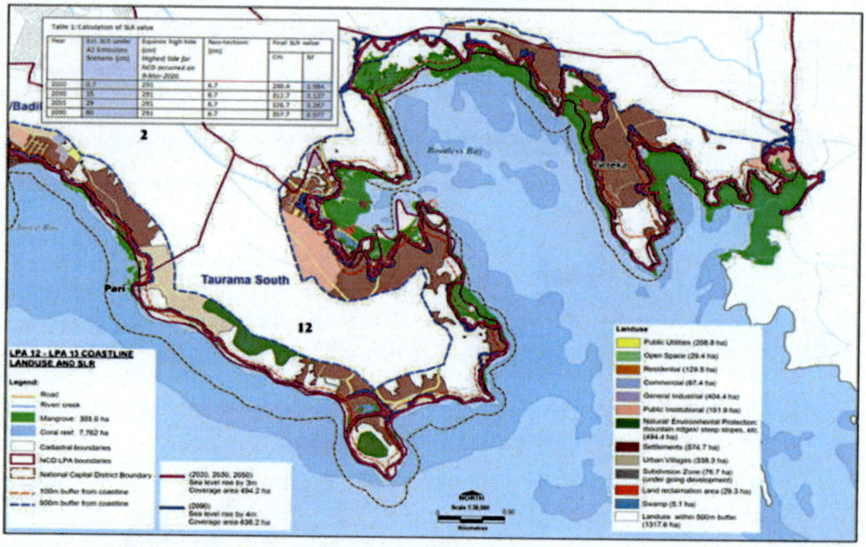

Fig. 4.10 Coastal land use and SLR in LPA 12 and 13

of the rainfall (78%) comes in the wet season that is influenced by the West Pacific Monsoons and the north-westerly winds. During the dry season, Port Moresby is exposed to the dry south-easterly winds, and the daily average temperature ranges from 28 to 32 °C (UN 2012).

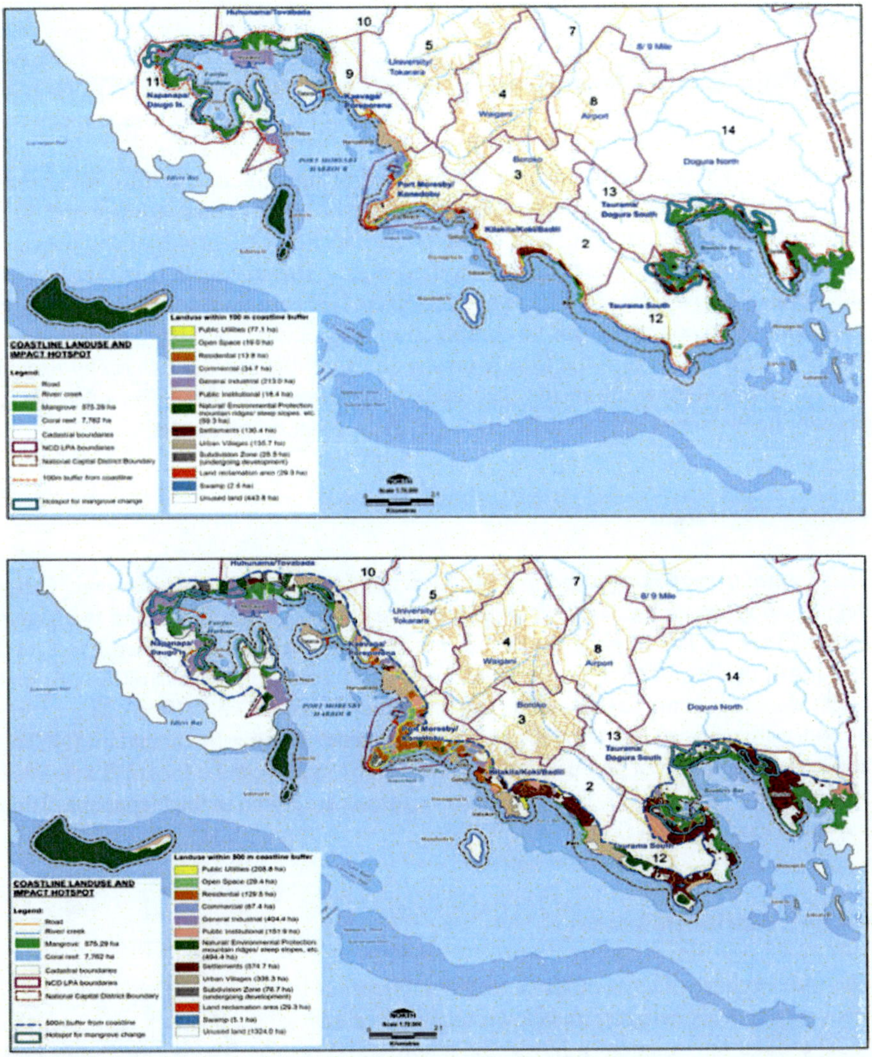

Fig. 4.11 Comparison of coastal land use developments within a 100 m and 500 m environmental protection buffer

The average yearly and monthly rainfall for 1990–2000 periods as indicated Fig. 4.2 is low as during the north-west season, rainfall can be twice as much as in the south-east season. The wet season had persisted longer from November to December to about April with a relatively shorter dry season happening between June and October (NCDC 2007).

4.4.1.2 Landform

Land above mean sea level ranges from <0.45 m from beaches to 3.5 m on mud flats to <50 m on low cliffs along the coastline, compared to the average mean sea level of 200 m above sea level for Port Moresby (Pain and Swadling 1980) and UN (2012). The landform of the city consisted of mountains and hills with weak or no structural control, hilly terrain with weak or no structural control and to a lesser extent, on undifferentiated coastal plain, and Homoclinal ridges and cuestas: inclined asymmetrical structurally controlled ridges. The soil depth is primarily very shallow to shallow soils (15–30 cm and 25–50 cm) (Shearman and Bryan 2007). The substrates are made up of one or combination of rocks, rocky cliff, boulders, sand and coral rubbles. Within the Port Moresby Harbour, the shoreline is rocky except for Ela Beach. The width of the tidal zone ranges from 1 to 3 m. The substrate is predominantly sandy and silty, but in some areas, coral rubbles exist.

4.4.1.3 Vegetation

The common vegetation type in Port Moresby is the savannah grassland. Other vegetation types include remnants of deciduous forest, beach forest and mangrove forest (Fig. 4.2). The original beach forest occurred around Ela Beach being replaced as a result of development on the waterfront. Remnant plant species of the original forest exist but scattered throughout the bay which make the vegetation composition of both native and exotic plant species with the latter being more abundant (NCDC 2007). The remnants of deciduous forest can be observed near the foot of Burns Peak and on well-drained areas. Dense mangrove forests are found in the Napanapa, Tuna bay and Bootless Inlet (Fig. 4.11).

4.4.1.4 Marine Resources

The wide expanse of bays, offshore reefs, its numerous islands, internationally significant wetlands of seagrass meadows, mangroves and sandy beaches make up the NCD coastal environment a haven for wildlife and people. The corals in the Port Moresby Bay are distributed in a patchy manner, and these coral reefs are part of the Papuan Barrier Reef, which is the third largest in the world (UN 2012). Maniwavie et al. (1998) sighted seven families of corals between Paga Point and Pyramid Point. Similarly, seagrass beds occur in patches of Port Moresby Bay and support a large array of marine life including macro algae (sea weeds). An indirect indication of the rich biodiversity can be classified from the distributions of corals, seagrass and mangrove ecosystems since these systems are highly correlated with biological diversity (Fig. 4.7).

4.4.2 *Socio-Economic Context of Port Moresby*

4.4.2.1 Population

The total population of Port Moresby from the 2011 census was 364,125 with a population density of 1189 persons per km^2. The growth pattern had been exponential with an annual growth rate of 3.3%, compared to PNG's 3.1% (NSO 2011; NCDC 2020). The population pattern from 1971 to 2011 is shown in Fig. 4.3. Port Moresby has the largest urban population in PNG compared to the other urban cities or towns. With the growth rate of 3.3%, the population of Port Moresby in 2019 would be 470,000 and is projected to be 680,000 in 2030 (NCDC 2020). A steady increase in population growth pattern confirms the increasing use of land developments and vulnerability risks involved within a confirmed city boundary.

4.4.2.2 Land Tenures

Land tenure of Port Moresby is both state and customary owned. Mobilizing customary land in NCD is a challenge as the traditional landowners continue to sell land through informal land dealings with developers in contravention to the current development plans/policies. This has caused rapid increase in informal developments depleting sensitive natural ecosystems (Fig. 4.4).

Land dealings relating to customary land are governed by the Land Groups Incorporation (Amendment) Act 2009 and Land Registration (Amendment) Act 2009. Land Groups Incorporation (Amendment) Act 2009 deals with the organization of the clans to manage their customary land, while the Land Registration (Amendment) Act 2009 is to facilitate voluntary registration of customary land for development through the ILGs. This process stimulates economic growth through appropriate land management and protection of valued environmental features potential for eco-tourism activities for traditional landowners or through joint partnership with government and private investors.

4.4.2.3 Land Use Developments

Being the capital city of PNG, Port Moresby is experiencing rapid urban growth (Fig. 4.5) in recent years due to high capital investment, cash flow and booming industries and employment opportunities. Its economic outlook is at promising levels with growing financial property investment and employment markets, stable economic conditions and rapidly private sector which is closely associated with the occurrence of massive infrastructure and commercial, industrial and residential developments throughout the city. These developments have caused massive destruction to the coastal resources including alterations to the physical appearances of coastal and terrestrial landforms and degradation of mangroves, reefs and seagrasses from land-based pollution.

4.5 Results

4.5.1 Coastal and Marine Resources

The biodiversity of the Bootless Bay is used as an indication of the rich biodiversity of the Port Moresby's coastal region (Table 4.2 and Fig. 4.7). The marine resources are compared with the PNG recorded data as indicated in Table 4.2. The Bootless Bay and Tuna Bay are habitats to 283 terrestrial plants, 23 species of mangroves, four species of seagrass, 512 fish species, 331 bird species, 281 species of coral reef, 20 species of macroalgae and 28 species of echinoderms (CEPA-JICA 2017). This rich marine biodiversity is very vulnerable to the increasing growing population, rapid land use developments and increasing sea level rise.

4.5.2 Urban Land Use Developments

The projected sea level rise of 3 m would slowly affect 64% of a total area of 930.37 hectares of land uses comprising built and coastal resources and environment from periods 2020–2055 according to Table 4.3 and Fig. 4.6. In contrast, by 2090, SLR would reach 4 m, and the land uses within the proximity of 513.28 hectares (36%) would be inundated or disappear forever. The mangrove forest makes up the highest area of 575.07 ha, highly vulnerable to the 3 m SLR from 2020 to 2055. The remaining mangrove area of 0.22 ha would be affected by 4 m SLR by 2090.

Table 4.2 Biodiversity of Bootless and Tuna Bay

Coastal/Marine resource	Bootless and Tuna Bay	PNG
Mangrove	23 species	44 species
Seagrass	4 species	12 species
Fish	512 species	1500 species
Bird	331 known bird species	831 species
Coral reef	281 species	514 species
Terrestrial plants	283 species	>200,000 species

Table 4.3 Existing land use developments in NCD vulnerable to projected 3 m and 4 m sea level rise from 2020–2090

Existing land uses	Land uses (ha) vulnerable to 3 m SLR for years 2020, 2030 and 2050	Land uses (ha) vulnerable to 4 m SLR for year 2090
Public utilities	26.6	47.6
Open space	7.9	18.3
Residential	0.7	1.5
Commercial	17.3	29.5
General industrial	56.7	68.6
Public institutional	14.2	22.6
Informal settlements	74.7	96.1
Urban villages	37.2	51.4
Subdivision zone (undergoing development)	6.6	11.25
Land reclamation	12.1	14.2
Swamp	0.3	2.91
Unused land	77.2	97.4
Natural hillslope areas	23.8	51.7
Mangrove forest	575.07	0.22

Figure 4.6 also indicated that formal land use developments within 123.3 ha are at high risks of 3 m SLR and 188.1 ha would be impacted by 4 m SLR. These formal developments include public infrastructure/utility, residential, commercial, general industrials and public institutional. It should be noted that 100 m and 500 m buffers do not present a more accurate assessment of coastal erosion or SLR, unlike projected 3 m modelling which present an accurate assessment based on microscale movement of fault lines and astronomical highest daily tides. However, both methods confirm that much of Port Moresby's coastal resources will be affected by increasing impacts of climate change phenomena on sea level rise. The reclaimed land with weak baseline structures and villages and settlements, including developmental areas and seashore natural resources, are vulnerable to coastal flooding and erosion. With time, these impacts will become more evident.

The maps in Figs. 4.8, 4.9, 4.10 and 4.11 are showing the land use types by sea level rise of 3 m and 4 m for three temporal periods 2020, 2030, 2055 and 2090. The map in Fig. 4.12 shows population figures within the 100 m and 500 m buffer. As noted, the 3 m SLR is within the 100 m buffer (Table 4.4).

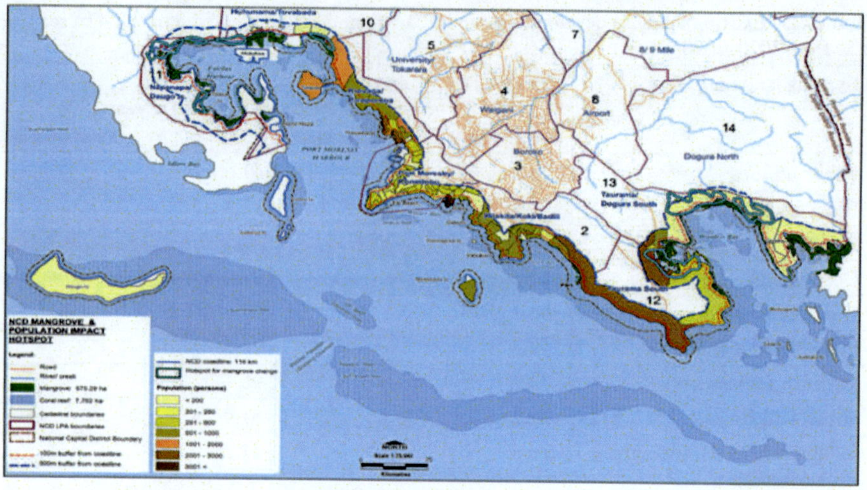

Fig. 4.12 Population within 100 m and 500 m buffer

Table 4.4 Existing land uses within 100 m and 500 m environmental buffer

	Land uses (ha) within 100 m environmental buffers	Land uses (ha) within a 500 m environmental buffer
Public utilities	77.1	208.8
Open space	19.0	29.4
Residential	13.8	129.5
Commercial	34.7	67.4
General industrial	213.0	404.4
Public institutional	18.4	151.9
Natural/environmental protection	59.3	494.4
Settlements	130.4	574.7
Urban villages	135.7	338.3
Subdivision zone (undergoing development)	25.5	76.7
Land reclamation	29.3	29.3
Swamp	2.5	5.1
Unused land	443.8	1324.0

4.6 Discussions

4.6.1 Sea Level Rise in PNG

The main effects of global warming due to greenhouse gas emissions in Port Moresby and PNG are sea level rise which will cause coastal inundation and erosion affecting coastal developments such as commercial, industrial, residential, road infrastructure and coastal communities. Sea level rise in PNG has been measured at 7 mm (1993–2010) and is predicted to rise between 40 and 150 mm by 2030. Port Moresby also experiences increases in storm surges, increase in the intensity and frequency of rainfall and increase in average daily temperatures. The Pacific Sea Level and Geodetic Monitoring Station for PNG was established in Lombrum, Manus Province, and records sea level for the entire country (Fig. 4.13).

As the sea level continues to rise into year 2030, more of our coastal areas development will be exposed to these natural forces as shown in Fig. 4.14. Rising sea level will also increase the mean depth of water, and so the speed of waves and energy they transmit across the reef and against the shore will cause changes in coastal sediment, erosion and deposition. Wave action and sea level rise will lead to the destruction of mangroves, coral reefs, seagrass and fish species, cultural and historical sites and other areas of special value on the coast or beaches. The sea level rise will also have social and economic impacts such as

- Saltwater contamination of drinking water.
- Economic costs for engineering structures and sewage systems will increase.
- Emigration and resettlement of communities.
- Land management and practices will have to change.

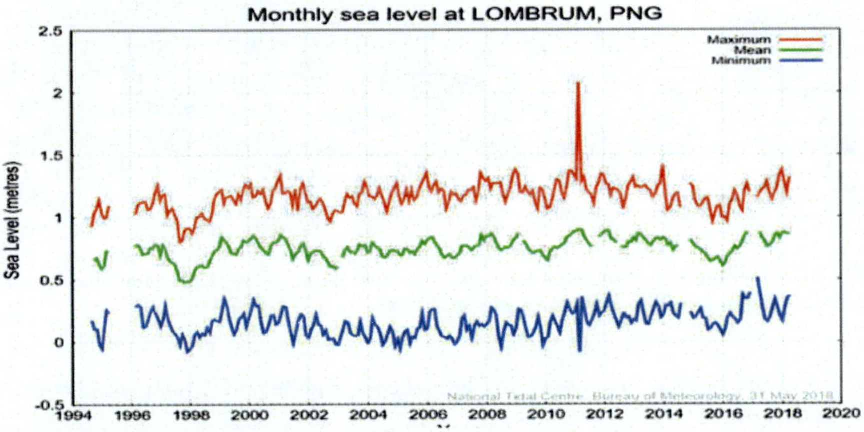

Fig. 4.13 Recorded monthly sea level rise for PNG

Fig. 4.14 Thermal energy warming sea currents to move faster and higher at Milne Bay, central and NCD provinces within the southern region of PNG

- Local and national economies will be reshaped with regard to marine and coastal resources.
- Loss of cultural heritage, psychologically and physically.
- Changes of land values will trigger major economic and social transformation.

- Some seafront species of mangroves will grow further inland as a result of seed dispersal from tidal currents (Fig. 4.17).

Coastal erosion is generally driven by the action of waves and currents (Fig. 4.14) and by mass wasting processes on slopes (UNISDR 2017); however, it is being accelerated along the Port Moresby's coastline due to storm surges and human activities influencing the tendency of landforms to erode. For example, the construction of coastal structures (e.g. seawalls) can lead to changes in coastal sediment transport pathways, resulting in erosion in some areas and deposit in other areas (Fig. 4.15). There had been undercutting of the coastal headlands, such the Paga Hills' cliffs and steep slopes areas and had contributed to mass wasting processes.

Fig. 4.15 Evidence of coastal flooding on infrastructure and seashore

Fig. 4.16 Marine biodiversity in Bootless Bay vulnerable to SLR and land use developments

Fig. 4.17 Seafront mangrove species Rhizophora apiculata (Ra) now grow further inland of LPA 13 where it never grew before

The current estimated 7 mm rise in sea level (NCDC 2020) would have significant adverse effects on the study area, particularly on low-lying shorelines and reclaimed areas. With a rise in sea level, beach sand will tend to wash offshore. As a result, the shoreline will not be as steady and will retreat in many places. Beaches that rely on obtaining sand and sediment from further alongshore will receive less—another factor increasing erosion and shoreline retreat. Coastal erosion would be even worse for Port Moresby because when rainfall is reduced, sediment from the land is likely to drop. The current rate of development in Port Moresby due to limited space is putting more pressure on natural resources (picture below) that will only enhance the effects of climate change on resources.

4.6.2 *Impacts of Sea Level Rise on Coastal and Marine Biodiversity*

The marine ecosystem of the Port Moresby Bay is progressively deteriorating due to the effects of climate change and of increasing developmental activities along the shoreline. The city's expansion and the over exploitation of coastal natural resources have reduced much of the original vegetation. Large portion of the city's landscape has been reshaped and reorganized including land (and beach) reclamations to cater for infrastructure, building material and fuel wood. Natural resources that are most vulnerable to the impacts of climate change (sea level rise and coastal erosion) and other anthropogenic activities and increasing population include shoreline, mangroves, sea grass, coral reefs and the rich biodiversity of fauna species habituating in the Port Moresby marine environment (Fig. 4.16).

Variations in ocean salinity, with levels becoming both fresher and saltier, would have a significant impact on coral reefs due to their low tolerance level, so large fluctuations in salinity would have adverse impacts on coral health. Coral reefs are under growing threat from ocean acidification, due to the reduction in ocean pH as CO_2 is taken up by ocean surface. Sea level rise over coral reefs provides corals with greater space to grow upward, however, with increased wave action, could also damage corals (Fig. 4.14). It was noted in Figs. 4.7, 4.8, 4.9 and 4.10 that the high diversity of marine resources within NCD would disappear and go extinct as projected 3 m SLR reached 2050. It was also evidence that some seafront species of mangroves are now growing further inland because of seed dispersal from strong and high tidal currents (Fig. 4.17).

Land use and climate change are recognized as two main drivers of future biodiversity change with potentially severe impacts on ecosystem services and human well-being. Habitat and land use changes, resulting from past, present and future human activities, have immediate impacts on biodiversity and ecosystem services, whereas the impacts of climate change have considerable lag times. Therefore, current and future land use projections are essential elements for assessing biodiversity and ecosystem change.

4.6.3 Impacts of Sea Level on Socio-Economic Development

Coastal seal level rise is among the most severe societal consequences of anthropogenic climate change. The mean sea level rise for PNG will continue as a consequence of anthropogenic climate warming and the impacts would impose an increase in exposure and sensitivity to the surrounding environment and communities. Societal concerns about sea level rise include potential shoreline recession, loss of coastal infrastructure, natural resources and biodiversity and in the worst case, displacement of communities and migration of environmental refugees. However, the impacts will vary strongly from region to region and coast to coast and therefore cannot be easily generalized, as changing mean and extreme coastal water levels depend on a combination of nearshore and offshore processes, related to climatic but also non-climatic anthropogenic factors.

It was noted that many land use developments in the city have greatly interfered and transformed the natural coastal landscape of Port Moresby without due consideration and standards to respond to the rising sea level. The highest recorded high tide in Port Moresby was on 9 March 2020 which reached a height of 2.91 m. Given this height, the extremes of waves in the future will greatly affect the built developments including road infrastructure by flooding over it and degrading them. The land reclamation for the construction of buildings adjacent to the Yacht club and Lancron Naval Base will have greater impacts from sea level rise because the highest height at which the land was reclaimed is less than the sea level rise predicted by the IPCC within the next thirty years unless adequate adaptation measures are created. As evidenced in Fig. 6, areas within 930.37 ha are at high risks of this 3 m sea level rise, and these include formal areas, informal settlements, urban villages, land reclamations, swamp wetlands, hillslope areas and mangrove forest. By 2030–2050, most of the low-lying

Fig. 4.18 Coastal developments at high risks of 3 m SLR

areas of Koki, Badili Ela Beach, Konedobu, Motukea, Pari, Baruni and Hanuabada would be greatly affected by the 3 m SLR (Fig. 4.18).

4.6.4 Policy and Social Options to Nature Base Solutions to Climate Emergency

The United Nation Framework Convention on Climate Change (UNFCCC) COP 25 in Madrid (2/12/19) recommended cities to integrate the role of nature-based solutions for climate emergency. In response to this call, certain policy frameworks have been developed at national and sub-national level such as the National Oceans Policy 2020–2030, National Protected Area Policy 2018, Bootless Bay National Marine Sanctuary Management Plan 2020–2025 and the NCD Urban Development Plan 2020–2030. These policies set out a framework to facilitate economic growth through appropriate land management, protecting valued environmental features, incorporating social issues into planning decisions, providing opportunities for efficient transport and mobility and safeguarding development from natural risks. Conserving and enhancing blue-green infrastructure is vital to support environmental services and amenity whilst reducing risks to climate change. Some localized solutions to climate actions along the coastal environment would include

- Use a land-based 250 m restricted development buffer from high watermark along the coastline.
- Use urban development plans/policies to restrict development of highly vulnerable uses such as hospitals and residential houses in high risk areas while permitting less vulnerable uses in lower-risk areas.
- Use hazard risk assessments based on a 100 years return period for coastal developments.
  - "The levels of habitable rooms must be raised above designated flood level by 300 mm" as shown in the diagram below adopted from Gold Coast City Council.

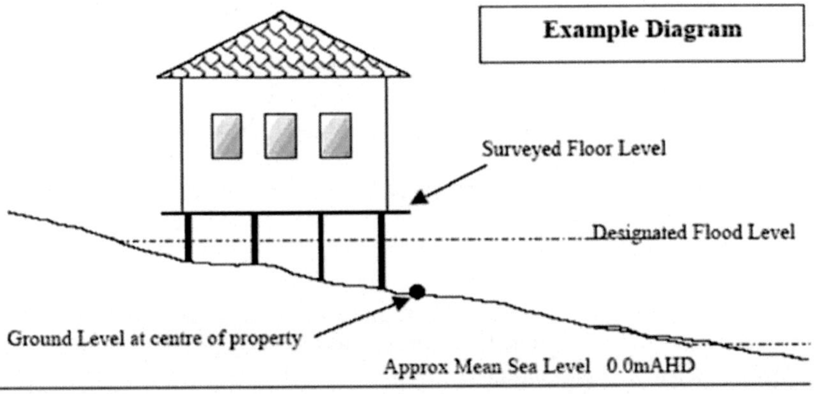

  - Building setbacks—Ensure front setbacks of buildings must be kept free from built structures because of the proximity of the surveyed property boundary to the seawall purposely to protect properties from extreme storm events and collages of the seawall.
  - Seawall design—All seawall structures along the coastline must comply with standard design of the NCDC purposely to ensure all seawall is uniform and provide good view to the waterfront. Rock boulders are part of the design to reduce strong wave breaks on the seawall.
  - Groyne—Groyne must be constructed perpendicular to the coast to collect sand deposited by longshore drift and protect sand from being washed away by waves and high tides. The effectiveness of this structure is to maintain the natural landscape by reducing erosions, wave currents and tides from destroying both the natural landscapes and built infrastructures.

- Flood warning systems and mapping—Systems to provide more accurate prediction of areas likely to flood and warning mechanisms to alert communities of impending threats could have the ability to limit loss of both property and life.

4.6.5 Adaptation Options to Reduce Coastal Vulnerability

A wide range of possible adaptation options are available to reduce vulnerability to negative impacts from sea level rise and other extreme weather conditions. However, ecosystem-based solution is the most preferred option as it is focusing on preserving key habitats such as mangroves, coral reefs and trees that provide natural protection and at the same time support ecosystem services, including fish and tourism. Some ecosystem-based option adopted from Rao et al. (2002) includes

- Replant mangroves—Mangroves provide an effective natural barrier to storm waves and stabilize sediment (limiting coastal erosion), help to maintain water quality and support coastal fisheries.
- Replant seashore buffer—Preserving intact vegetation as well as replanting where necessary along the coastline could reduce coastal erosion and assist in slowing intensity of waves.
- Reduce upslope vegetation and land clearing—Preserving remaining trees and grassland by reducing clearing and re-vegetation where possible can assist in reducing hillslope erosion and flash flooding especially during an extreme storm/rainfall event.
- Reduce coral depletion—Coral reefs provide local protection and provide multiple ecosystem resources including sustaining fisheries and influencing local sediment process (including beach nourishment). Limiting overfishing and using proper fishing methods can assist in maintaining these services.
- Monitoring and enforcement—Once areas have been preserved and damaging practices have been curtailed, monitoring and enforcement are required to ensure these positive actions continue.

4.7 Conclusion

The study of the Port Moresby coastal region's physical environment has raised a series of issues relating to the vulnerability status of the area to the effects of climate change. The most sensitive areas are the marine environments and the developed areas within the 100 m buffer of the seashore. The development and reclamation activity appears to be having negative nearshore impacts on marine life, water quality and tidal flushing cycles and flooding effects of population. These areas include Ela Beach, the Paga Point foreshore and offshore reef areas are extremely vulnerable sites to the effects of sea level rise. Most natural vegetation in the area has been cleared; however, large areas of savannah grass remain, and their retention warrants further consideration.

Port Moresby is rapidly developing, and it is also experiencing environment, sociocultural and economic challenges in managing its urban growth and providing equal opportunities for its rapidly growing population. Port Moresby as a coastal city faces climatic risks of flooding, coastal inundation and impacts from strong winds and extreme temperatures. Lack of sound management strategies of resource use coupled with challenging impacts of climate change could result in coastal ecosystems depletion and degradation.

References

Boori MS (2010) Coastal vulnerability, adaptation and risk assessment due to environmental change in Apodi-Mossoro estuary, Northeast Brazil. Int J Geomat Geosci 1(3). Integrated Publishing Services

Bukvic A, Rohat G, Apotsos A, de Sherbinin A (2020) A systematic review of coastal vulnerability mapping. Sustainability 12, 2822. https://doi.org/10.3390/su12072822. https://www.mdpi.com/journal/sustainability

Commonwealth of Australia (2014) Papua New Guinea—Port Moresby 2019 tide predictions calendar—climate and oceans support program in the Pacific. Australian Government

Conservation and Environment Protection Authority and Japan International Cooperation Agency (2018) Biodiversity project 2015–2020 Annual Report 2017, Issue March 2018

Dolan AH, Walker IJ (2006) "Understanding vulnerability of coastal communities to climate change related risks". In: Journal of coastal research, winter. Special Issues No. 39

Downing T, Patwardhan (2020) Vulnerable assessment for climate change adaptation. Adaptation and planning framework technical paper 3. Habana/Oxford

Government of Papua New Guinea (2015) National marine conservation assessment for PNG, conservation and environment protection authority, p 51

Gallina V, Torresan S, Zabeo A, Critto A, Glade T, Marcomini A (2020) A multi-risk methodology for the assessment of climate change impacts in coastal zones. Sustainability, EISSN 2071-1050, Published by MDPI https://www.mdpi.com/2071-1050/12/9/3697/htm

Handayani W, Rudiarto I, Setyono JS, Chigbu UE, Sukmawati (2017) Vulnerability assessment: a comparison of three different city sizes in the coastal area of Central Java, Indonesia. Adv Clim Chang Res 8, 286–296. Available on line at http://www.sciencedirectcom

Kiele R, Vuvu J (2008) Addressing an innovative approach to determine change in Mangrove forest in the bootless bay, Central province by using GIS/RS techniques. In: Addressing national and global issues through scientific research proceedings. UPNG, SNPS, pp 90–101

Lang L (2003) Managing natural resources with GIS. ESRI, USA

Maniwavie T, Aitsi J, Federizon (1998) Rapid ecological assessment of the walter and joyce bays of Port Moresby. Report prepared for JICA

Mukhopadhyay A, Dasgupta R, Hazra S, Mitra D (Jan–April 2012) "Coastal hazards and vulnerability: a review" Int J Geol, Earth Environ Sci ISSN: 2277-2081. Online Int J 2(1):57–69

National Statistics Office. Census data. Port Moresby https://www.nso.gov.pg/statistics/population

National Capital District Commission (2007) Port Moresby town local development plan, vol 2. SMEC PNG Limited. NCD

Nguyen TTX, Bonetti J, Roges K, Woodroffe CD (2016) Indicator-based assessment of climate-change impacts on coasts: a review of concepts, methodological approaches and vulnerability indices. Oceans Coast Manag 123:18–43

National Capital District Commission (2020) Port Moresby towards 2030. National district development plan revive 2020. NCDC

Nelson AS (2018) Coastal zones. Tulane university. https://www.tulane.edu/~sanelson/Natural_Disasters/coastalzones.htm

Pain C, Swadling P (1980) Sea level changes, coastal landforms and human occupation near Port Moresby—a pilot study. In: Science in New Guinea Journal 1980. NCD. https://www.researchgate.net/publication/275037322

Papua New Guinea—Port Moresby (2019) Tide predictions calendar. Climate and oceans support program in the Pacific. Australian Government

Proceedings of the 8th International Coastal Symposium (ICS 2004), III(Winter 2006):1316–1323. Coastal Education & Research Foundation, Inc.

Rao NS, Carruthers TJB, Anderson P, Saxby T, Durbin T, Jungblut V, Hills T, Chape S (2002) A comparative analysis of ecosystem-based adaption and engineering options for Lami Town, Fiji. A synthesis report by the secretariat of the Pacific regional programme

Shearman PL (2010) Recent change in the extent of Mangroves in the Northern Gulf of Papua, Papua New Guinea. AMBIO J Hum Environ 39(2):181–189

Shearman PL, Bryan JE (2007) Papua New Guinea resource information system handbook. University of PNG. Papua New Guinea

Tragaki A, Gallousi C, Karymbalis E (2018) Coastal hazard vulnerability assessment based on geomorphic, oceanographic and demographic parameters: the case of the Peloponnese (southern Greece). Land 7, 56. https://doi.org/10.3390/land7020056. https://www.mdpi.com/journal/land

UN Habitat (2012) Port Moresby, Papua New Guinea—climate change vulnerability assessment. UN Habitat

United Nations Office for Disaster Risk Reduction (2017) "7. Coastal erosion hazard and risk assessment" words into action guidelines: national disaster risk assessment hazard specific risk assessment. UNISR

United Nations Office for Disaster Risk Reduction (2017) "Coastal erosion hazard and risk assessment" UNISDR

Vivien MG, Daniels, Richard C, White, Tammy W, Birdwell, Kevin R (1994) "The development of a coastal risk assessment database: vulnerability to sea-level risk in the US Southeast" In: Journal of coastal research, 1994, Special Issue No. 12, Coastal hazards perception, susceptibility and mitigation. Coastal Education & research Foundation, Inc. pp 327–338

Chapter 5
Assessment of Mangrove Distribution and Drivers of Mangrove Cover Change from 2010 to 2020 of the National Capital District (Port Moresby) Papua New Guinea

Gertrude Miria, Regina Kiele, and Eddly Joel

Abstract Mangroves are a very diverse group of specialized coastal trees or shrubs that are adapted to the marine environment and inhabit highly productive inter-tidal ecosystems in the tropics and subtropics providing critical goods and services to coastal communities and contributing to global climate change mitigation owing to substantial carbon stocks. The inter-tidal areas are harsh environments between land and sea, subject to both high and low rainfall, hot and cold temperatures, variable amounts of freshwater oxygen and salt, and soils that can be pure clay or mixed with sand, silt, coral rubble or rock. These variations in chemical and physical factors influence mangrove ecosystems to display extreme diversity in plant composition, forest structure and growth rates (Manivawie (1998) An introduction manual on the biology and restoration of mangrove ecosystem. MIRC).

Despite their importance, the mangrove distribution in National Capital District (NCD) continues to decline primarily due to anthropogenic drivers. According to Kiele, 2008, mangrove cover in the Bootless Bay has decreased from approximately 2.768 km^2 in 1985 to about 1.031 km^2 in 2004 with a decrease in mangrove area of about 1.7 km^2 within a space of about 18 years.

Conserving and enhancing mangrove forest is vital to maintain environmental services and amenity and also increase our adaptation measures to climate change. Mangrove forests have survived a number of catastrophic climate events and it is highly likely that they will survive into the foreseeable future as sea level, global temperatures, and atmospheric CO_2 concentrations continue to rise. The United Nation Framework Convention on Climate Change (UNFCCC) COP 25 in Madrid (2/12/19) recommended cities to integrate the role of nature-based solutions for climate emergency. This chapter will assess mangrove cover change over a 10-year

G. Miria (✉) · E. Joel
National Capital District Commission, NCD, Port Moresby, Papua New Guinea
e-mail: GertrudeM@ncdc.gov.pg

R. Kiele
School of Natural and Physical Sciences, University of PNG, NCD, Port Moresby, Papua New Guinea
e-mail: kielerh@upng.ac.pg

K.-G. Kim and M. Thioye (eds.), *Planning Climate Smart and Wise Cities*,
The Urban Book Series, https://doi.org/10.1007/978-3-030-80165-6_5

period and identify several drivers responsible for mangrove change, mangrove destruction and mangrove degradation along the NCD coastline through remote sensing and GIS techniques and literature analysis.

Keywords Coastal region · Mangrove cover change · National capital district · Climate change · Sea level rise · Urbanization · Papua New Guinea

5.1 Introduction

Mangroves are an essential part of the marine park's ecosystem, providing food and shelter for a variety of marine species. In Papua New Guinea (PNG), many coastal communities depend on mangroves for their livelihood and have been harvesting mangrove trees over generations for construction materials and firewood. Mangroves also provide a natural defense against storm surges, coastal erosion, and coastal flooding. An analysis conducted by government's Climate Change and Development Authority (CCDA) has highlighted community-based mangrove planting as a cost-effective measure for coastal communities. Unfortunately, overexploitation and degradation of mangrove areas means communities are losing this protection.

This study describes the findings of mangrove cover changes using the Geographical Information System (GIS) in mapping the time-serious of 2010–2020 and the different drivers of this changes across the coastal region of the National Capital District (NCD), an administrative and capital city of PNG. This study will provide an understanding that rapid losses of mangrove forests have had negative consequences on the environment, the climate, and humanity, through diminished benefits such as carbon storage, coastal protection and fish production. Restoration of mangrove forests is possible, and has already been undertaken in many settings, but such efforts have been piecemeal, and many have failed due to poor regulatory and enforcement measures by authorities.

5.2 Background

Studies showed that mangrove losses around the world has been severe over the last several decades, including PNG. Mangroves cover approximately 137,760 km^2–152,360 km^2 of the world's surface (Wilson 2017), however, the total area of mangrove forests in the world have drastically declined from approximately 16.5 million ha in 1980s to 15.2 million ha in 2000s (Thanh et al.2017). Mangroves have been heavily impacted by degradation and deforestation with 20–35% of global mangrove extent lost over the last 50 years. The twentieth century mangrove losses were largely dominated by forest clearing and exploitation for timber production and raw materials as well as rapid coastal population growth and urban expansion (Goldberg et al. 2020).

According to the United Nations Food and Agriculture Organization (UN FAO), mangrove losses over the periods 1980–2005 were estimated at approximately 3 million hectares with the rate of degradation over the period 1990–2000 estimated at 1% per year, which is twice the rate of terrestrial rainforest. Mangrove degradation has resulted to the release of carbon into the atmosphere estimated at 0.02–0.12 Pg per year, representing 10% of total emissions resulting from deforestation (Thomas et al. 2017). In addition, a recent global research found that more than 25% of the world's original mangrove cover have been destroyed and lost forever (Wilson 2017).

In contrast, PNG mangrove forest covered an area of 574,867 ha in 2002 with 66% found in the Gulf and Western Provinces along the coast of the Trans-Fly River delta towards West Papua in Indonesia (Shearman et al. 2010). The most recent land-cover study of PNG estimated that the country possesses 592,900 ha of mangroves distributed along all the coastal provinces. The largest area of contiguous mangrove system is found in the Purari and Kikori area of the Gulf of Papua and covered an area of 260,822 ha (Shearman 2010).

Shearman (2010), noted that no large scale clearance of mangroves for aquaculture farming and land reclamation have occurred in PNG since 2002. The majority of the mangrove estate is only utilised by local people for subsistence use, firewood gathering and fishing, both of which are unlikely to have had any significant impact on mangrove area. However, mangrove cover distribution in the NCD from 1985 to 2004 have seen a great declined by approximately 2.768 km^2–1.031 km^2 reducing mangrove area to 1.7 km^2 within a space of 18 years (Kiele and Vuvu 2008). The destruction of mangrove forests has reduced their viability and the quality of socioeconomic and ecological services, consequently creating environmental problems, such as land degradation, soil erosion, and loss of habitats of animal and plant species. Thus, information of the status of spatial distributions of mangrove forests is deemed critical for decision-makers and scientists to evaluate current land-use management policies.

Mangrove forests are globally recognized as highly carbon-rich tropical ecosystems that provide a range of critical economic and ecological services to surrounding coastal population (Goldberg et al. 2020). Mangroves like other plants, convert energy from the sun into organic matter through the process of photosynthesis and contain very high levels of above ground biomass as well as high quantities of below ground biomass which equates to carbon storage levels equivalent to those measured in dense Amazonian rainforests. This makes mangroves amongst the most carbon rich ecosystems in the tropics (Thomas et al. 2017). Mangrove productivity is important because it has direct impact on the health and function of the marine food chain and being an evergreen forest makes mangroves a highly productive ecosystem year round.

Despite their importance for local livelihoods, biodiversity and carbon sequestration mangrove forests are greatly threatened by both anthropogenic activities and natural threats such as sea level rise. Global mangrove loss has been attributed primarily to human activity. Rapid urbanization, inefficient land use planning and management, and the increasing impacts of climate change have become pressing environmental, socio-economic and cultural issues in NCD. The effects of storm surges and sea level rise are heavily impacting NCD's beaches, ecosystems, coastal infrastructure and communities. With these realities, come the tasks of mitigating, adapting to and recovering from such impacts. Despite the challenges, PNG mangroves are among the most species rich and productive ecosystems in the region and holds approximately 600,000 ha, which are also amongst the most biologically diverse in the world. Of the total 44 mangrove species in PNG, 23 species are found in the Bootless Bay area in the NCD and Central Provinces (CEPA-JICA 2017).

5.3 Purpose of the Study

This study investigates the distribution of observed drivers of change in mangrove forest extent, over the period 2010–2020 using time-series google images. This study uses both spatial and temporal information to achieve an understanding of the changes in mangrove forest extent across the 88.14 km coastline of the NCD (Port Moresby city), in PNG. This study will generate time series mangrove cover changes mapping, which provides a critical tool for encouraging restoration and enabling robust, data-driven policy changes and investments in driving climate change adaptation projects in mangrove restoration and rehabilitation projects.

5.4 Aim of the Study

This study attempts to present a comprehensive assessment of the mangrove ecosystem within the NCD Coastal region. The techniques of GIS tool will be applied to achieve the following goals:

i. To map out mangrove cover changes and assess the different drivers of change over a 10-year period.
ii. Identify threats associated with mangrove cover changes across local planning areas of the NCD.
iii. Identify planning measures to help reduce mangrove degradation and depletion.

5.5 Materials and Methodology

5.5.1 *Study Area*

Port Moresby is a coastal city within the foothills located on the shores of the Gulf of Papua, at latitude 9 degrees 29′ South of the equator. NCD is the largest and most developed urban centre in PNG in terms of population size and function. Its total area is about 265 km^2 and a total population of approximately 1 million people. The

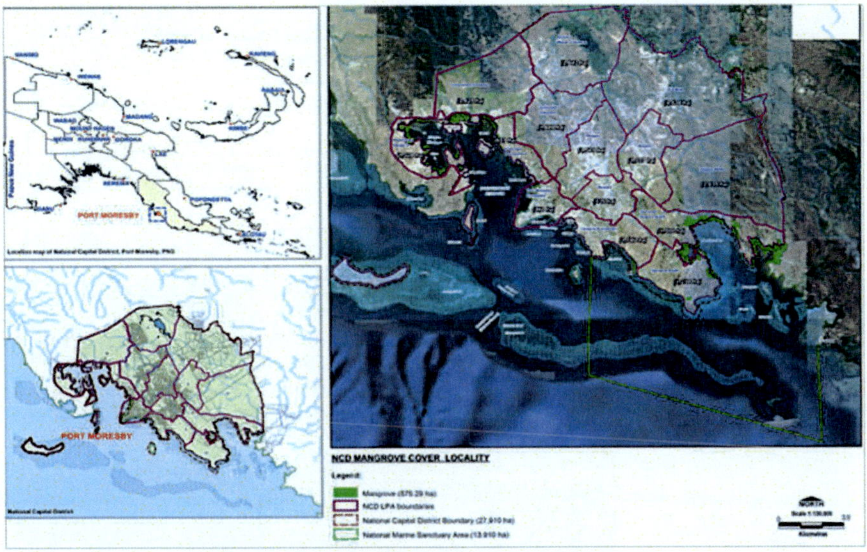

Fig. 5.1 Location map of the national capital district, Port Moresby, Papua New Guinea

National Capital District Commission (NCDC) is responsible for providing basic municipal services including the delegated responsibility of urban land use planning and management in the city on both alienated and customary land.

The NCD's extensive coastline of 88.14 km (NCDC, 2006) offers a great opportunity for development and at the same time contains significant and ecologically importance of mangroves, seagrasses and reef areas. The composition of the coastline starts in the north-west of Fairfax harbor and running south through Walta Bay, Joyce Bay and to the south-east of Bootless Bay bounded with Central Province.

It is home to 24 mangrove species in 11 families of the world's mangrove species. The dominant species include Rhizophoraceae family (Bruguiera, Ceriops and Rhizophora) as according to the CEPA-JICA Project, 2020. The location of the study area is shown in Fig. 5.1.

5.5.2 *Application of GIS Tool*

Mapping and monitoring the changing extent of mangroves has proved difficult, with most attempts relying on a combination of remotely sensed data and existing field maps to catalogue mangrove extent for a single point in time. Remote Sensing and GIS when used in combination, these datasets provide an opportunity to detect changes in mangrove extent over time. The Cartographic Model (Fig. 5.2) shows major steps involved in the collection of data and mapping of NCD Mangrove Cover Change. Existing spatial databases including satellite images (Google Earth with

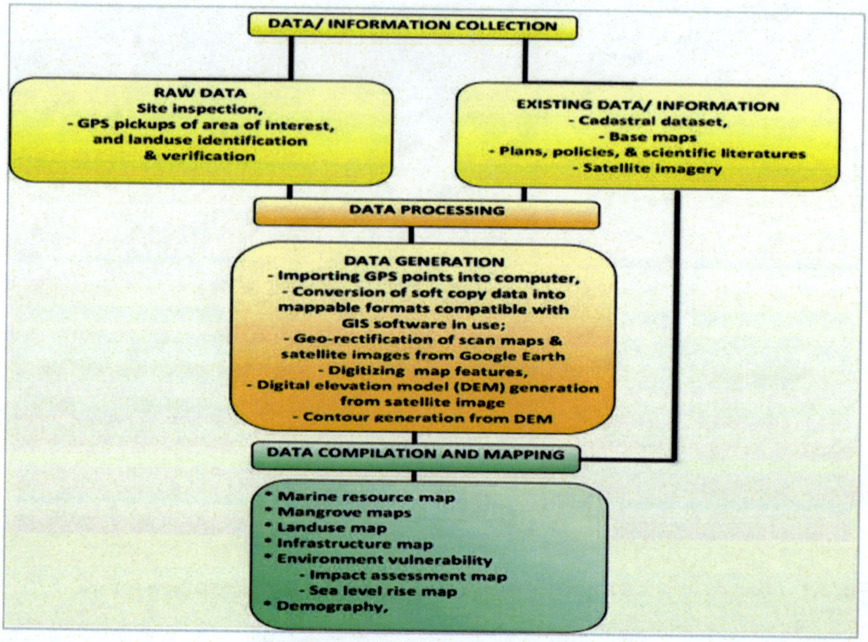

Fig. 5.2 Cartographic model showing the general process involved in the data collection and mapping for NCD mangrove cover assessment

spatial resolutions <5 m) and site inspection assessments were used in this analysis. A Magellan explorist 510 GPS was use to collect locational points of mangroves. ArcGIS, a GIS software was used in the assessments of the GIS datasets.

5.5.2.1 Data Collection

Data collection is the prerequisite step prior to undertaking any mapping task. In this study, two main datasets were used which involves collection of raw data and use of existing data. Field inspection was undertaken using Global Positioning Satellite (GPS) device to obtain the coordinates (latitude and longitude)/locations of mangrove loss, gain and intact and other important features which were used to create new data set.

Digital Elevation Model (DEM) was also generated from satellite imagery for some parts without contour information. Areas were contour information were present, the DEMs were generated directly from the contours. All existing datasets were collected from authentic sources, such as, statutory bodies, urban development plans and policies, and other reputable scientific literatures.

5.5.2.2 Data Processing

In the data processing phase, raw datasets were edited and converted into mappable format by manually using digitization and keyboard data entry, or converting them using PDF or image formats and georeferenced into mapping software before using them as base maps or reference maps. DEMs for Napa Napa area and Daugo Island were generated from satellite image as the existing contour data was not available for the north-west part of the study area. Ancillary sources such as Google Earth images, land use/land cover maps, forest cover change classification maps, GIS datasets and past research were also used to aid the study, particularly in the areas where quality images were problematic. the images acquired were not enhanced prior to the classification.

5.5.2.3 Data Compilation and Mapping

Data compilation and mapping was the final stage of different datasets compiled together to portray a desired representation of reality in the form of maps. Thus, the mapping output of mangrove cover change from the years 2010–2020 aid this assessment.

5.6 Results

The results from the assessments are presented in Tables 5.1, 5.2, 5.3 and Figs. 5.3, 5.4 and 5.5. The results have confirmed that mangroves along the NCD coastline are declining in its spatial context. There was a total of 715.56 ha of mangroves in 2010, and was reduced by 36.5%, 34% in 2015 and 29.4% in 2020, respectively. Thus, the remaining mangrove ecosystem in 2020 is 575.29 ha (Table 5.1, Figs. 5.6, 5.7 and 5.8).

Mangrove loss in NCD from 2010 to 2020 account for 344.79 ha and the highest hotspot for these losses were in LPA 11 with 131.6 ha, LPA 13 with 101.35 ha and LPA 12 with 80.19 ha (Table 5.2 and Fig. 5.9). In contrast, mangrove gain in NCD from 2010 to 2020 stands at 166.85 ha, the highest gain of mangroves was at LPA 13

Table 5.1 Mangrove cover change trend from 2010 to 2020

NCD mangrove cover changes 2010–2020 and different drivers of change					
Year	Mangrove cover (ha)	Mangrove cover change (ha)	Mangrove loss (ha)	Mangrove gain (ha)	Intact mangrove (ha)
2010	715.56	148.14	68.85	79.29	636.21
2015	667.29	152.99	100.60	52.39	614.92
2020	575.29	210.52	175.34	35.17	518.87

Table 5.2 Comparison of mangrove forests areas in the national capital district coastal region

	Distribution by LPA of different drivers of mangrove cover change over a 10-year period from 2010-2020.					
Coastal Region	Mangrove cover (Hectares)	Mangrove cover change (Hectares)	Mangrove loss (Hectares)	Mangrove gain (Hectares)	Mangrove Change hotspot (Hectares)	Intact mangrove (Hectares)
LPA 11	684.8	182.49	131.62	50.87	182.49	627.2
LPA 10	5.06	6.49	4.53	1.96		2.12
LPA 9	54.76	19.45	11.96	7.03		45.95
LPA 1	0.34	0.68	0.34	0.34		0
LPA 2	42.77	22.36	14.35	8.01		34.21
LPA 12	303.91	126.55	80.19	46.36	126.55	253.32
LPA 13	866.5	153.63	101.35	52.28	153.63	807.2

with a gain of 52.28 ha, followed by LPA 11 (50.87 ha) and LPA 12 (46.63 ha). Port Moresby town (LPA1) and Badili–Koke (LPA 2) have the least mangrove coverage areas because developments were concentrated in these areas from first colonial administration to World War 2 and after the war, developments continued to grow and spread northward and eastward of NCD (NCDC 2007).

Observation analysis in Table 5.3 and Fig. 5.3 summarized the types of species identified and their zonation and substrate that determined their growth. Field observation showed that healthy and intact mangroves can be found within Taurama and Bootless Inlets in the south-east of Port Moresby and in Fairfax harbor in the north-west area, while unhealthy and poor growths of mangroves can be found along the coastlines of Dogura Inlet, Gereka settlement, Joyce Bay, Konebada-Idubada and Baruni village. According to the CEPA-JICA, 2017, Bootless Bay is habitat to 283 terrestrial plants, 23 mangrove species, 4 seagrass species, 81 bird species, 4 species of marine mammals, 1 species of saltwater crocodile, 2 species of sea snakes, 3 species of turtles, 512 species of fishes, 284 species of reef coral, 20 species of macro algae, and 28 species of echinoderms.

Table 5.3 Types of species identified and their zonation and substrate during field observation

Location	Mangrove species observed	Zonation	Condition	Substrate	Reason
Taurama inlet	Bruguiera gymnorrhiza (Bg), Rhizophora stylosa (Rs), Bruguiera sexangula (Bs)	A-Rs B-Bg, Rs, Bs C-Bg. Bs	Healthy and well grown	Mud and silt	Shore is soft and muddy enough for these species to take root. Other factors include right amount of salinity, temperature and tidal fluctuations that enable good growth of mangroves
Dogura inlet	Avicennia eucalyptifolia (Ae), Rhizophora stylosa (Rs), Osbornia octodonta (Oo), Ceriops Tagal (Ct)	A-Ae, Ct, Rs, Oo	Poor growth	Gravel, sand and rocky	Being at the sea front, they are exposed to strong wind and wave action that can easily erode them. Other possible effects maybe tidal movements and rocky substrate that prevent them from regeneration
Bootless inlet	Rhizophora stylosa (Rs), Osbornia octodonta (Oo), Ceriops Tagal (Ct), Bruguiera gymnorrhiza (Bg), Rhizophora apiculata (Ra), Rhizophora mucronata (Rm), Bruguiera sexangula (Bs)	A-Rs, B-Ct, Ra, Rm, Bs, Bg C-Oo, Ceriops	Healthy and good growth	Mud, silt, gravel and rocky gravel	Shore is soft and muddy enough for these species to take root. Other factors include right amount of salinity, temperature and tidal fluctuations that enable good growth of mangroves

(continued)

Table 5.3 (continued)

Location	Mangrove species observed	Zonation	Condition	Substrate	Reason
Konebada–idubada	Rhizophora stylosa (Rs), Osbornia octodonta (Oo), Ceriops Tagal (Ct)	A-Rs, Oo	Poor growth	Gravelly sand, rocky	Fragmented mangroves due to build developments. The natural coastal landform has been transformed and the shorefront is a reclaimed land that reduces mangrove growth
Baruni village	Rhizophora stylosa (Rs), Osbornia octodonta (Oo), Ceriops Tagal (Ct)	A-Rs B-Ra, Ct C-Oo, Ct		Mud, gravel, sand	Regeneration is prolific in Baruni village but seedling mortality is high due to the relatively closed canopy under the mature stand. The understory is therefore fairly open and there is very little growth of common ferns like Acrosticum nor that of Acanthus species

(continued)

Table 5.3 (continued)

Location	Mangrove species observed	Zonation	Condition	Substrate	Reason
Fairfax habor	Rhizophora stylosa (Rs), Osbornia octodonta (Oo), Ceriops Tagal (Ct), Bruguiera gymnorrhiza (Bg), Rhizophora apiculata (Ra)	A-Rs, B-Ct, Ra, Rm, Bs, Bg C-Oo, Ct		Gravel, sand and rocky	Forest zonation is distinguishable with Heriteria, Avicennia and Ceriops occupying the landward end, Bruguiera and Rhizophora apiculata in the middle, and R. stylosa at the seaward end

5.7 Discussion

5.7.1 Mangroves in the National Capital District

The results of 2020 mangrove forests of NCD cover a total area of 575.29 ha (Fig. 5.4) representing 0.1% of the total Mangrove forests in PNG.[1] The mangroves are concentrated disproportionately towards a fewer to none in the central coastal region of Port Moresby Town, Koke and Babili area in the Local Planning Area (LPA) 1 and 2 while the remaining forest are found in the north-west and south-east of Port Moresby.

Mangrove community is the most important vegetation within the Bootless Inlet and the Fairfax harbor with a good stand of mangrove forests reaching 10 m in height. The larger diameter trees include Avicennia marina, Rhizophora apiculata, Bruguiera gymnorhiza and B. sexagula. Average tree girth in the areas observed is around 30 cm in diameter (CEPA-JICA 2017). These mangroves serve as nursery grounds for fisheries and buffers both marine and terrestrial environment from severe erosion and natural catastrophe.

The distribution or zonation by mangrove species depend on a number of factors: the level and duration of tides and floods; the amount of salt and oxygen in the water and soil; the type of soil or substrate, including its pH[2]; the amount of fresh water;

[1] Papua New Guinea mangrove forest is estimated at 592,900 ha distributed along all the coastal provinces of PNG (Shearman 2010).

[2] PH (hydrogen ion concentration) is one of the methods that deal with the physical and chemical properties of soil. The acidity influences the chemical transformation of most nutrients available to plants. Good soils have a pH vale of between 6 and 7, which is neutral. Lower than 6 are acidic. Acidic soils do not readily allow the transformation chemical nutrients to plants. Soil characteristics directly influence mangrove productivity and structure (English et al. 1994).

Fig. 5.3 Mangrove distribution from the shoreline, mudflats and landward from bootless bay in South-east to Napa-Napa in North-west of NCD, 2020

the climate; and the geomorphology. The inter-tidal areas are harsh environments, subject to both high and low rainfall, hot and cold temperatures, variable amounts of freshwater oxygen and salt, and soils that can be pure clay or mixed with sand, silt, coral rubble or rock and are constantly influenced by the tidal regime of the area. These variations in chemical and physical factors influence mangrove ecosystems to display extreme diversity in plant composition, forest structure and growth rates (Maniwavie 2009). The intertidal zone can be divided into three main zones or categories as shown in Fig. 5.6.

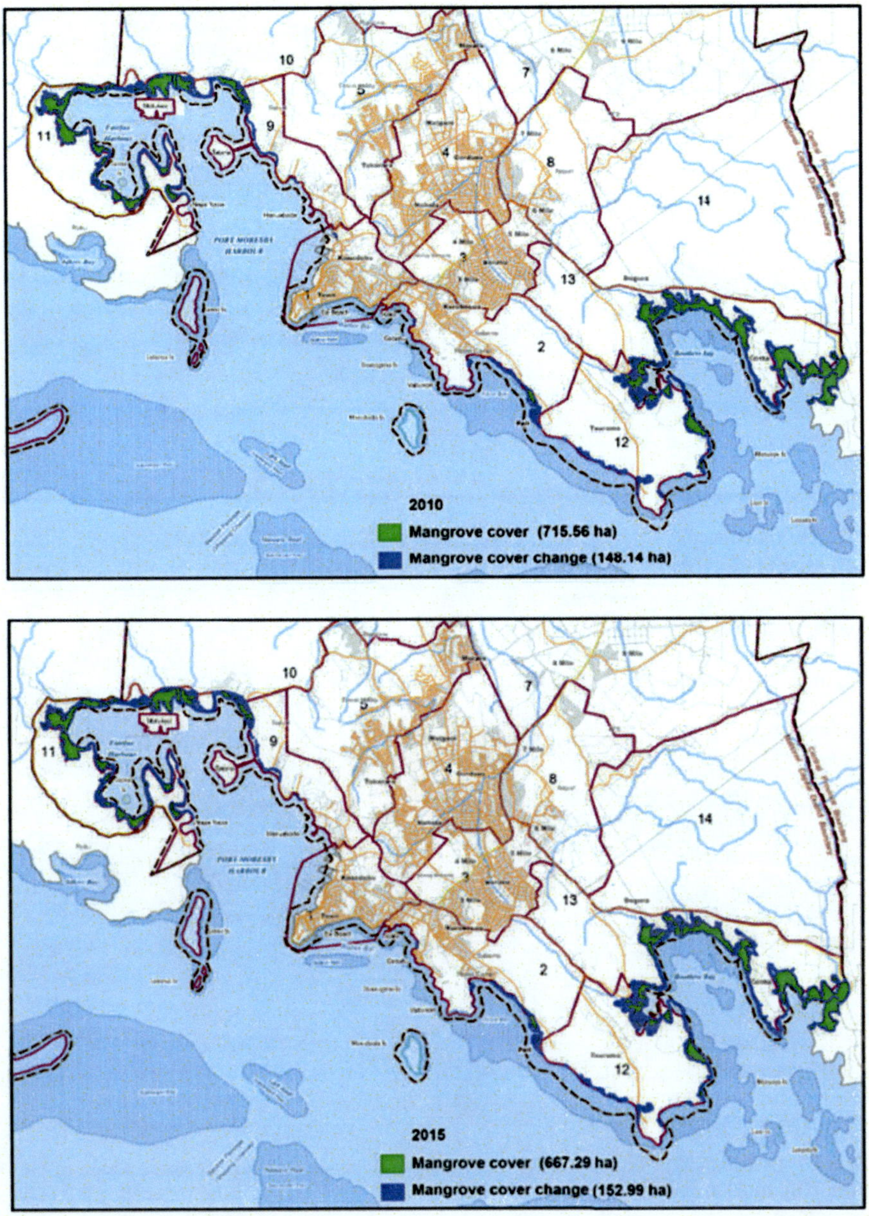

Fig. 5.4 Mangrove distribution pattern of 10-year trend from 2010 to 2020 in NCD coastal region

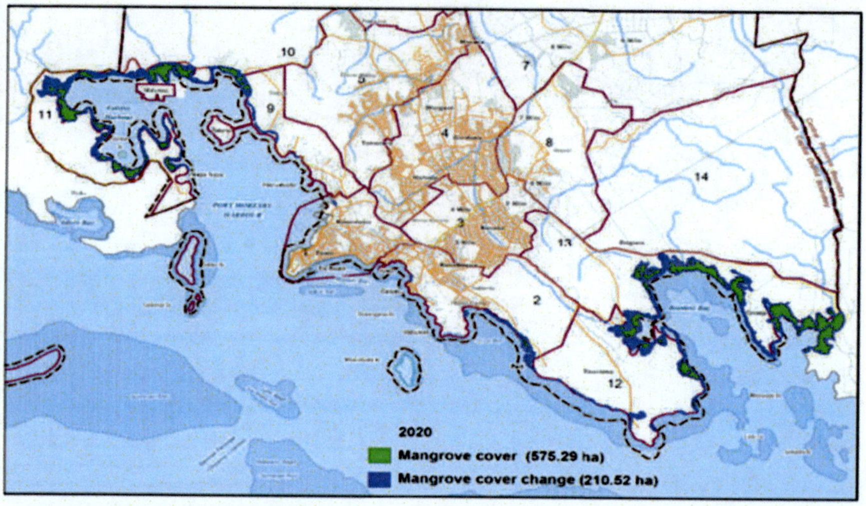

Fig. 5.4 (continued)

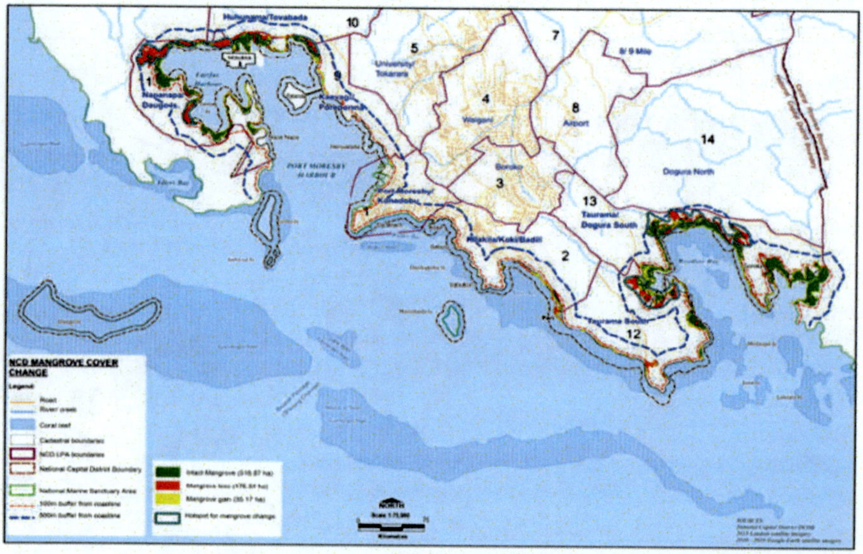

Fig. 5.5 Distribution of different drivers of mangrove change across the coastal region of NCD from 2010 to 2020

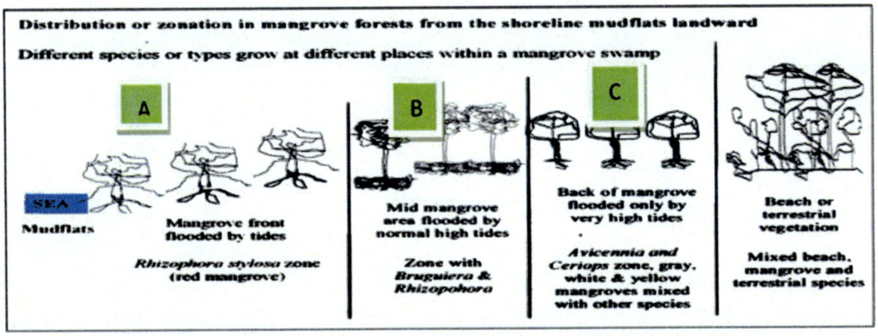

Fig. 5.6 Zonation in species distribution

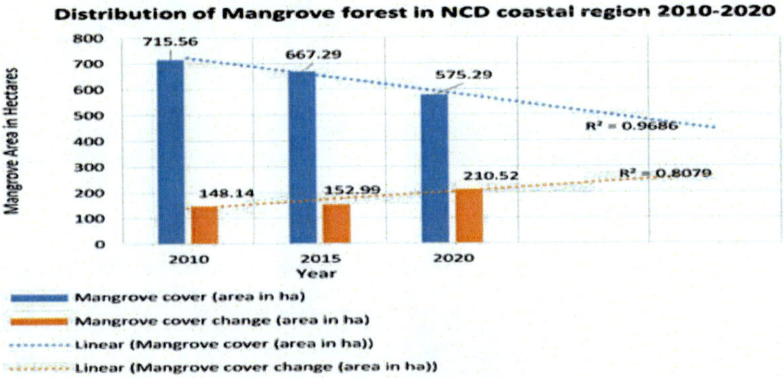

Fig. 5.7 Distribution of mangrove forest in NCD 2010–2020

- Sea front (A)—submerged on most tides.
- Middle(B)—between the mud flat/shoreline and land, only submerged by medium and high tides.
- Mangrove back(C)—only covered by spring or very high tides.

According to Fig. 5.4, the mangroves around Bootless bay (Dogura Taurama area) and Fairfax harbor (Poreporena Napa-Napa area) are the remaining intact natural mangroves in NCD. This remaining mangroves are currently under increased pressure from rapid coastal developments and if not managed and conserved well would result to another degraded site in the city.

A linear trend line in Fig. 5.7 describes a continuous rise or fall over time. For example, a blue linear trend line shows a steady decline in area of mangrove cover change annually over a 10-year period and orange linear trend line indicate a steady increase in area of mangrove cover change annually over a 10-year period. The R2 value of 0.9686 (mangrove cover) and 0.8079 (mangrove cover change) indicates a good fit of the estimated trend line values to the actual data which means the R-squared value is most reliable as it is at or near 1.

Fig. 5.8 Conservation easement at Bautama under the bootless bay national marine sanctuary established by national protected area policy through the GoPNG commitment to the 2017 international oceans conference at New York

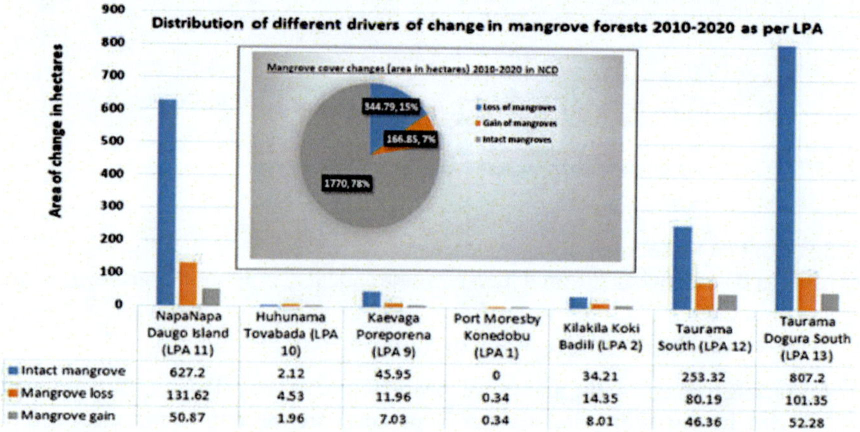

Fig. 5.9 Different drivers of change in mangrove distribution 2010–2020 in NCD coastal region

5.7.2 Mangroves at the Receiving End of Development and Climate Change

The NCD is rapidly developing, and it is experiencing environment, socio-cultural and economic challenges in managing urban growth and providing equal opportunities for its rapidly growing population. Global statistics of Environmental Performance Index 2020 show that PNG ranks 125 (28.4%) out of 180 countries of the world for Environmental Health and rated 152 (35.1%) in Ecosystem vitality. Furthermore, its urbanization rate is rising at 13.35%. This situation has a significant bearing on the general health status of our rich biodiversity. With lack of sound management strategies of resource use coupled with challenging impacts of climate change, NCD as a coastal city is facing climatic risks of flooding, coastal inundation and impacts from strong winds and extreme temperatures.

The degradation of mangroves diminishes its function of regulating the flow of sediments and mud, which impacts corals and reef habitat. Furthermore, spawning grounds for fish diminishes with the degradation of mangroves. The number of mud crabs are said to have decreased in population. The quality of sea water, run-offs and inlets into certain parts of the Bootless Bay, especially Tuna Bay is of grave health concern to the local communities. The color of the water in and around the mangroves and shorelines is murky and mostly not clear. Children along the bay area have developed painful rashes and skin diseases which are likely linked to the declining water quality.

Past work by TNC and partners in PNG has demonstrated that mangrove forests provide critical goods and services to local communities such as food and firewood, supporting water quality, and stabilizing shorelines to provide a buffer against flooding from storm surge, sea-level rise, and coastal erosion. The annual economic value of mangroves, based on the products and services they provide, has been estimated to be between USD$200, 000 and USD$900, 000 per hectare. Given this rate, the 575.29 ha of mangrove in NCD would cost between US$115,058,000 and US$517,761,000.

PNG's mangrove forests also hold vast stores of "blue carbon", and therefore can play an important role in helping PNG to contribute to global efforts to mitigate climate change. Importantly, mangrove forests also have an important role in helping PNG adapt to changing climate conditions including sea-level rise. Indeed, coastal flooding and erosion from sea-level rise and storms are identified as key threats in PNG's Nationally Determined Contribution under the Paris Agreement.

Progress has been made to integrate climate change adaptation and risk reduction both at the national policy level and the local level. What has emerged is that, while drawing on international expertise and support, most initiatives and support are targeted at local needs and problems and can be addressed from within as is often quoted, "think globally but act locally" (Fig. 5.8).

5.7.3 Anthropogenic Threats and Impacts on Mangrove Cover Change

Degradation of mangrove ecosystem from overharvesting for firewood, settlements, urbanization and industrial developments along the coastal zones of NCD are evident, and they have significant negative impacts on biodiversity and other ecosystem services. The Tuna Bay, once known as the "spawning ground of the yellow fin tuna" has lost its significance as the catch of yellow fin tuna has dropped from between 500 and 700 tuna in 2012 to 200 −300 tuna in 2018 and the stock of tuna catch has reduced to less than 100 tuna in 2020. With these realities, come the tasks of mitigating, adapting to and recovering from such impacts. These impacts are compounded by various urban land use activities encroaching into mangrove ecosystems now putting them under continuous pressure of depletion forever.

The composition of Intact mangroves in NCD that still remains in their natural state from 2010 to 2020 stands at 1770 ha which make-up 78% of the total area of mangrove cover. LPA 13 has the highest area of 807.2 ha intact mangroves which is protected under the Bootless Bay National Marine Sanctuary. Despite conservation efforts by the Government for the protection of the mangroves at Tuna Bay, uncontrolled land use activities continue to occur and the authority is lagging behind its enforcement measures. Figure 5.10a shows national and local authorities taking charge in enforcement effort to stop illegal construction works at Tuna Bay.

The results in Fig. 5.9 indicate that current hotspots for mangrove loss are in LPAs 11, 12 and 13 as much of the land in these localities are under customary ownership and uncontrolled informal land sales outside the formal land registration process. This has resulted to recent informal developments and urbanization projects have caused ecological and social problems for the surrounding villages and attempts by the government and other relevant authorities to solve these problems have been piecemeal.

(a)

(b)

Fig. 5.10 **a** Authorities served *Stop-work notice* and *unauthorized use notice* using both physical planning act 1989 and environment act 2002 to developer known as customary landowners. **b** Various uses of mangrove due to increasing urbanization

5.7.4 Mangroves in the Context of Changing Climate

Changes in climate and other factors affect mangroves in complex ways. Mangroves are also sensitive to climate change impacts, associated with the Anthropocene enhanced greenhouse effect and ocean warming. Sea level rise in the last few decades has largely resulted from ocean warming and thermal expansion. In PNG, mean sea level rise is measured at 7 mm between 1993 and 2010, and is projected to rise to 150 mm by 2030 according to the NCD Urban Development Plan 2030.

The effects of storm surges and sea level rise are heavily impacting NCD's beaches, ecosystems, coastal infrastructure and communities. With these realities, come the tasks of mitigating, adapting to and recovering from such impacts. Since these impacts are compounded by anthropogenic activities, NCD's mangrove ecosystems are continuously under pressure as there appeared to be a lot of encroachment into the mangroves by built developments. Field observations showed that retreat of mangroves from the seaward edge was a key sign of sea level rise impacts (Fig. 5.11). There is now a concern that sea levels are rising more quickly and higher than originally expected and altering the flooding patterns of astronomical tides as demonstrated in Fig. 5.12. This means that changes in sea level rise will enhance daily low and high tides.

The study of the vulnerability assessment of Sea Level Rise projects the PNG mean sea level to rise by 3 m between 2020, 2030 and 2050 (Fig. 5.14). Based on this projected 3 m SLR, about 575.07 hectares of mangrove forests in NCD could

Fig. 5.11 Mangrove distribution and expansion into terrestrial land at Napa-Napa area

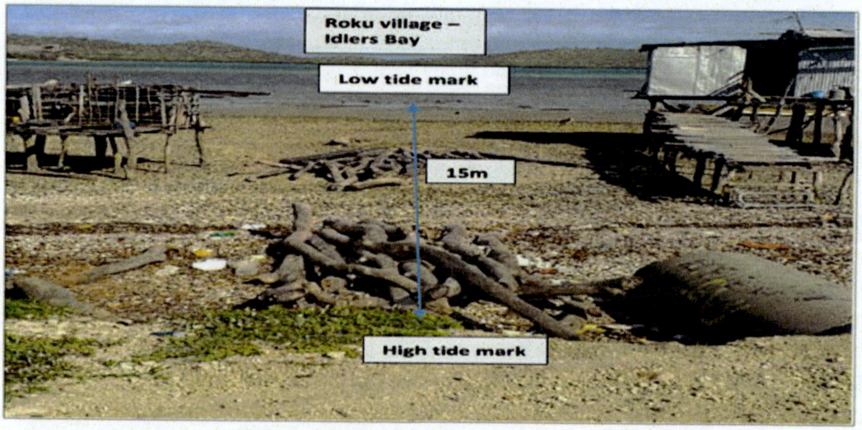

Fig. 5.12 High tide event in 2020, reaching a height of 15 m at Roku village, Idlers Bay

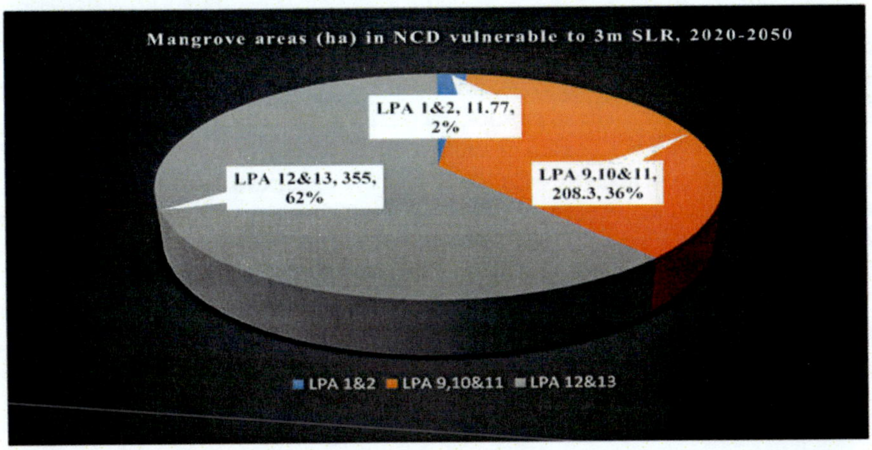

Fig. 5.13 Mangrove areas vulnerable to 3m sea level rise projected to occur from 2020 to 2050

potentially disappear in this period leading up to 2050 (Fig. 5.13). The level of sea water is very crucial to the functioning of mangrove ecology and sea level rise will affect every aspect of the ecosystem.

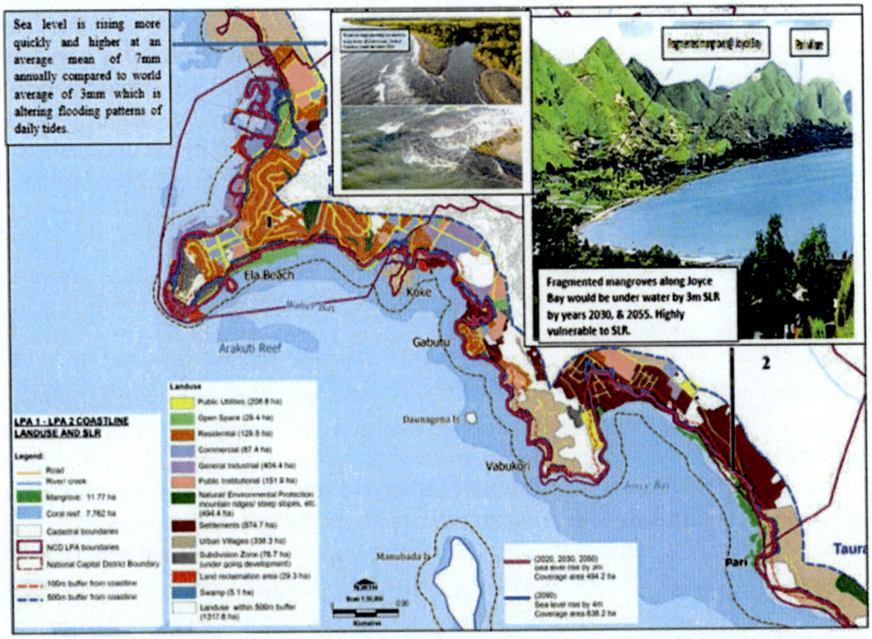

Fig. 5.14 Sea Level Rise a major threat to mangrove survival in the coastal region of NCD

Sea level rise could affect mangrove health and composition due to changes in salinity, recruitment, inundation, and changes in the wetland sediment budget. The change in the local salinity regimes will create a competitive interactions of mangroves with other wetland species. Increases in frequency or intensity of tropical storms and cyclones in combination with sea-level rise may alter erosion and sedimentation rates in mangrove forests and lead to mangrove destruction and depletion. The increased atmospheric carbon dioxide (CO_2) concentration may increase mangrove growth by stimulating photosynthesis or improving water use efficiency, but the consequences of this growth enhancement for the ecosystem are unknown. Various studies have revealed that flooding caused by increased precipitation, storms and sea-level rise may result in decreased productivity, photosynthesis and survival. Inundation of lenticels in the aerial roots can cause the oxygen concentrations in the mangrove to decrease, resulting in death. Inundation is also projected to decrease the ability of mangrove leaves to conduct water and to photosynthesize.

5.8 Conclusion

This study provides a more consistent and reliable mangrove inventory of change and their drivers in mangrove forest extent over the period 2010–2020 using remote sensing and GIS techniques along with other ancillary sources such as existing GIS datasets, high resolution imagery, Google Earth and other published literatures. The results reinforce previous claims that mangroves have been historically degraded across their range. Other than sea level rise, changes in air, and sea surface temperature, stresses from storms, variations in precipitation as well as anthropogenic disturbances, have adversely affected the mangrove ecosystem in this decade.

Although the distribution of mangroves in NCD account for only 0.1% of the total 592,900 ha in PNG, emphasis should be placed on quantifying ecosystem services in monetary terms thereby forging an economic counterbalance in the face of heavy pressure from land use developments. Regardless of size, mangrove forests are valuable resources for Papua New Guineans and this study can aid in the decision making process for the conservation and rehabilitation of the mangrove forests on a local scale.

Understanding the role of mangrove ecosystem and their response to climate change should help in making an informed decision and management plan for these vulnerable mangroves that are at high risks of the projected 3 m sea level rise from period 2020–2050. How mangroves will respond to such changes is a complex issue and much research needs to be done to understand and mitigate the impacts of climate change. Further research into change detection and accurate assessment of the causes and consequences of changes based on ground truth are needed to comprehend the true state of mangrove forests in NCD.

References

CEPA-JICA (2017) Biodiversity project 2015-2020. Report for second workshop for bootless bay marine conservation initiative

Ellison JC (1993) Mangrove retreat with rising sea-level, Bermuda. Estuarine, coastal and shelf science. 37:75–87. https://doi.org/10.1006/ecss.1993.1042. [CrossRef] [Google Scholar]

Ellison JC (1998) Holocene stratigraphic records of vegetation change and sedimentation from the Ajkwa river estuary mangroves, Irian Jaya. Report presented to P.T. freeport Indonesia company, 300 k expansion project, Tembagapura, Irian Jaya, Indonesia

Ellison JC (1997) Mangrove ecosystems of the western and Gulf provinces, Papua New Guinea, a review. Science in New Guinea, vol 23. pp 1–15

Ellison JC (1997a) Mangrove community characteristics and litter production in Bermuda. In: Kjerfve B, Lacerda LD, Diop ES (eds) Mangrove forests of the Latin America and Africa regions, vol 5. UNESCO, Paris

Ellison JC (2000) How south pacific mangroves may respond to predicted climate change and sea level rise. In: Gillespie A, Burns WCG (eds) Climate change in the south pacific: impacts and responses in Australia, New Zealand, and Small Islands states. Kluwer Academic Publishers, Dordrecht, p 289

Government of Papua New Guinea (2015) National marine conservation assessment for PNG, Conservation and environment protection authority. p 51

Goldberg L, Lagomasino D, Thomas N, Fatoyindo T (2020) Global declines in human-driven mangrove loss. pp 5844–5855. Retrieved 2021, from https://onlinelibrary.wiley.com

Kiele R, Vuvu J (2008) Addressing an innovative approach to determine change in mangrove forest in the bootless bay, central province by using GIS/RS techniques. In: Addressing national and global issues through scientific research proceedings. pp 90–101. UPNG, SNPS

Mowbray D et al (1996) From Rio to Rai. Environment and development in Papua New Guinea, vol 5 & 2. UPNG Press, Port Moresby

Maniwavie T, Aitsi J, Federizon (1998) Rapid ecological assessment of the Walter and Joyce Bays of Port Moresby. Report prepared for JICA

Miller GT (1998) Living in the Environment, 10th edn. Wadsworth Publishing Company, USA

Marine Conservation and Management Training Course notes (2008), MIRC

Manivawie T (2008) An introduction manual on the biology and restoration of mangrove ecosystem. MIRC

National Statistics Office. Census Data, Port Moresby. https://www.nso.gov.pg/statistics/population

National Capital District Urban Development Plan (2006) NCDC

Nita A (2007) Resources and adaptation measures course notes. UPNG Press, Port Moresby

National Capital District Commission (2007) Port Moresby town local development plan, vol 2. SMEC PNG Limited. NCD

National Capital District Commission (2020) Port moresby towards 2030. National district development plan revive 2020. NCDC

Papua New Guinea—Port Moresby (2019) Tide predictions calendar. Climate and oceans support program in the Pacific. Australian government

Samuel P (2008) Oceans and coastal resource management. Semester 2 lecture notes. UPNG, Port Moresby, PNG

Shearman PL (2010) Recent change in the extent of mangroves in the Northern Gulf of Papua, Papua New Guinea. Retrieved 2021, from https://www.ncbi.nlm.nih.gov/pmc/articles/PMC3357694/, pp 181–189

Son N-T, Chen C-F, Chen C-R (2017) Mapping mangrove density from rapideye data in Central America. www.degruyter.com/document/, https://doi.org/10.1515/geo-2017-0018/html

Thomas N, Lucas R, Bunting P, Hardy A, Rosenqvist A, Simard M (2017) Distribution and drivers of global mangrove forest change 1996–2010. PLoS One 12(6):e0179302. Retrieved 2021, from https://www.researchgate.net/publication/317422133_Distribution_and_drivers_of_global_mangrove_forest_change_1996-2010

UN Habitat (2013) Port Moresby, Papua New Guinea—climate change vulnerability assessment. UN Habitat

United Nations Office for Disaster Risk Reduction (2017) "7. Coastal erosion hazard and risk assessment" Words into action guidelines: national disaster risk assessment hazard specific risk assessment. UNISR

Weber P (1993) "Abandoned seas: reversing the decline of the oceans". World watch paper 116. World watch institute. p 66

Wilson R (2017) Impacts of climate change on mangrove ecosystem in the coastal and marine environments of Caribbean small Island developing states (SIDS). Retrieved 2021, from www.gov.uk/guidance/commonwealth-marine-economies-programme, pp 60–82

Part III
Fostering Climate Action in Cities

Chapter 6
Possible Climate Actions in Cities

Gajanana Hegde

Abstract This chapter discusses trends in greenhouse gas emissions in cities primarily in the building sector and priorities for climate action for mitigation in the light of global temperature goals. Possible technology/measures for climate action sourced from IPCC and IEA literature, carbon market mechanisms and those indicated in recently prepared Nationally Determined Contributions (NDCs) of countries are included.

Keywords NDCs · IPCC assessment report · IPCC guidelines · CDM · Carbon pricing · Indoor air pollution · Energy efficiency · Behavioral change

Disclaimer The content of this chapter has no relationship whatsoever with the UNFCCC process or the work of the secretariat. Its findings, interpretations and conclusions do not reflect the views of the secretariat. The secretariat does not vouch for the accuracy, currency or completeness of the data or information contained in the article and excludes all liability for any loss or damage arising (including through negligence) in connection with and as a result of any reliance on the data and information contained in the article and/or its use. Questions concerning the analysis and views expressed in the article shall be addressed to the author.

6.1 Introduction

A good understanding of the sources of emissions in a city should precede the discussion on possible climate actions in the city. As emphasized in Chap. 1, the latter should ideally encompass holistic planning and deployment of clean technologies augmented by digital technologies. This chapter includes general information on trends and sources of emissions in cities, possible measures for climate action

G. Hegde (✉)
UNFCCC secretariat, Haus Carstanjen Office, Martin Luther King Strasse 8, Bonn 53175, Germany
e-mail: GHegde@unfccc.int

K.-G. Kim and M. Thioye (eds.), *Planning Climate Smart and Wise Cities*,
The Urban Book Series, https://doi.org/10.1007/978-3-030-80165-6_6

together with the breadth and depth of emissions reduction that could be achieved with those measures. A reference is made regarding the measures explored by stakeholders under carbon market mechanisms and those indicated in recently prepared Nationally Determined Contributions (NDCs) of countries.

Among sectors such as energy generation, buildings, transport and waste relevant for the cities, buildings stand out due to the diverse, distributed and significant nature of the emission sources they include with consequent opportunities to mitigate them. The buildings sector—including residences, offices, shops, hotels, schools, and other public and commercial premises—today accounts directly and indirectly for approximately 30% of the final energy consumed around the world including almost 55% of global electricity consumption. When both the construction and use phases are taken into consideration, it accounts for about 37% of today's global CO_2 emissions (IEA 2020a). In this context, it is important to note that half of the buildings that will be used in the second half of this century are yet to be built, often in regions where mandatory building codes do not exist yet. GHG emissions from the residential building sector more than doubled between 1970 and 2010. Majority of the growth in emissions in the sector came from indirect CO_2 emissions from electricity use. Move from rural to urban areas and from informal to formal housing in developing nations, increasing wealth and lifestyle changes that influence the living space, type and number of appliances often contribute to increasing residential emissions (UNFCCC Secretariat 2020). Contrary to the conventional thinking that urbanization results in higher aggregate residential energy consumption and emissions, it is not always the case and it depends on stage of urbanization. For example, most Sub-Saharan African countries are in an accelerating stage of urbanization that outpaces the economic growth leading to a decrease in aggregate energy consumption. This is possibly due to investment in energy infrastructure lagging resulting in limited access to modern energy and low ownership rates of electrical appliances; biomass is still the primary cooking fuel in most of the urban populations in Africa. In contrast, Asia and the Middle East and North Africa regions are urbanizing in phase with thriving economies with an increase in aggregate residential energy consumption. Developing regions in Asia have the highest annual rates of urbanization and consume more than 32% of the total residential energy use worldwide. Continuing rapid urbanization is expected to increase the residential energy demand in this region beyond any other region in the world (Wang et al. 2020).

Causes of housing-related emissions are manifold. These include geographic location and climate, type of building and its physical characteristics and the number, type and efficiency of energy-consuming devices in the building and the amount of time they are used, and occupant behavior and sociodemographics, encompassing the number of occupants, their choices and interaction with their surroundings (UNFCCC Secretariat 2020). The use of energy for space cooling typically by means of an electric powered fan or air conditioning (AC) system is growing faster than for any other end use in buildings, more than tripling between 1990 and 2016. Global sales of ACs have been growing rapidly; since 1990, annual sales of ACs nearly quadrupled to 135 million units. There are now about 1.6 billion in use. Almost a fifth of all the electricity used in buildings is for cooling. Local air pollutants caused by cooling energy demand have similarly grown. The lion's share of the projected growth in energy use for space cooling by 2050 comes from the emerging economies. By 2050, it is projected that two-thirds of the households in the world will have air conditioners and global energy demand from air conditioners is expected to triple by 2050. The efficiency of ACs varies widely—in all major markets today, people are typically buying air conditioners whose average efficiencies are less than half of what is available (IEA 2018).

The need for climate action and clean energy technology in cities is not just about dealing with the climate crisis. It is equally about achieving sustainable development in areas such as health and energy access. Energy production and use is the leading source of air pollution: nine-out-of-ten people around the world breathe polluted air every day causing several premature deaths each year. Outdoor (ambient) pollution accounts for 3 million premature deaths and around 2.5 million die prematurely from indoor air pollution caused by combustion of fuel for cooking and heating. The biggest impact is in emerging economies, where more than 2.6 billion people still do not have access to clean cooking (IEA 2020a).

6.2 Sources of Emissions

2006 IPCC guidelines provide internationally agreed methodologies to estimate and report GHG emissions and removals and are often used by national governments for their GHG inventories. These guidelines classify emissions across five sectors, i.e., Energy, Industry, Agriculture, Forestry and Other Land Use (AFOLU), Waste and Other/Cross-sectoral. Figures 6.1, 6.2 and 6.3 below contextualize these emission sources for the case of cities based on the work of Program for Energy Efficiency in Buildings (PEEB), GIZ. In addition to direct emissions (e.g., use of fossil fuels for

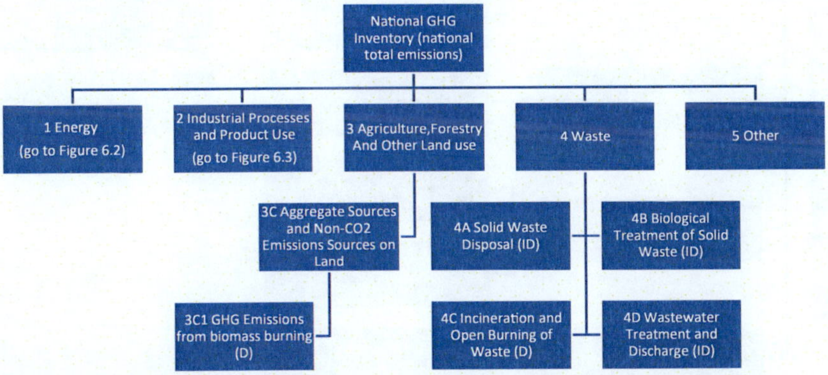

Fig. 6.1 Main categories of emissions in 2006 IPCC guidelines relevant for cities (D–direct emissions, ID–indirect emissions)

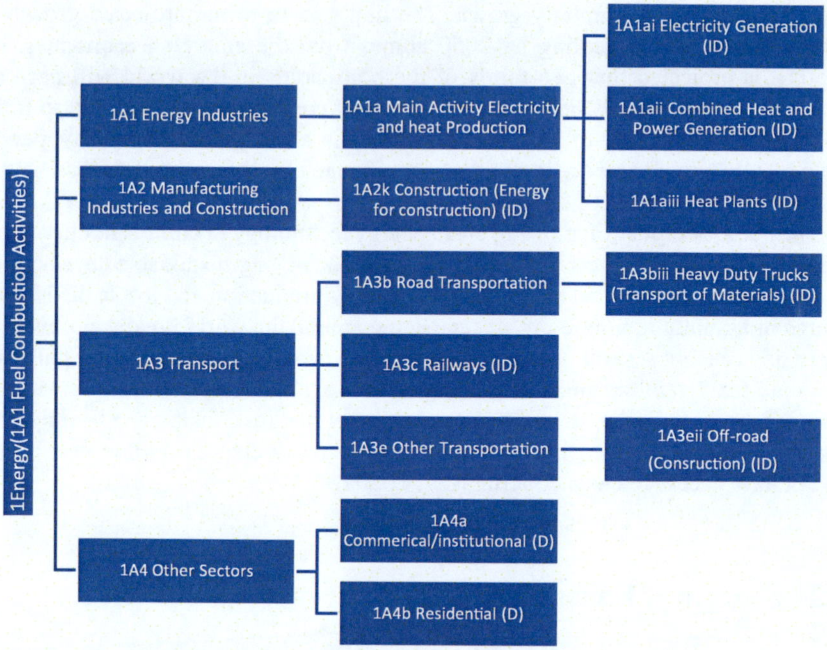

Fig. 6.2 Main categories of energy-related emissions in 2006 IPCC guidelines relevant for cities

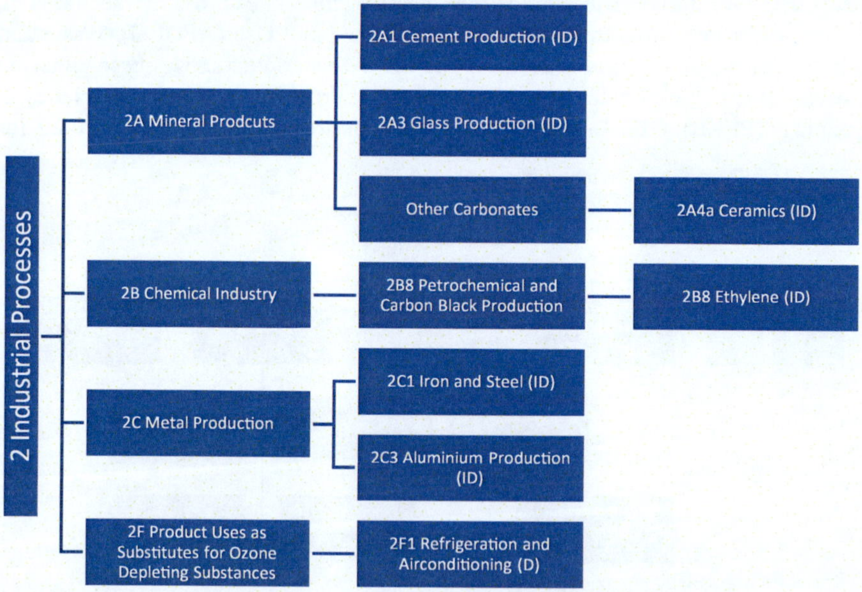

Fig. 6.3 Main categories of industrial processes-related emissions in 2006 IPCC guidelines relevant for cities

heat or electricity in the buildings) and indirect emissions (e.g., emission embedded in materials used in buildings or electricity supplied from the grid), the construction of new buildings leads to emissions from land use change under certain circumstances. Therefore, as emphasized under Chap. 1, recognizing the linkage between buildings, city services such as transport and waste management and land use change will help develop effective mitigation strategies, policies and actions such as urban planning.

An understanding of relative contribution of different sources of emissions in the city will help prioritize the climate action in the cities. In that regard, considering that emissions related to the operation or use of the existing buildings represent the majority (many times 70–90%) of the total lifecycle emissions in a given year, it may be useful to estimate, direct emissions due to fuel combustion in buildings (1A4a and 1A4b) and indirect emissions due to power/heat generation outside of buildings (1A1a), where relevant use of biogas/biomass for cooking subcategory 3C1 should be included.

Emissions associated with the embodied energy of construction materials including maintenance, repair and renovation would also be relevant to consider. Typically, these would include indirect emissions due to the production of cement (2A1), glass (2A3), chemicals (2B) and metal (2C).For an accurate estimation, indirect emissions due to construction activities (1A2k), transport of production materials (1A3), end of life activities of buildings—on-site deconstruction (1A2k), waste disposal (4) and transport (1A3) should also be considered.

Tools such as the Google Environmental Insights Explorer (https://insights.sustainability.google/) and the CURB tool from the World Bank (https://www.worldbank.org/en/topic/urbandevelopment/brief/the-curb-tool-climate-action-for-urban-sustainability) could also be useful to consider in this regard.

6.3 Strategies for Climate Action

There are many opportunities for climate action in the cities. Increasingly innovative cost-effective tools and technologies are becoming available that enable such climate action. For example, as stated before, heating and cooling energy use in residential and commercial buildings is expected to grow by 79% and 84%, respectively, over the period 2010–2050. Strategic and planning level of the technological options, design practices and behavioral changes can achieve large reductions in building energy use (50–90% in new buildings, 50–75% in existing buildings) (IPCC 2014). This could be achieved for example through adopting essential steps in the design of low-energy buildings are: (1) building orientation, thermal mass, and shape; (2) high-performance envelope specification; (3) maximization of passive features (daylighting, heating, cooling, and ventilation); (4) efficient systems meeting remaining loads; (5) highest possible efficiencies and adequate sizing of individual energy-using devices; and (6) proper commissioning of systems and devices. Similar conclusions are drawn in recent IEA reports. For instances, IEA Sustainable Development Scenario requires a global annual reduction in energy intensity per m2 of at least 2.5%. According to IEA, this could be achieved by 2030 with more stringent building energy codes, deep energy renovations, a tripling of heat pumping technology uptake and a 50% improvement in the average seasonal performance of air conditioners, along with other energy efficiency measures (IEA 2020b).

6.4 IPCC Assessment Reports and Buildings

Working Group III to the Fifth Assessment Report of the Intergovernmental Panel on Climate Change, in Chap. 9 Buildings, notes that in 2010, buildings accounted for, besides high amount of energy consumption, approximately one-third of black carbon emissions and a significant amount of refrigerant emissions. While the business as usual energy use and related emissions may double or potentially even triple by mid-century due to several key trends, it could reduce drastically if today's cost-effective best practices and technologies are broadly diffused (*robust evidence, high agreement*) the report noted (IPCC 2014). Recent advances in technology, design practices and know-how, coupled with behavioral changes, can achieve a two to tenfold reduction in energy requirements of individual new buildings and a two to fourfold reduction for individual existing buildings largely cost-effectively or sometimes even at net negative costs. Recent developments in technology and know-how enable construction and retrofit of very low- and zero-energy buildings, often at little marginal investment cost, typically paying back well within the building lifetime (*robust evidence, high agreement*). In existing buildings, 50–90% energy savings have been achieved throughout the world through deep retrofits (*medium evidence, high agreement*). Energy efficient appliances, lighting, information communication (ICT) and media technologies can reduce the substantial increases in electricity use that are expected due to the proliferation of equipment types used and their increased ownership and use (*robust evidence, high agreement*). Building codes and appliance standards with strong energy efficiency requirements that are well enforced, tightened over time and made appropriate to local climate and other conditions have been among the most environmentally and cost-effective. Experience shows that pricing is less effective than programs and regulation (*medium evidence, medium agreement*). In addition to technologies and architecture, behavior, lifestyle and culture have a major effect on buildings' energy use; three to fivefold difference in energy use has been shown for provision of similar building-related energy service levels (*limited evidence, high agreement*). In developed countries, evidence indicates that behavior informed by awareness of energy and climate issues can reduce demand by up to 20% in the short term and 50% of present levels by 2050. These findings of IPCC are further illustrated in Table 6.1.

Table 6.1 Mitigation strategies, energy-savings potential and enabling policies

Mitigation strategies	Specific measures/Potential reduction[1]		Key Enabling Policies
Carbon efficiency – Building-integrated renewable energy – Switch to low-carbon fuel such as electricity – Use of natural refrigerants – Advanced biomass stoves	Solar electricity generation through rooftop PV installations	15–58%	Carbon tax, feed-in tariffs extended for small capacity; soft loans for renewable technologies
Energy efficiency of Technology Building envelope performance, efficient appliances, daylighting, lighting, heating, ventilation, air conditioning, automation, smart controls	High-performance thermal envelope with efficient heating, ventilation and air conditioning (see Table 6.2 for potential for individual appliances)	9.5% to 68%	Public procurement, appliance standards, tax exemptions, soft loans
System (infrastructure) efficiency Housing standard, integrated design, urban planning, district heating/cooling, building controls, energy storage, distributed energy generation, cogeneration, trigeneration	Overall system efficiency[2]	30–70%	Building codes, preferential loans, subsidized financing schemes, Energy Service Companies (ESCOs), Energy-Saving Performance Contract (EPCs), suppliers' obligations, white certificates, integrated design process (IDP) into urban planning, importance of policy packages rather than single instruments
	Deep retrofits	40–80%	
	Passive house standard and nearly/net zero and energy plus energy buildings	30–70%[3]	
	Integrated design process	Upto 70%	
Energy service demand reduction	Behavior and lifestyle changes of users, smart metering	20–40%[4]	Awareness raising, education, energy audits, energy labeling, building certificates and ratings, energy or carbon tax, personal carbon allowance

Source adopted from IPCC (2014) Table 9.1 on page 676

[1] Versus Business as Usual (BAU) baseline

[2] Energy savings by building type: (i) Detached single-family homes: 50–75% (total energy use); (ii) Multi-family housing: 80–90% (space heating requirements); (iii) Multi-family housing in developing countries: 30% (cooling energy use), 60% (heating energy); (iv) Commercial buildings: 25–50% (total HVAC), 30–60% (lighting retrofits)

[3] 83% residential heating energy and commercial heating and cooling energy, 50% commercial heating and cooling energy

[4] 20–40% for behavioral change, 40% for lifestyle change

Table 6.2 Reduction potential in buildings for various end uses due to on-site RE generation, efficiency improvements or behavioral changes

End use	On-site emissions free energy supply (%)	Device efficiency (%)	System efficiency (%)	Behavior change (%)
Heating	20–95%	30–80%	90%	10–30%
Hot water	50–100%	60–75%	40%	50%
Cooling	50–80%	50–75%	67%	50–67%
Cooking	0–30%	25–75%		50%
Lighting	10–30%	75–99%	80–93%	70%
Refrigerators		40%		30–50%
Dishwashers		17%		75%
Clothes washers		30%		60–85%
Clothes dryers		50%		10–100%
Office computers and monitors		40%		
General electrical loads	10–120%			

Source Adopted from IPCC (2014) Table 9.2 on page 687

6.5 C40 Cities Strategic Approach to Climate Action

C40 partnered with the McKinsey Center for Business and Environment to quantitatively assess the biggest opportunities for emissions reduction in cities. Assessment revealed 12 opportunities across four action areas with the greatest potential in most global cities to curb emissions (C40 2017). The study concluded that more progress can be made by concentrating on a small number of high value opportunities than by spreading effort over hundreds of potential actions. Success will require cities to find creative ways to tackle operational challenges including aligning stakeholders, supply chains, procurement practices and financing. According to the study, cities could achieve significant of emissions reduction by 2030 and build the knowledge and foundational capabilities needed to reach net zero carbon by 2050. Incremental investment required would be in the order of \$50 to \$200 per metric ton of CO_2 equivalent. However, all opportunities provide a positive return on investment in the mid to long term, whether through direct cash flow for investors (e.g., in the case of renewables and efficiency improvements) or broader boosts to economic activity in the city (e.g., transit-oriented development). For many opportunities, up-front investments are paid back within five to ten years.

More over carbon pricing instruments could be an important enabler here. World-wide use of carbon pricing instruments[1] is growing and to date have covered over a fifth of global emissions. By 2019, 14,500 projects were implemented to result in issuance of cumulative 4 billion tCO_2e of credits (World Bank 2020). Clean Development Mechanism (CDM) accounted for 50% of credits issued so far. However, in 2019, two-third of credits were issued by independent crediting mechanisms such as Verra (VCS), Californian compliance offset system, Gold Standard and several other systems. Companies offsetting emissions for compliance or voluntary commitments, besides the requirements of national governments for compliance commitments, were the key drivers for the demand growth. In 2019, about US$45 billion was raised through carbon pricing systems which included 31 emission trading systems (ETS) in operation (World Bank 2020). Prices ranged from US$1/$tCO_2e$ to US$119/$tCO_2e$, whereas the average price was $2/$tCO_2e$ (IMF 2019) and 50% of the credits were valued under US$10/$tCO_2e$. In 2019, about 1600 companies disclosed that they currently use internal carbon pricing or that they anticipate doing so within two years (CDP 2019). The High-Level Commission on Carbon Prices estimated that carbon prices of at least US$40–80/$tCO_2$ by 2020 and US$50–100/$tCO_2$ by 2030 are required to cost-effectively reduce emissions in line with the temperature goals of the Paris Agreement (CPLC 2017). As of today, less than 5% of GHG emissions currently covered by a carbon price are within this range (World Bank 2020) but they are expected to increase and are already increasing in several jurisdictions.

Four action areas identified by above cited C40 study included:

Decarbonizing the electricity grid: Utilities and regulators must play a central role in increasing the share of renewables together with energy storage to ensure grid reliability. Measures may include setting clear decarbonization goals, aggregating demand for renewables, promoting energy efficiency and shifting more urban energy consumption to electricity (especially in transportation and heating). Potentially cities could achieve a grid mix of 50–70% renewables (specifically, solar and wind, balanced with other zero-emission generation source such as hydro) by 2030 depending on local resource characteristics, and market and regulatory structure. This level would capture 35–45% of the total emissions reductions needed in that time frame at a cost as low as $40 to $80 per megawatt hour.

Optimizing energy efficiency in buildings: In buildings around the world, heating and cooling account for 35–60% of total energy demand and, on average, produce nearly 40% of emissions. Several opportunities based on widely available technologies offer the potential to significantly reduce emissions from buildings. These include raising building standards for new construction, retrofitting building envelopes, upgrading heating, ventilation and air conditioning (HVAC) and water heating technology, and implementing lighting, appliance and automation improvements. While cities generally have more influence over this area than many others,

[1] Typically covers carbon tax, emission trading system (ETS), carbon crediting mechanisms, result-based finance, internal carbon pricing in companies, policies that directly or indirectly impact price of carbon.

progress will still require city leaders to work closely with building owners, both residential and commercial, real estate developers and building occupants. This action area is particularly important since building stock tends to turn over only every 30 to 50 years, getting it wrong will lock in emissions, and potential costs, for decades. In contrast, getting it right will reduce energy costs—as well as provide more resilient, comfortable spaces to live, work, and play—for city residents. Focused acceleration in this action area can close 20 to 55% of the gap between current emissions trends and 2030 abatement targets, depending on the local climate and population growth of the city, at an average cost of $20 to $100 per metric ton of CO_2 equivalent.

Enabling next-generation mobility: An unprecedented range of mobility options is now available to cities including multiple, reinforcing trends in mobility and land use planning. The key to reducing emissions through these trends is to ensure that all residents have access to a variety of attractive, affordable low-carbon mobility options. The development of complete, compact communities that meet the mobility needs of residents and business is foundational to building stronger cities and enabling next-generation mobility. Transit-oriented development implemented today promotes smart densification through better land use planning, lays the foundation for more multimodal transport and reduced carbon emissions in the long term. Initiatives to encourage walking and cycling within cities' existing land use patterns as well as targeted enhancement of mass transit, such as the introduction of bus rapid transit (BRT) on main arteries, can collectively start to lower emissions in the short term. In addition, cities can accelerate emissions reductions by enabling uptake of next-generation vehicles, which take advantage of new electric, shared, connected, and autonomous technologies, and by optimizing freight transport and delivery. Focused acceleration in this action area can contribute emissions reductions equal to 20–45% of 2030 targets, depending on urban income levels and population density. In the process, these efforts can increase national income by reducing congestion and transforming the quality of life for residents by alleviating local air pollution and improving equitable access to mobility options.

Improving waste management: Cities can tackle waste emissions in a resource-effective way first reducing waste upstream, then repurposing as much useful finished product as possible; then recycling, composting, and otherwise recovering materials for use; and finally, managing disposal to minimize emissions of any remaining organic matter. Innovative models for waste management can help cities rethink their need for traditional collection and disposal infrastructure, and forward-looking cities are already going further and planning the transition to a fully "circular economy," shifting resource consumption from linear flows to continuous reuse. Depending on the starting point of existing waste management services, as well as composition of waste, focused acceleration can achieve up to 10% of the emissions reductions needed by 2030, as well as numerous benefits to local resource resilience and health.

6.6 Climate Action Pledges in Cities in Global Climate Action (NAZCA) Portal

Considering that cities represent a significant portion of the world economy and population, it comes as no surprise that Global Climate Action (NAZCA) portal[2] of the UN Climate Change listed over 10,693 actors from cities representing over 11,915 climate actions as of March 2021 (Global Climate action, NAZCA 2021). Pledges covered intended climate actions in thematic areas of energy, industry, transport, human settlement, land use, water, ocean and coastal zones. Integrated spatial planning, renewable energy generation, wide spread use of efficient appliance and low-carbon intensity materials, incentives and infrastructure for the use of non-motorized transport and public transport, electromobility are included. Bridging fragmented value chains for zero-emission, efficient, and resilient buildings and knowledge exchange are also proposed. Proposed measures include strengthening integrated mandatory energy codes, increased use of the design tools for efficiency, use of data and information for integrated urban planning and risk zoning and development of adaptation construction techniques to resist wind, water, heat and humidity. Considering that climate actions in cities often involve multiple stakeholders, some of the proposed measures envisaged subnational and non-state actors cooperating across borders to mitigate or adapt to climate change, usually under the patronage of the national governments.

6.7 Climate Action in Cities and NDCs

Countries communicate their NDCs and update them as part of the five-year cycle of reviewing their targets and commitments under the Paris Agreement. NDCs are very relevant to the building sector, with many countries identifying the sector as an important source of GHG emissions and an area for action in their NDCs. Overall, 136 countries mentioned buildings in their first NDC submission, 53 countries mention building energy efficiency and 38 specifically call out building energy codes, indicating the importance of building energy efficiency to our climate future (Global ABC 2020). Tables 6.6, 6.7, 6.8 and 6.9 (under section 6.10) include a compilation of a sample of climate action measures indicated in NDCs updated toward end of 2020, segregated into energy generation, buildings, transport and waste sectors.

[2] Climate actions, initiatives and associated data registered on the portal are provided by CDP, Carbon Climate Registry, The Climate Group, Global Investor Coalition on Climate Change, UN Global Compact, Global Covenant of Mayors, Climate Bonds Initiative and the UN Environment's Climate Initiatives Platform.

6.8 Climate Action in CDM Methodologies

As the possible role of market mechanisms is one of the focus areas of this book and this chapter, Tables 6.3, 6.4 and 6.5 include a compilation of measures included in the CDM methodologies developed to date. As noted, CDM accounts for 50% or more of all carbon credits issued to date and has been an important open source of information for possible climate action in project-based approaches. Under the CDM, numerous methodologies were developed over the last two decades, mostly through a bottom up process, where stakeholders worked with regulatory bodies of the CDM to continuously update and refine the possible climate action and associated monitoring, reporting and verification approaches based on the context on the ground. While each of such a methodology unleashed new areas of climate action, not all of them resulted in an implemented project due to various barriers for project implementation. Furthermore, being a project-based mechanism, CDM does not cover measurement of impacts of implemented policies. Despite these limitations in relation to a holistic and integrated approach, CDM methodologies represent an important repository of experience gained through implementation of climate action in thousands of projects in hundreds of countries over the last two decades (CDM, 2019, 2018).

Table 6.3 Climate action relevant to buildings sector in CDM methodologies

Subsector	Measures covered
Renewable electricity (captive power)	• Renewable electricity generation for captive use and mini-grid
Thermal energy for cooking	• Switch from non-renewable (unsustainable) biomass for thermal applications such as residential cooking including with electricity • Biogas/biomass thermal applications for households/small users • Solar cookers for households • Energy efficiency measures in thermal applications of non-renewable biomass such as cooking
Solar water heating	• Solar water heating systems
Energy efficiency in water delivery	• Water pumping efficiency improvements • Demand-side energy efficiency activities for specific technologies such as lamps, ballasts, refrigerators, motors, fans, air conditioners, pumping systems and chillers
Water purifier	• Distribution of zero-energy water purification systems for safe drinking water • Low greenhouse gas emitting safe drinking water production systems
Water saving	• Demand-side energy efficiency activities for installation of low-flow hot water savings devices

(continued)

Table 6.3 (continued)

Subsector	Measures covered
Refrigerators/chillers	• Power saving through energy efficient chillers • Dissemination of energy efficient household appliances such as refrigerators, air conditioners, washing machines • Energy efficiency and HFC-134a recovery in residential refrigerators
Lighting	• Demand-side activities for efficient lighting technologies such as light-emitting diode (LED) lamps to households • Demand-side energy efficiency activities for installation of energy efficient lighting and/or controls in buildings • Substituting fossil fuel-based lighting with LED lighting systems
Street lighting	• Demand-side activities for efficient outdoor and street lighting technologies
Whole building	• Energy efficiency technologies and fuel switching in new and existing buildings • Energy efficiency and fuel switching measures for buildings • Installation of cogeneration or trigeneration systems supplying energy to commercial building • Energy efficiency and/or energy supply projects in commercial buildings • Energy efficiency space heating measures for residential buildings • Energy efficiency and renewable energy measures in new residential buildings
District heating/cooling	• Energy efficiency improvement projects—boiler rehabilitation or replacement in industrial and district heating sectors • Introduction of an efficient district heating system • Fossil fuel displacement by geothermal resources for space heating • Introduction of a new district cooling system • Supply-side energy efficiency improvements—generation

Table 6.4 Climate action relevant to waste sector in CDM methodologies

Subsector	Measures covered
Alternative waste—composting	• Avoidance of methane emissions through composting • Avoidance of methane emissions through excavating and composting of partially decayed municipal solid waste (MSW)
Alternative waste treatment—other technologies	• Alternative waste treatment processes • Less carbon intensive power generation through continuous reductive distillation of waste • Avoidance of methane production from decay of biomass through controlled combustion, gasification or mechanical/thermal treatment • Avoidance of methane production from biomass decay through controlled pyrolysis • Methane avoidance through separation of solids from wastewater or manure treatment systems • Destruction of hazardous waste using plasma technology including energy recovery

(continued)

Table 6.4 (continued)

Subsector	Measures covered
Alternative waste treatment—aerobic	• Avoidance of landfill gas emissions by in situ aeration of landfills • Avoidance of landfill gas emissions by passive aeration of landfills • Methane oxidation layer (MOL) for solid waste disposal sites
Landfill gas recovery	• Flaring or use of landfill gas • Landfill methane recovery
Lagoons and biodigester—biogas	• Treatment of wastewater • Methane recovery in wastewater treatment • Methane recovery through controlled anaerobic digestion
Manure treatment	• GHG emission reductions through multi-site manure collection and treatment in a central plant • GHG emission reductions from manure management systems • Methane recovery in animal manure management systems • Methane recovery in agricultural activities at household/small farm level
Aerobic wastewater treatment	• Mitigation of greenhouse gases emissions with treatment of wastewater in aerobic wastewater treatment plants • Avoidance of methane production in wastewater treatment through replacement of anaerobic systems by aerobic systems
Utilization of biogenic methane	• Natural gas substitution by biogenic methane produced from the anaerobic digestion of organic waste • Biogenic methane injection to a natural gas distribution grid • Biogenic methane use as feedstock and fuel for town gas production • Methodology for collection, processing and supply of biogas to end-users for production of heat • Hydrogen production using methane extracted from biogas
Recycling	• Recovery and recycling of materials from solid wastes • Recovery and recycling of materials from e-waste

Table 6.5 Climate action relevant to transport sector in CDM methodologies

Subsector	Measures covered
Bicycles, tricycles, e-bikes or e-tricycles	• Lightweight two and three wheeled personal transportation
Bus systems	• Bus rapid transit projects
Mass rapid transit systems	• Mass rapid transit projects • Cable cars for mass rapid transit system (MRTS)
Energy efficiency	• Emission reductions by electric and hybrid vehicles • Transportation energy efficiency activities using retrofit technologies • Transport energy efficiency activities using post—fit Idling Stop device • Emission reductions through improved efficiency of vehicle fleets

(continued)

Table 6.5 (continued)

Subsector	Measures covered
Fuel switch	• Introduction of low-emission vehicles/technologies to commercial vehicle fleets • Plant oil production and use for transport applications • Biodiesel production and use for transport applications • Introduction of Bio-CNG in transportation applications • Introduction of LNG buses to existing and new bus routes
Transportation of cargo	• Modal shift in transportation of cargo from road transportation to water or rail transportation
Transportation of liquid fuels	• Modal shift in transportation of liquid fuels
Technology for improved driving	• Transportation energy efficiency activities installing digital tachograph systems to commercial freight transport fleets • Emission reductions through improved efficiency of vehicle fleets

6.9 Compilation of Climate Action Measures in Cities

Section below includes a simplified listing of climate action categories for the ease of reference drawn from several sources mentioned above. Within each category of action (e.g.,energy efficiency in appliances), there are several opportunities for intervention such as policies, infrastructure upgrades and awareness programs that could be implemented. This list represents the most common and impactful categories of action that cities can take to reduce GHG emissions, but it is not meant to be exhaustive. Diversity in regulatory and economic practices, social and cultural practices requires that policy measures and packages are unique to cities. Some measures such as energy labeling schemes have a reinforcing impact on other policy types assessed, such as loan, grants and energy efficiency obligations. Information and feedback policies can help facilitate effective implementation of other instruments. Finally, most policy instruments would be unable to function without policies enforcing standards for measuring efficiency in systems or buildings.

Measures in energy generation
Subcategory: district heating/cooling
• Establish district heating/cooling
• Efficiency improvements in district heating/cooling
• Fuel switching in district heating/cooling
Subcategory: municipal renewable energy
• Municipal renewable energy installation (solar, wind, biomass, etc.)
• Incentive program for residential and commercial building-integrated renewable energy
• Municipal clean energy procurement

Measures in transport
Subcategory: mode shift
• Build bus rapid transit system
• Operational upgrades for existing transit (mapping, route optimization, ticketing upgrades)
• Travel demand management (car sharing, congestion charging zones, parking restrictions and cycle hire program)
• Bicycle infrastructure upgrades
• Pedestrian infrastructure upgrades
• Parking infrastructure management to disincentivize private car use
Subcategory: fuel switch and fuel efficiency
• Public vehicle electrification
• Taxi electrification
• Electric vehicle infrastructure upgrades
• Transit electrification
• Bus electrification
• Bus fuel switching
• Private electric vehicle incentives (reduced fees, parking, tolls, taxes)
• Low-emission truck incentives (reduced fees, taxes)
• Freight (delivery) system improvement (freight consolidation centers and improved logistic info, last mile bike delivery)

Measures in urban planning
• Transit-oriented development
• Infill/densification
• Urban forestry and green space programs

Measures in waste
Subcategory: waste management (sorting, recycling, composting, treatment)
Residential/commercial/industrial recyclables collection
Subcategory: waste collection and transportation energy
Subcategory: waste to energy

Besides the above several cross-cutting climate action measures may be implemented in cities[3] including:

- Grants/subsidies for cost reduction, preferential (no interest or lower interest) loans;
- Tax incentives/exemptions;
- Taxation/energy/carbon taxation based on carbon footprint;

[3] Adapted from UNFCCC Secretariat (2020).

- Energy efficiency obligations (obligated parties, generally energy utilities meet energy-saving targets by delivering or procuring energy savings at the customer end of the energy system);
- Feed-in tariffs (guaranteed payments to incentivize renewable energy in the grid);
- Obligation to companies/utilities to allocate money to energy efficiency, renewable energy and/or low-income assistance programs;
- Cooperative procurement, e.g., cohousing build/community shared renewable systems (building as a single unit to building as a system and stock of materials with value and usability at end of life to promote circular economy);
- Risk assessment of standard materials to incorporate externalities so that the full value of circular economy alternatives can be assessed;
- Database and market venue (real-time or digital) for reusable material supply and demand;
- Energy labeling—as a real estate marketing tool;
- Development of rules and guidelines on the labeling of recycled building materials;
- Institutional financial support of innovative solutions;
- Support for the research and development (R&D), innovation and production of innovative technologies;
- Circular economy solutions: Reducing, reusing, or recycling materials to reduce embodied emissions in the construction of new or refurbished buildings;
  - Circular supplies (prioritize renovation, minimize use of virgin material, design for ease of future renovation and reuse of space);
  - Products as a service (user pays for materials or products that remain owned by the developer or producer);
  - Product lifetime extension: Maintain and extend life of product through smart maintenance, repairs, upgrades and renovation;
  - Resource recovery: More innovation in material recycling.
- Urban green infrastructure through integration of green spaces in urban planning at various scales to alter urban temperatures (forests, parks, street trees and grasslands but can include mangroves, coral reefs, sea grasses, sand beaches and dunes in the coast environment);
- Urban blue infrastructure (surface temperature of water is lower compared to vegetated areas which in turn are markedly cooler than streets and roofs, includes ponds, wetlands, rivers, lakes and streams, as well as estuaries, seas and oceans);
- Awareness raising, information provision and labeling;
  - Capacity building, training, technical assistance and one-stop shops;
  - Benchmarking;
  - Mandatory energy audits;
  - Voluntary or mandatory energy-saving targets;
  - Voluntary or mandatory energy performance standards for products and technologies;
  - Banning inefficient products;
  - Mandatory renovation.

6.10 Climate Action Measures in NDCs

This section includes extracts from a sample of NDCs relevant to climate action in cities, based on 2020 updates from countries available on NDC Registry (interim) via https://www4.unfccc.int/sites/ndcstaging/Pages/LatestSubmissions.aspx accessed on March 1, 2021. Tables 6.6, 6.7, 6.8 and 6.9 include compilations of measures in buildings, waste, energy and transport sectors. Information is for illustration purposes, is summarized and translations are approximate. Furthermore, some of the measures proposed are conditional subject to availability of finance, technology, etc. For accurate and detailed information, reader is advised to access the original NDC document available on the above site.

Table 6.6 Examples of mitigation measures included in the area of buildings in a sample of NDCs

Andorra	• Any new building constructed from January 1, 2020 should be of almost zero-energy consumption (2030) • Administrative buildings shall have renewable sources of electricity to reduce the consumption of electricity and heating in public establishments (2050) • Carry out energy audits before 2022 in all the administration's heated buildings (2030) • To improve the energy efficiency of public lighting with regulation through a regulation (2030)
Monaco	• Improving the energy and environmental performance of all existing and future buildings (The best energy is the energy we do not consume) • Policies and measures simultaneously target the renovation of the already built heritage (envelopes and energy systems), uses and changes in behavior and sustainable construction methods for new buildings (adapted to the Mediterranean climate and to the specific features of Monaco) with the aim of a high level of optimization of the energy efficiency of all buildings • This optimization requires a gradual reinforcement of the regulatory thermal requirements for new buildings and renovations, as well as prioritization and an increase in the annual rate of renovation, supported by financial mechanisms • The government is supporting the adaptation of construction methods to local climatic conditions through Monaco's Mediterranean Sustainable Building approach and the training of construction players in new • It should be noted that the Principality of Monaco has a strong interest in "blue energy" and in particular in solar thermal energy as a substitute for fossil fuels
Norway	• Ban on the use of mineral oil for heating of buildings from 2020 (The ban covers the use of mineral oil for both main heating (base load) and additional heating (peak load), in residential buildings, public buildings and commercial buildings) • Energy requirements in the building code (energy requirements in buildings to ensure more energy efficient buildings)
Republic of Maldova	• Improving the infrastructure of hospitals for operationalization to "green" standards • Improving the energy performance of buildings is a priority of public policies in the Republic of Moldova, and the draft Government Decision (No. 1103 of 14.11.2018) for the approval of the regulation regarding the periodic inspection of the air conditioning systems in buildings. It was enacted on: 21.12.2018

(continued)

Table 6.6 (continued)

Switzerland	• New buildings are better insulated than in the past (Recently, the existing legal framework has again been revised and existing policies and measures strengthened in view of Switzerland's commitment under the Paris Agreement for the period from 2021 to 2030 (third CO_2 Act) • Heating oil is increasingly replaced by natural gas and electricity (e.g., for heat pumps) (Recently, the existing legal framework has again been revised and existing policies and measures strengthened in view of Switzerland's commitment under the Paris Agreement for the period from 2021 to 2030 (third CO_2 Act))
United Kingdom of Great Britain and Northern Ireland	• 600,000 heat pump installations per year by 2028, creating a market led incentive framework to drive growth • Homes built to the Future Homes Standard will be "zero carbon ready" and have 75–80% lower carbon dioxide emissions than those built to current standards • Green Homes Grant to improve the energy efficiency of homes and replace fossil fuel heating, green home finance initiatives to improve the energy efficiency of around 2.8 million homes, improving around 1.5 million to EPC C standard by 2030 • Lower carbon heating and cooking with no change in experience for domestic consumers through hydrogen blends and reducing the emissions of the gas used by up to 7% • Public Sector Decarbonization Scheme • For homeowners who live off the gas grid, particularly in rural areas, upgrades to their heating systems through the Homes Upgrade Grant • Social Housing Decarbonization Fund to continue upgrading the least efficient social housing • Strengthen energy efficiency requirements for private sector landlords • To support those least able-to-pay, extend the Energy Company Obligation to 2026, so suppliers can help improve the draughtiest and coldest homes
EU 27	• Directive (EU) 2018/844 on energy performance of buildings
Columbia	• Sustainable construction: Guidelines for the design and construction of sustainable buildings, achieving reductions in energy consumption, which allows for the reduction of GHGs linked to development of housing and other buildings, through (1) reduction of natural gas and electrical energy consumption in new housing not including Social and Priority Housing (VIS and VIP) and (2) reduction of natural gas and energy consumption in new buildings other than housing • Goal: Gradual implementation until 100% of the buildings new nationwide by 2026, maintained until 2030
Costa Rica	• In 2030, 100% of new buildings will be designed and built by adopting systems and low emission and low resilience technologies bioclimatic parameters
Cuba	• The installation of 833,333 solar heating units (1 million m2 of surface area) in the residential and industrial sectors • The installation of 15.25 million LED lamps in the residential and public sectors • The replacement of 2 million electric resistance stoves with induction stoves

(continued)

Table 6.6 (continued)

Dominican Republic	• Air conditioning replacement program for all consumer and service sectors based on new, more efficient standards • Programs for the reconversion of domestic refrigerators for all consumer and service sectors based on new, more efficient standards • New standards to introduce efficient lighting for public and residential areas • Introduction of energy efficiency standards in new buildings • High PCG (HFC) gas destruction program in low efficiency refrigeration and air conditioning equipment according to the Kigali Amendment protocol. (Qualitative)
Ecuador	• Efficient cooking program (replacement of liquid petroleum gas (LPG) stoves with induction stoves)
Panama	• By 2025, Panama will have developed the "Technical Guide to Climate Change for Human Settlements" with mitigation and adaptation components • By 2025, Panama will have developed a "Climate Change Plan for the Infrastructure Sector" that includes adaptation and mitigation components • By 2025, Panama will integrate the climate change dimension into public investment projects through the implementation of the "Climate Change Technical Guide for Public Investment Infrastructure Projects"
Uruguay	• Use of solar collectors for domestic hot water in large users, industrial and residential users: 50 MWth of installed capacity for 2025 • Implementation of a residential area pilot plan of smart grids, replacing 100,000 electricity meters by smart meters by 2025 • "Implementation of the 2024 Energy Efficiency Plan," which includes, among others, the following measures: – Replacement of current equipment with efficient equipment: 4 million incandescent light bulbs replaced in the residential sector, and 30% of LED lights in public lighting by 2025 – Mandatory energy efficiency labeling in household devices by 2025: lamps, water heaters, air conditioners and refrigerators – Regulation of energy efficiency labeling in new homes and buildings by 2025 – Implementation of the Pilot Program for the Improvement of Energy Efficiency in housing, in Montevideo • Implementation of Energy Efficiency Certificates in all consumer sectors. Its monetary value will be determined by annual targets and funding available, which includes at least 0.13% of the previous year total energy sales • Wider use of solar collectors for hot water in large users, industrial and residential users: 100 MWth of installed capacity for 2025 • Extension of the Pilot Program for the Improvement of Energy Efficiency in 5% of homes throughout the country by 2025 • Renewal and consolidation of the energy efficiency plan, which includes, among others, the following measures: – Replacement of current equipment with efficient equipment: 80% of built-in LED luminaires in public lighting by 2025 – Mandatory labeling of energy efficiency in household devices: other household appliances, gas-burning appliances and wood-burning appliances by 2025

(continued)

Table 6.6 (continued)

	– Implementation of energy efficiency labeling program in used and non-residential buildings by 2025 – Consolidation of smart grids including household appliances and smart meters in two neighborhoods or towns by 2025
Marshall Islands	• Ensure building code focus on energy efficiency (Marshall Islands electricity roadmap) • Mandate the purchase of energy star (or similar energy rated labeling) appliances for all government agencies ("soft" internal government policy). (Marshall Islands electricity roadmap) • Minimum energy performance standard (MEPS) and appliance labeling scheme. (Marshall Islands electricity roadmap) • A ban on incandescent light bulbs. (Marshall Islands electricity roadmap) • Mandate the purchase of energy star (or similar energy rated labeling) appliances for all government agencies ("soft" internal government policy)
Rwanda	• Efficient lighting in buildings: further dissemination of CFL and LED lamps in residential, commercial and institutional buildings. Supported by government subsidies and VAT exemptions on energy-saving lamps. Reduction of grid-based GHG emissions • Solar water heater (SWH) program: installation of solar thermal water heaters within urban residential buildings supported by use of loans and grants to subsidize purchase costs, as part of the national green building code minimum compliance system • Solar street lighting: installation of solar lighting and LED systems to replace high-pressure sodium (HPS) lamps for street lighting and public spaces within populated areas and main roads by 2024. Reduction of grid-based GHG emissions • Efficient cook stoves: dissemination of modern efficient cook stoves to 80% of the rural population and 50% of the urban population by 2030, achieving a more sustainable balance between supply and demand of biomass, and reducing firewood and fossil energy consumption for cooking • Solar street lighting: installation of solar lighting and LED systems to replace high-pressure sodium (HPS) lamps for street lighting and public spaces within populated areas and main roads by 2024. Reduction of grid-based GHG emissions
Senegal	• Standardization and labeling of household appliances and office equipment • Distribution of 800 000 improves cook stoves by 2030 compared to 350 K in 2016 • 27,000 biodigesters by 2030 • 48 K biodigesters by 2030 • Promotion of biochar

6.11 Conclusions

Cities include a variety of sources that cause GHG emissions including the construction and operation of the buildings, electricity generation, transport services and waste management. For example, the buildings sector today accounts for almost 55% of global electricity consumption, and when both the construction and use phases are taken into consideration, it accounts for about 37% of today's global CO_2 emissions.

Table 6.7 Examples of mitigation measures included in the area of waste in a sample of NDCs

Argentina	• Toward 2030, and with the aim of guaranteeing the minimization of waste generation, icmprovements will be adopted in of sustainable consumption and production • Toward 2030, and with the aim of guaranteeing the minimization of waste generation, the circular economy will be promoted as one of the tools available • The circular economy will be promoted as one of the tools available, among others, to achieve sustainable development, proper waste management—including separation at source and the promotion of the recycling industry
Bangladesh	• Already two city corporations adopted the incineration technology to reduce waste and other two city corporation will also follow them • Around 100 municipalities will establish biogas plant to reduce emission
Brunei Darussalam	• Minimizing the amount of waste that needs to be disposed of through the adoption of best practices and innovative technologies
Combodia	• New sanitary landfills with LFG extraction at the Dangkor landfill • Increase the share of waste disposed at sanitary landfills with LFG extraction from 0% in 2020 to 50% by 2030 and extract LFG from the Dangkor landfill (Estimated accrued income in the period of 2020–2030: 1. Electricity sales: USD 69.8 million, 2. Carbon revenues: USD 25.97 million based on USD 5/carbon credit) • Composting of biodegradable organic fraction of MSW supplemented with separation of organic waste (at source) • If 10% of all MSW generated is composted by 2030 then up to 0.5 $MtCO_2e$/year of GHG emissions can be avoided by 2030 • Production of refuse-derived fuel (RDF) from either (a) fresh MSW or (b) old MSW mined from the Dangkor landfill (GHG ER from RDF + anaerobic digestion up to 0.2 $MtCO_2e$/year) • Implementation of National 3R strategy • Centralized recycling facility for industrial waste from the garment sector (Reduce 0.11$MtCO_2e$ from 2021 to 2030 at an average of 0.01 $MtCO_2e$/year) • Better management of industrial wastewater in the food & beverage sector (5–10% of total CH4 emissions)
Chile	• In 2020, develop a national organic waste strategy, aimed at increasing the recovery of organic waste generated at municipal level by recovering nutrients, organic materials or substrates into the productive process and, thereby, contributing to climate change adaptation and mitigation
Costa Rica	• By 2025, at least 10 municipalities implement the National Composting Plan • By 2030, at least 50% coverage will be achieved of sanitary sewage in the areas of high population density, incorporating criteria of resilience to climate change • By 2030, at least 50% of the water residuals in areas of high population density will receive treatment • During the first two years of implementation of this NDC, Costa Rica will have published and implementation of its policy instruments such as the Action Plan for the Management Integral Solid Waste 2021–2026 and the National Composting Plan 2020–2050, articulating emission reduction efforts, with a transformation approach to the economy and the bioeconomy

(continued)

Table 6.7 (continued)

Dominican Republic	• Small-scale energy generation plants based on biomass (agricultural and forestry) and solid waste • Develop a national strategy for organic waste, to increase the recovery of this type of waste generated in municipalities, in order to reduce CH_4 emissions • Capture and direct use of methane from landfills for energy purposes • Recycling of (new) wastes with value (by-products) for energy purposes, composting (organic fertilizers) and others • Introduction of circular economy (CE) processes in the Waste sector. Establish a short-, medium- and long-term CE roadmap that generates metrics and indicators for an MRV system at the subsector level • Use of municipal, industrial and biological waste for clinker kilns in cement plants • Recycling of (new) wastes with value (by-products) for energy purposes, composting (organic fertilizers) and others
Ecuador	• Active methane capture and electricity generation at the El Inga (Quito) landfill • Active methane capture and electricity generation from the Pichacay (Cuenca) landfill • Active capture of methane in Santo Domingo de los Tsáchilas and Ambato landfills • Composting with forced aeration
Jamaica	• The country is currently developing proposals for a country-wide Integrated Waste Management (IWM) PPP concession which is expected to improve the management and reduce the emissions of the waste sector • Pilot projects to explore increasing the use of biodigestors • Pilot projects to explore biodiesel from cooking oil
Japan	• Reduction of municipal solid waste incineration • Reduction of municipal solid waste disposed of by direct landfill • Introduction of semi-aerobic landfill system for final disposal site of municipal solid waste • Promote the advanced technologies in the sewage sludge incineration facilities • Promotion of measures for energy efficiency and conservation of/energy generation from sewerage systems • Promotion of sorted collection and recycling of plastic containers and packaging
Maldives	• Waste to energy. The planned installation of 8 MW in Thilafushi and 1.5 MW in Addu City will be completed. These systems will be optimized for grid connection and electricity production
Mexico	• Treatment of wastewater from both municipal and industrial sources • Recycling • Reuse • Composting • Biodigestion
Monaco	• Since 2016, the principality has been deploying an ambitious strategy to limit the amount of waste produced/reduce all types of waste at source (market or distribution ban for some products) • Reduce plastic waste, the incineration of which is responsible for the majority of emissions in this sector • Zero single-use plastic waste by 2030 policy, prohibit the marketing or distribution of certain products

(continued)

Table 6.7 (continued)

	• Direct waste to material recovery • Improve their recovery (wastes) (measures to simplify sorting and to increase the quantities collected)
Mongolia	• Reduce the waste volume for landfill through the improved waste treatment and recycling process • Increase the share of the population with access to improved sanitation and hygiene facilities • In addition, if conditional mitigation measures such as the carbon capture and storage and waste-to-energy technology are implemented, then Mongolia could achieve a 27.2% reduction in total national GHG emissions
Nepal	• By 2030, create an enabling environment for both public and private sector to treat industrial and municipal waste, including fecal sludge • By 2030, adopt and implement waste segregation, recycling and waste-to-energy programs in at least 100 municipalities • By 2030, the burning of healthcare waste in 1400 healthcare facilities will be prohibited by proper management of healthcare waste through the application of non-burn technologies • Promote the 3Rs (Reduce, Reuse, Recycle) approach to waste management, along with source segregation and management of degradable and non-degradable waste • Focus on coproduction of energy and organic fertilizer from solid waste, wastewater and fecal sludge
Nicaragua	• The comprehensive development project of La Chureca has included not only the sealing of the landfill of the same name, considered the largest in Latin America, but also the construction of a recycling plant where garbage collectors work, as well as the construction of houses, a school for the more than 250 families that inhabit the place and the reduction of the gases produced by the decomposing garbage itself that before the intervention produced spontaneous combustions on the entire surface of the landfill, now they are led outside by a circuit of pipes and gasification chimneys through which methane gas comes out. These gases are planned to be used in electricity generation by the Mayor of Managua • Since 2007, the wastewater treatment systems have increased significantly, by 2010, 13 departmental capitals provide treatment to wastewater. Since the start of operations of the Managua Wastewater Treatment Plant, the percentage of treatment of wastewater collected in the city has improved significantly from 35.22% in 2007 to 98.19% for 2011 and the treatment index went from 19.66% to 57.63% nationwide
Norway	• Requirement to collect landfill gas (landfill directive incorporated into national law requires all landfills with biodegradable waste to have a system for extracting landfill gas) • Prohibition of depositing biodegradable waste (landfilling of easy degradable organic waste was prohibited in 2002 and was replaced by the wider prohibition of depositing from 2009 that applies to all biodegradable waste.) • Agreement with industry to minimize waste (agreements primarily to ensure that waste is collected and sent to approved treatment) • Measures to increase waste recycling (waste regulations for a number of waste fractions and a tax on beverage packaging) • Tax on final disposal of waste (tax on incineration up to 2010 and for landfills up to 2015)

(continued)

Table 6.7 (continued)

Republic of Korea	• Measures to minimize wastes, such as single-use items in particular, will be implemented
Rwanda	• Landfill gas utilization: extraction and utilization of landfill gas (LFG) for power generation in connection to semi- or fully controlled landfills for urban areas. Reduced CH4 emissions from landfill sites and avoided CO_2 emissions from displacement of fossil-based electricity use • Waste-to-energy (WtE) plants: development of WtE plants in Kigali and other urban areas through energy recovery options other than LFG. Avoided CO_2 emissions from displacement of fossil-based electricity use • Aerobic composting: development of commercial-scale aerobic composting systems for agricultural and forestry residue, manure, food processing, household kitchen and garden waste, and biosolids (organic solids from treated sewage). Reduction in CH_4 emissions, since methane-producing microbes are not active in the presence of oxygen • Wastewater treatment plants (WWTP): investment in wastewater treatment and reuse technology, reducing methane emissions from wastewater and providing a nutrient-rich digestate that can be used as a fertilizer. Reduction of CH_4, and CO_2 emissions
Senegal	• Achieving 85% access rate to the sewerage network by 2030 (i.e., a treatment rate of nearly 70% and a depollution rate of over 55%) • Rehabilitation or closure of departmental and uncontrolled landfills by 2030 • Construction of standardized grouping points and integrated waste management centers • Promulgation of regulatory texts on solid waste management
United Arab Emirates	• The Emirate of Abu Dhabi has adopted the single-use plastic policy 2020 that aims to make Abu Dhabi free of single-use plastic bags by end of 2021 • Diversion of waste away from landfills is a key waste management objective and related federal and emirate-level targets have been put in place. For instance, the emirate of Sharjah has set an ambitious target to reduce waste to landfill to zero by 2021, and processing and recycling is being prioritized to enable the achievement of the target • In keeping with the approach of reducing waste diverted to landfills and deriving value from municipal solid waste, waste-to-energy projects are being developed across the UAE. The under-development Dubai center for waste processing, with the capacity to process 1.9 million tons of municipal solid waste and an installed power capacity of 200 MW, is expected to be operational by 2024 • The wastewater treatment and sanitation sector in UAE has witnessed significant advances over the last few years with the expansion of wastewater treatment capacity. The use of treated water, for purposes such as irrigation and district cooling, is also steadily increasing • In conjunction with technology-based enhancement in domestic agriculture, the UAE has taken a comprehensive approach to reducing food waste by engaging local residents, government organizations and businesses in initiatives to reduce, and encourage treatment of, food waste. The nationwide food waste pledge launched in 2018 encourages the UAE's hospitality sector to adopt efficient food production practices. The UAE aims to cut food waste by half by 2030

(continued)

Table 6.7 (continued)

United Kingdom of Great Britain and Northern Ireland	• The resources and waste strategy sets out England's plans to move away from a linear economy, toward a more circular and sustainable economy in which natural resources are used efficiently and waste is minimized • Scotland's Making Things Last strategy sets out priorities to achieving its commitment to moving toward a more circular economy—where products and materials are kept in high value use for as long as possible • The welsh government has consulted on a new circular economy strategy. This proposed a range of actions which seek to keep resources in use for longer and avoid waste. The final strategy will be published in the coming months • Northern Ireland published the second iteration of the national waste prevention program stopping waste in its Tracks in 2019. The program aims to decouple waste from growth and the actions contained within are designed to help Northern Ireland transition to a low-carbon circular economy • Explore new and innovative ways to manage emissions from landfill • We will work with the waste sector to ensure that different waste materials going into energy recovery processes are treated in the best possible way, to minimize environmental impact and maximize their potential as a resource • There is an opportunity to undertake further research to accelerate methane production in the early life of a landfill site to reduce the length of aftercare required. This could help to reduce emissions from landfill further and in a sustainable way • We will work toward no food waste entering landfill by 2030. Many local authorities have introduced separate collection of food waste and we will work to support more so that the amount of food waste sent to landfill continues to decline • We will take action through the Courtauld 2025 Agreement to reduce the amount of food that is wasted in the UK. This could deliver up to £20 billion worth of savings to the UK economy between 2015 and 2025. We will also divert more food from landfill to support resource productivity and avoid further emissions
Uruguay	• Introduction of CH_4 capture and burning in final disposal systems of solid urban waste (with and without electric power generation): 60% of urban solid waste generated placed in final disposal sites with this technology • Development of solid urban waste management systems, including improvements in the treatment and final disposal systems. This development includes the extension of capture and burning of CH_4 systems and/or the introduction of technologies to reduce CH_4 generation to new final disposal sites, so that 90% of the urban solid waste is disposed of in sites with these technologies • Improvement in the industrial wastewater treatment systems with technologies that reduce CH_4 emissions. This includes the implementation of new CH_4 capture and burning systems in anaerobic treatments, reaching the treatment systems where 30% of emissions are generated
Vietnam	• Taking measures to manage, develop and apply technology for treatment of domestic solid waste, ordinary industrial solid waste and special solid waste to limit GHG emissions
Zambia	• Improved solid waste handling and disposal practices • Reduction in waste generation • Reduction in generation of biomethane from waste • Improved wastewater treatment practices

Table 6.8 Examples of mitigation measures included in the area of energy generation and supply in a sample of NDCs

Bangladesh	• The government has set a target of generating 250 MW from solar home system by 2030 • Bangladesh has installed around 5.8 million solar home systems (SHSs) across the country, meaning that almost 18 million beneficiaries are getting solar electricity which is around 11% of the total population of Bangladesh • Around 60.6 MW of solar has been installed on the vacant rooftops of government and private buildings • The government has extended a refinancing scheme to finance alternative energy generation projects like small-scale solar and micro grids, to improve energy access for those living in off-grid areas
Maldives	• Increase of electricity production by renewable energy (RE) with storage and grid stabilization • Increase of efficiency of generators—significant upgrading of the existing power production infrastructure needs to be done via routine scheduled maintenance, synchronization and optimization of power production (increase supply and demand-side efficiency) • Upgrading the grids to minimize grid loss—reducing grid loss to at least 5% (increase supply and demand-side efficiency) • Use of liquefied natural gas (LNG) for electricity generation within greater Malé region. The diesel used for power production could be replaced with LNG for the greater Malé region with the proposed LNG plant in Thilafushi and the interconnectivity bridge
Marshall Islands	• Increase reliability of the grid, and build public confidence in the grid, so people are less likely to seek alternatives as backup (Marshall Islands electricity roadmap) • A loan scheme is being introduced by the Marshall Islands development bank in late 2018, funded by the Taiwan international cooperation and development fund, to provide householders with loans to purchase both energy efficient appliances and private rooftop solar (Marshall Islands electricity roadmap) • Signal that there could be a tariff review to ensure partially self-sufficient users are paying a fair price for being able to access the network, to better inform their future investment decisions (Marshall Islands electricity roadmap) • Continue the import tax exemption for solar PV panels but do not actively promote it. (Marshall Islands electricity roadmap) • Consider standards for imported equipment, and the licensing of installers. (Marshall Islands electricity roadmap) • Include the integrity of roof structures in the revised building code, with a view to supporting PV installations, especially on larger roofs. (Marshall Islands electricity roadmap)
Mexico	• Actions that increase the participation of clean energy in the national electric network • Strengthening and optimization of existing infrastructure • Promotion of innovative technologies in the areas of storage • Smart grids • Actions that will promote the optimization of the processes of the refining and processing systems have been identified, including the implementation of the methane emissions reduction policy

(continued)

Table 6.8 (continued)

Mongolia	• Hydropower plants, wind power plants, solar power plants • Heat pumps for heating utilities • Improved efficiency of energy production • Reduce electricity and heat transmission and distribution grid losses • Improved efficiency of energy production • Improve the heat supply in cities and towns (improving the efficiency of heat only boilers) • In addition, if conditional mitigation measures such as the carbon capture and storage and waste-to-energy technology are implemented, then Mongolia could achieve a 27.2% reduction in total national GHG emissions • Use coal bed methane
Nepal	• By 2030, increase the reliable supply of clean energy, ensuring access to all • Strengthen transmission and distribution links to support upscaling of e-cooking, e-heating, e-transport and charging stations • Increase the quantity (kWh), quality, reliability and affordability of electricity access from renewable sources
Republic of Korea	• Ban on constructing new coal-fired power plants • Additional reduction of existing coal power plants • Conversion to eco-friendly fuels such as LNG (from coal) • Scale-up of renewable energy (to increase the share of renewable energy up to 20% by 2030 and 30–35% by 2040) • The solar and wind power capacity is expected to increase threefold from the 2019 level by 2025 (from 12.7 GW to 42.7 GW)
Rwanda	• Grid-connected hydropower generation: development of 56.75 MW large hydrocapacity (capacity > 5 MW), 24.5 MW small and mini hydroprojects (capacity < 5 MW) and 75 MW regional projects by 2030. Displacement of GHG emissions from fossil fuel power generation (peat, diesel oil) • Solar mini-grids: 68 MWp of solar mini-grids to be installed in off-grid rural areas by 2030, as reflected in the rural electrification strategy. Displacement of traditional biomass fuels, diesel and kerosene for domestic energy use • Solar mini-grids: 68 MWp of solar mini-grids to be installed in off-grid rural areas by 2030, as reflected in the rural electrification strategy. Displacement of traditional biomass fuels, diesel and kerosene for domestic energy use • Off-grid rooftop solar electrification: penetration of off-grid solar and rooftop solar PV panels consistent with the ESSP targets of around 1,500,000 households to be electrified through, equivalent to 250,000 connections per year. Displacement of grid power and diesel consumption and associated GHG emissions
Singapore	• Singapore is pushing ahead to spur the deployment of solar PV through continued investment in research, development and demonstration (RD&D) to reduce cost, improve efficiency and enable innovative modes of deployment such as floating, offshore, and building-integrated PV
Suriname	• Further investments for expansion in both grid-connected and off-grid capacity, as well as efficiency programs. The government is currently reviewing expansion plan options (Energy efficiency standards: introducing EE standards by developing legislation (amendment of electricity act) and set up a dedicated organization for implementation.)

(continued)

Table 6.8 (continued)

	• Development of sustainable and scalable innovative business model for installing, operating and maintaining mini-grids, including a payment system (demonstrate sustainable housing model: promote renewable energy (RE) access by move to the sustainable electrification of +200 villages in the interior by the replacement of existing use of diesel by solar supply and solar/hybrid systems.) • Facilitate commercial finance and investment for mini-grid projects are structured and applications evaluated by commercial banks. Seeking technical and institutional support to introduce new modalities, which includes public–private (investment) partnerships (PPPs), for rural electrification in order to facilitate investment (PPPs: provide incentives for investors by encouraging an investment-friendly environment through risk mitigation by a guarantee fund) • Implementation of fiscal sustainability measures to promote energy efficiency. The current tariff is a disincentive to private investment in more efficient technology. The impact of such reforms on citizens, a study of impacts and compensation measures will be conducted and a just transition plan prepared. (Energy efficiency—Subsidy and fiscal reform: Promote energy efficiency (EE) and energy conservation through energy-savings equipment (energy efficient appliances) by providing them to customers at reduced prices; including equipment labeling and performance standards) • Adoption of renewable energy act to provide the legal, economic and institutional basis for the promotion of the use of renewable energy resources focusing on methods of electrification serving the interior, such as grid expansion (connecting off-grid systems to the national grid), solar PV systems, mini-grids, and micro- and small-scale hydropower development. (Policy and regulatory framework: implementation of the electricity authority suriname (sector regulator) and the development of the renewable energy act and the rural electricity act)
United Arab Emirates	• Development of both photovoltaic and concentrated solar power projects with record-breaking low electricity prices • The UAE is also investing in nuclear energy with the development of the four-reactor Barakah nuclear power plant • In addition to grid-based clean power, distributed electricity generation, rooftop solar photovoltaic in particular, is being promoted in the emirate of Dubai through the adoption of a net-metering program • In order to facilitate grid connection of diverse sources of supply and enhance grid stability, UAE utilities have also been making advances in deployment of energy storage technologies. The emirate of Abu Dhabi is home to the world's largest virtual battery plant with a capacity of 108 MW. The batteries are distributed across 10 sites but can be controlled as a single plant, securing supply for emergencies and allowing for peak shaving • A 250 MW pumped storage hydropower project is also being developed in the Hatta mountains in Dubai wherein water will be pumped to an upper reservoir when surplus solar power is available; the stored water will then be used to generate electricity to meet demand during evening peak hours or when needed • The UAE has developed the region's first commercial-scale network for carbon capture, use and storage. Inaugurated in 2016, the Al Reyadah project captures CO_2 at an emirates steel facility, and compresses, dehydrates and transports it through a pipeline for injection into oil wells for enhanced oil recovery applications. The project's current installed capacity to capture, transport and inject CO_2 is up to 800,000 tons per year. Building on the success of Al Reyadah, plans are underway to expand carbon capture capacity in the country

(continued)

Table 6.8 (continued)

	• Having adopted a zero routine flaring policy in its operations, the country's oil and gas companies are regional leaders in flaring avoidance design and operation. ADNOC has designed and operated its facilities to utilize the recovered gas in normal operations through recovery compressors or by adding value to its downstream chain • The use of state-of-the-art techniques like optical imagery, infra-red capabilities and Leak Detection and Repairs (LDAR) has allowed the UAE oil and gas sector to monitor and manage fugitive emissions across the value chain, and the sector continues to build on these successes • The UAE's oil and gas sector is also undertaking initiatives to power production facilities with cleaner energy, increase the efficiency of energy • The Dubai electricity and water authority has developed a world-class research and development center that focuses on solar testing and certification, energy efficiency and smart grids • In 2018, the UAE government launched the artificial intelligence laboratory in collaboration with Khalifa university of science and technology and the international renewable energy agency (IRENA). The laboratory uses artificial intelligence to optimize the use of renewables in the UAE by mapping optimal solar power locations across the country—in addition to tracking air pollutants and monitoring water quality
United Kingdom of Great Britain and Northern Ireland	• By 2030, we aim to produce 40 GW of offshore wind, including 1 GW of innovative floating offshore wind in the windiest parts of our seas • Our aim is for the UK to develop 5 GW of low-carbon hydrogen production capacity by 2030. This will be supported by a range of measures, including a £240 million net zero hydrogen fund, and setting out next year, our hydrogen business models and a revenue mechanism for them to bring through private sector investment • Lower carbon heating and cooking with no change in experience for domestic consumers through hydrogen blends and reducing the emissions of the gas used by up to 7% • We are pursuing large-scale new nuclear projects, subject to value-for-money • Phase out the use of unabated coal to produce electricity by 2025 • We are announcing up to £385 million in an advanced nuclear fund. This will enable investment of up to £215 million into small modular reactors to develop a domestic smaller-scale power plant technology design that could potentially be built-in factories and then assembled on site • We are also committing up to £170 million for a research and development program on advanced modular reactors • We are doubling down on our ambition to be the first country in the world to commercialize fusion energy technology, enabling low carbon and continuous power generation. We are already providing £222 million for the visionary STEP program (Spherical Tokamak for energy production), which aims to build the world's first commercially viable fusion power plant in the UK by 2040, and £184 million for new fusion facilities, infrastructure and apprenticeships to lay the foundations of a global hub for fusion innovation in the UK • We will provide £100 million for energy storage and flexibility innovation challenges—essential technology as we move toward an increasingly renewables heavy system to allow us to store energy over hours, days and even months

Table 6.9 Examples of mitigation measures included in the area of transport in a sample of NDCs

Andorra	• Increase the share of electric vehicles for passenger transportation by 20% (2030) • Increase the share of electric vehicles for passenger transportation by 50% (2050)
European Union	• Regulations (EU) 2019/631 setting CO_2 emission standards for new passenger cars and light commercial: new, binding targets will reduce CO_2 emissions from road transport. CO_2 emissions per kilometer from passenger cars sold in the EU must be reduced, on average by 37.5%
Monaco	• Creation of park-and-ride facilities at border crossings and a wide range of alternative mobility solutions (gondolas, support for walking and cycling by strengthening equipment and e-services) • Public transportation will be gradually replaced to reach zero CO_2 emissions in 2030 • Support the replacement of internal combustion vehicles with electric vehicles • Electric helicopters could be brought into service by 2030 to provide regular services between Monaco and Nice (France) • Principality has banned the use of heavy fuel oil for navigation in its territorial waters and is deploying devices to supply electricity to ships in ports
Norway	• Tax exemptions and other advantages (advantage as free or low charges for toll roads, ferries and public parking) for electric vehicle • CO_2 dependent registration tax for new passenger cars including special rules for plug-in hybrid cars
Republic of Maldova	• Resilient urban infrastructure to reduce exposure to climate risks • Increasing climate resilience of transport infrastructure (roads, bridges, viaducts, railways, tracks)
Switzerland	• Strengthening existing fuel efficiency policies and measures for car
United Kingdom of Great Britain and Northern Ireland	• By 2030: ban sales of new petrol and diesel cars and vans • From 2030 onwards: all sold vehicles have a significant zero-emissions capability (e.g., plug-in and full hybrids) • From 2035 onwards: all sold vehicles have 100% zero emissions • Accelerate the roll out of charging infrastructure (rapid charge points on motorways and major roads and installing more on-street charge points near homes and workplaces) • Build first hundreds, then thousands, of miles of segregated cycle lane • Create more low-traffic neighborhoods to stop rat-running and allow people to walk and cycle • Financial support the electrification of UK vehicles and their supply chains (including developing "Gigafactories" to produce the batteries needed at scale) • Extend plug-in car, van, taxi and motorcycle grants to 2022–23 • Phase-out the sale of new diesel heavy goods vehicles (HGVs) • Freight trials to pioneer hydrogen and other zero-emission lorries • Introduce of at least 4000 more British built zero-emission buses
Cambodia	• 30 vehicle inspection centers in operation by 2030 • Promote integrated public transport systems in main cities • Shift long distance freight movement from trucks to train • Develop national road construction and maintenance design standards • Repair and rehabilitate existing road infrastructure and ensure effective operation and maintenance systems

(continued)

Table 6.9 (continued)

Japan	• Improvement of fuel efficiency • Promotion of next-generation automobile • Traffic flow improvement • Promotion of public transport • Modal shift to railway • Comprehensive measure for eco-friendly ship transportation • Reduction of land transportation distance by selecting nearest port • Comprehensive low-carbonization at ports • Optimization of truck transport • Energy consumption efficiency improvement of railways • Energy consumption efficiency improvement of aviation • Accelerated promotion of energy-saving ships • Making vehicle transport business more eco-friendly by eco-driving • Promotion of collective shipment • Promotion of intelligent transport systems ITS (centralized control of traffic signals) • Development of traffic safety facilities (improvement of traffic signals, and promotion of the use of LED traffic lights) • Promotion of automatic driving, eco-driving • Utilization of the special zones system for structural reform for global warming measures • Promotion of inter-ministry collaborative measures following roadmap of global warming measures, etc. • Promotion of car sharing
Maldives	• Establishment of vehicle/vessels emissions standard • Establishment of efficient transport management system • Promotion of hybrid vehicles
Mongolia	• Switch to Euro-5 standard fuel • Switch the transport of coal form road to rail • Switch the heating of passenger train to electric heating
Nepal	Electric vehicles: • By 2025: – Sales of e-vehicles will represent 25% of all the sales of passenger private vehicles – Sales of four-wheeler public passenger vehicles will represent 20% of all the sales of passenger private vehicles (not including electric rickshaws and electric-tempos) • By 2030: – Sales of e-vehicles will represent 90% of all the sales of passenger private vehicles – Sales of four-wheeler public passenger vehicles will represent 60% of all the sales of passenger private vehicles (not including electric rickshaws and electric-tempos) Other measures • By 2030: develop 200 km of the electric rail network to support public commuting and mass transportation of goods

(continued)

Table 6.9 (continued)

Republic of Korea	Zero-emitting vehicles: • Deploy 3 million EVs • Deploy 850,000 hydrogen vehicles Freight transport: • Modal shift in freight transport from road to rail and 12 shipping • Expand ships fueled by LNG
Singapore	• Promote public and shared transport and active mobility • Phase out internal combustion engine vehicles • Promote the adoption of LEV and ZEV • Enhance the environmental friendliness of the transport infrastructure
Vietnam	• Apply energy efficiency measures in transport • Change freight transportation models • Restructuring the transportation market • Shifting from private to public means of transport • Shifting from conventional fuels to biofuel, natural gas and electricity • Improving the energy efficiency of transport vehicles
Marshall Islands	• Changes to ship design by implementing energy efficiency measures (efficient propellers, sail assist, low-carbon fuels • Changes to ship operations by means of sea efficiency improvements • Changes ship fuel to sea coconut diesel blends Changes to ship docking facilities to reduce the amount of time (and fuel) required to load and unload ships
Papua New Guinea	• Promote clean fuel technology regulations to set standards for the GHG emissions, and economic incentives for fuel-efficient vehicles: – Encourage the introduction of fuel-efficient transport equipment – Encourage sustainable substitution of fossil fuels with biofuels – Monitor vehicle fleet-weighted fuel and CO_2 efficiency – Eliminate high emission vehicles – Establish low-carbon fuel standards • Encourage the introduction of hybrid and electric vehicles
Tonga	• 30% improvement in fuel efficiency for newly registered light-duty vehicles (through registration fees, import tariffs, or fuel economy standards) • 10% of all newly registered light-duty vehicles to be electric or hybrid vehicles by 2030
Rwanda	• Vehicle emissions standards introduced to increase vehicle emissions performance (including tax incentives and scrappage of older vehicles, and inspection) • Public transport infrastructure, including bus rapid transport (BRT), bus lanes, non-motorized transport lanes, and other modal shift projects Introduction of electric vehicles (EVs) from 2020 onwards (buses, passenger cars, motorcycles)
Senegal	• Diversification of transport modes, with recourse to rail and maritime transport • Increasing the use of sustainable public transit (bus rapid transit, regional express train) • Promotion of hybrid cars

(continued)

Table 6.9 (continued)

Colombia	• To introduce 600,000 electric vehicles (cab, vehicles of passengers, light vehicles, light trucks, official vehicles) • To achieve the use of the performance-based navigation system (PBN) in the 100% of airports and flights of the country • Motor freight modernization program: renew 100% of the heavyweight fleet by 2023 • Promote waterborne transport of 8.0 million tons in the Magdalena river by 2030 • Increase the modal share of transportation by bicycles to 5.5% in selected cities • Finance the development and implementation of transit-oriented development initiatives in selected cities to decrease the use of motorized vehicles
Costa Rica	• 2025: – Implementation of a sustainable logistics model in ports, cities and logistics centers – Adoption of standards to migrate the motorcycle fleet to zero emissions • 2030: – 8% of the public transport fleet will be zero emissions – The electric passenger train will enter into operation in the greater metropolitan area of San Jose, driven by renewable electric energy • Infrastructure will be improved to increase non-motorized trips (bicycle and walking) by at least 5%
Cuba	• 2030: introduction of more than 55 000 electric vehicles and the installation of around 25 thousand recharge stations
Dominican Republic	• New and extension of existing lines from the Santo Domingo metro • New cable car line • Implementation and adequation of BRT systems in Santo Domingo and in Santiago de los Caballeros • Replacement of the diesel bus fleet by electric buses • Definition and application of a policy of renovation of taxis and conchos. Modernization of the public vehicle fleet by electric and hybrid units • Adequation of the bicycle lane network by promoting the use of bicycles in the big cities • Creation of express bus lanes in big cities • Implementation of a vehicle inspection program
Ecuador	• Operation of the Quito metro (22 km) • Operation of the Cuenca Tramway (12 km)
Uruguay	• Mix of 5% of bioethanol in naphta, and 5% biodiesel in diesel • Implement mandatory energy efficiency labeling for light combustion vehicles until 2025 • Introduction of public transportation electric vehicles by 2025: 15 buses and 150 taxis • Introduction of 150 utility electric vehicles by 2025 • Installation of the first electrical route of Latin America, installing power systems for electric vehicles in the national routes that connect Colonia-Montevideo-Chuy • Mix of 10% bioethanol in naphta and 10% biodiesel in diesel • Introduce 110 electric buses and 550 taxis by 2025 • Introduce 900 utility electric vehicles by 2025 • Replacement of 5% of the light private vehicle fleet by electric vehicles • Introduce a network of fast-charging stations for electric vehicles in DC

While there is a steep increase of emissions from cities in recent years due mainly to rapid urbanization in many parts of the world, particularly in the developing regions in Asia coupled with increased delivery of energy and mobility services, numerous solutions to curtail the emissions have been conceived and road tested. These include climate action to optimize energy efficiency in buildings, decarbonizing the electricity grid, enabling next-generation mobility and improving waste management. Measures such as more stringent building energy codes, deep energy renovations including renewable energy generation, high performance of air conditioners and heat pumps have shown that it is possible to drastically reduce emissions from buildings to accomplish net zero-energy consumption or even be energy positive. Besides technological, policy solutions, behavior change and sufficiency considerations (e.g., sharing of services to meet needs while reducing collective emissions) are beginning to take shape. This chapter has listed a number of such technologies/measures that the cities could implement drawn from literature sources related to carbon market mechanisms, NDCs of countries besides IPCC and IEA literature. Enabling role that carbon markets and climate finance could potentially play to bridge the gap in meeting financing, technology and capacity development needs is highlighted. While the listing of climate actions represents the most common and impactful categories of action that cities can take to reduce GHG emissions, it is not meant to be exhaustive. Diversity in regulatory and economic practices, social and cultural practices require that policy measures and packages are unique to cities. Some measures such as energy labeling schemes have a reinforcing impact on other policy types assessed, such as loan, grants and energy efficiency obligations. Information and feedback policies can help facilitate effective implementation of other instruments. Finally, most policy instruments would be unable to function without policies enforcing standards for measuring efficiency in systems or buildings.

References

C40 Cities (2017) Focused acceleration: a strategic approach to climate action in cities to 2030

CDM (2018) Achievements of the clean development mechanism- harnessing incentive for climate action (2001–2018) Available via. https://unfccc.int/sites/default/files/resource/UNFCCC_CDM_report_2018.pdf. Accessed 1 Feb 2021

CDM (2019) Methodology booklet 11th edn. Available via https://cdm.unfccc.int/methodologies/documentation/meth_booklet.pdf. Accessed 1 Feb 2021

CDP (2019) CDP disclosure 2019. Available via: https://www.cdp.net/en/climate/carbon-pricing/carbon-pricing-connect. Accessed 1 Feb 2021

CPLC (2017) Report of the high-level commission on carbon prices, May 29, 2017.

GCA (2021) Global climate action NAZCA. Available via: https://climateaction.unfccc.int/. Accessed 5 March 2021

GlobalABC (2020) 2020 Global status report for buildings and construction: towards a zero-emission, efficient and resilient buildings and construction sector. Available via: https://globalabc.org/resources/publications. Accessed 1 March 2021

UNFCCC Secretariat (2020) Human settlements: sustainable low emission housing and building solutions. Available via: https://unfccc.int/sites/default/files/resource/Technical_Paper_2020-Low_Emission_Report.pdf. Accessed 10 September 2021

IEA (2018) The future of cooling: opportunities for energy-efficient air conditioning, OECD publishing. Available via: https://www.iea.org/reports/the-future-of-cooling. Accessed 10 March 2021

IEA (2020a) Energy technology perspectives. Available via: https://www.iea.org/reports/energy-technology-perspectives-2020. Accessed 15 Feb 2021

IEA (2020b) Tracking buildings 2020. Available via: https://www.iea.org/reports/tracking-buildings-2020. Accessed 5 March 2021

IMF (Dec 2019) Putting a price on pollution, finance and development 56(4)

IPCC (2014) Climate change 2014. Mitigation of climate change. Contribution of working group III to the fifth assessment report of the intergovernmental panel on climate change. Chapter 9, "Buildings". Available at: www.ipcc.ch/site/assets/uploads/2018/02/ipcc_wg3_ar5_full.pdf

NDC Registry (interim) 2021. Available via: https://www4.unfccc.int/sites/ndcstaging/Pages/LatestSubmissions.aspx. Accessed 1 March 2021

Wang Q, Lin J, Zhou K, Fan J, Kwan M-P (2020) "Does urbanization lead to less residential energy consumption? A comparative study of 136 countries". Energy 117765

World Bank (2020) State and trends of carbon pricing 2020. Washington, DC

Chapter 7
Innovative New Market Mechanisms from Project to Mitigation Activities in the Urban Context: A New Paradigm

Kwi-Gon Kim and Hee-Sun Choi

Abstract Market mechanisms are important instruments to achieve cost-effective mitigation and help to finance more green infrastructure projects, and thus a key topic of international negotiations on a new international agreement under the United Nations Framework Convention on Climate Change (UNFCCC). City-wide urban approaches to market mechanisms hold much promise and other interesting opportunities.

This chapter explores how existing CDM mechanisms can be reformed to assure transparency and environmental integrity market mechanisms for mitigation activities in the urban context. Following the introduction (Sect. 7.1), the second section of the chapter describes a framework for developing urban carbon markets in terms of procedure, methods and techniques, and tools. To make the second section vivid, the third section addresses an urban model of market mechanism in terms of (1) locally governed mechanism, (2) spatial dimension of the carbon market, (3) scoping and integration of sectors, and (4) baseline, additionality, and monitoring/verification. A new post-2020 can be an expansion of the CDM toward aggregated level mechanisms to build on a market mechanism for mitigation, through the strengthening of programmatic approaches. Different levels of mechanisms for scaling up of mitigation can be achieved by appropriate additionality and baseline determination procedures. The fourth section covers a holistic step-by-step guide for setting baseline, additionality determination, and aggregation procedures. Big data and blockchain technology can best complement other instruments which allow for robust additionality and baseline setting. The fifth section deals with a testbed approach under the urban carbon market mechanism.

Keywords Market mechanism · CDM mechanisms · International emission trading · Urban carbon markets · Carbon finance · Urban platform business · Carbon emission rights · DLT/blockchain technology · CDM program of

K.-G. Kim (✉)
Seoul National University, 1 Gwanak-ro, Gwanak-gu, Seoul, South Korea
e-mail: kwigon@snu.ac.kr

H.-S. Choi
Korea Environment Institute, Bldg B, 370 Sichenong-daero, Sejong, South Korea
e-mail: choihs@kei.re.kr

K.-G. Kim and M. Thioye (eds.), *Planning Climate Smart and Wise Cities*, The Urban Book Series, https://doi.org/10.1007/978-3-030-80165-6_7

activities · ICT infrastructure · Cross-effects · Transparency · Policy options · Scoping · Blockchain model · Cross-sector approach · Project design document · Additionality · Testbeds

7.1 Introduction: Urban Approach to Carbon Finance

Carbon markets are included in climate policy proposals at the national level. Yet, they are not included at the city level. Cities are important for new carbon market mechanisms. The urban approach to carbon finance is one of the several key components that the UNFCCC, World Bank, ICLEI, and their partners are working in response to the urgent need to develop new carbon market mechanisms.

We need local approaches to urban markets. The various types of planning platform models described in Chap. 1 could provide us with a wide range of opportunities for new urban carbon markets and carbon financing according to the Paris Agreement. The urban platform businesses are growing rapidly.

The rapid expansion in urbanization and urban dwellers whereby in about 30 years close to 79% of the world population will be living in cities. 5 billion people have mobile devices, the usage is increasing, and the ICT infrastructures have been capable of sharing information on every front: together, they can help us monitor energy, resources efficiency, and impact of GHG emissions. Urban areas currently use about 67% of the world's energy and account for over 71% of global GHG emissions.

It is now recognized that city is the main cause of global warming. However, at present, less than 1% of carbon is credited with carbon emission rights by greenhouse gas reduction activities in cities.

Perhaps this is the principal reason why reform of existing CDM is needed at this stage.

7.2 Framework for Developing Urban Carbon Markets

This framework helps to develop urban carbon markets for enacting green economic stimulus package that will create jobs, while reducing greenhouse gas emissions, expediting energy transition, green transition, and COVID-19 recovery transition, and enhancing climate resilience.

A framework for developing urban carbon markets should include:

- Standardized carbon crediting frameworks have the potential to substantially streamline CDM procedures for highly replicable projects including urban land use (e.g., transit-oriented development, infill/densification, urban forestry, and green space programs)
- As standardized baselines move beyond a pure-project-by-project approach to the definition of baseline, and additionality, they need to open the perspective to

extend the CDM to the crediting of policies, plans, and programs, and with that to include sectors and cross-sectors not yet really reached by the CDM.

- Crediting frameworks that define the scope of a methodological assessment of urban market mechanisms and overall volume of eligible carbon credits in advance of project implementation could facilitate the testing of new approaches for crediting activities in sectors that can be integrated into sector cohesion.
- The similarity between standardized baselines frameworks and the improved PoAs—in particular those dealing with microscale activities (and these are most of all PoAs)—is striking and shows the overall logic and consistency of the CDM reform process (World Bank 2010).

Some researchers connected to World Bank, World Resources Institute, Perspectives, and ICLEI are interested in a city-wide approach to carbon finance. The current chapter is in that respect building upon earlier project with Urban Environment Accords (UEA).

The future of urban carbon markets can be characterized by level of localization, spatial dimension of markets, scoping and integration of sectors, baseline, additionality and transparency, procedure of a new carbon market methodology, empirical analysis through testbed developments, registration of a new urban methodology and direct local benefits: co-benefits of mitigation and adaptation.

A conceptual framework is shown for developing urban carbon markets in Fig. 7.1.

7.3 Methodology: An Urban Model of Market Mechanism for an Integrated Holistic Approach

Within the framework described in Sect. 7.2, methodologies to assess the GHG mitigation potential of land use and land use-related sectoral mitigation activities in urban contexts can be developed under the new market mechanisms in the form of either the mandatory or voluntary carbon markets or maybe both based on a step-by-step approach.

An urban model of market mechanism is a mechanism which can be designed and blended with climate finance in order to mobilize for NDC implementation.

7.3.1 Localizing SDGs and Climate Actions for the Market Mechanisms: The Rationale for Urban Carbon Market Mechanisms

New urban carbon market mechanisms should use methodological approaches better aligned with national/regional circumstances. It should be better-integrated mechanisms in policy, planning, and program framework and with other sources of financing.

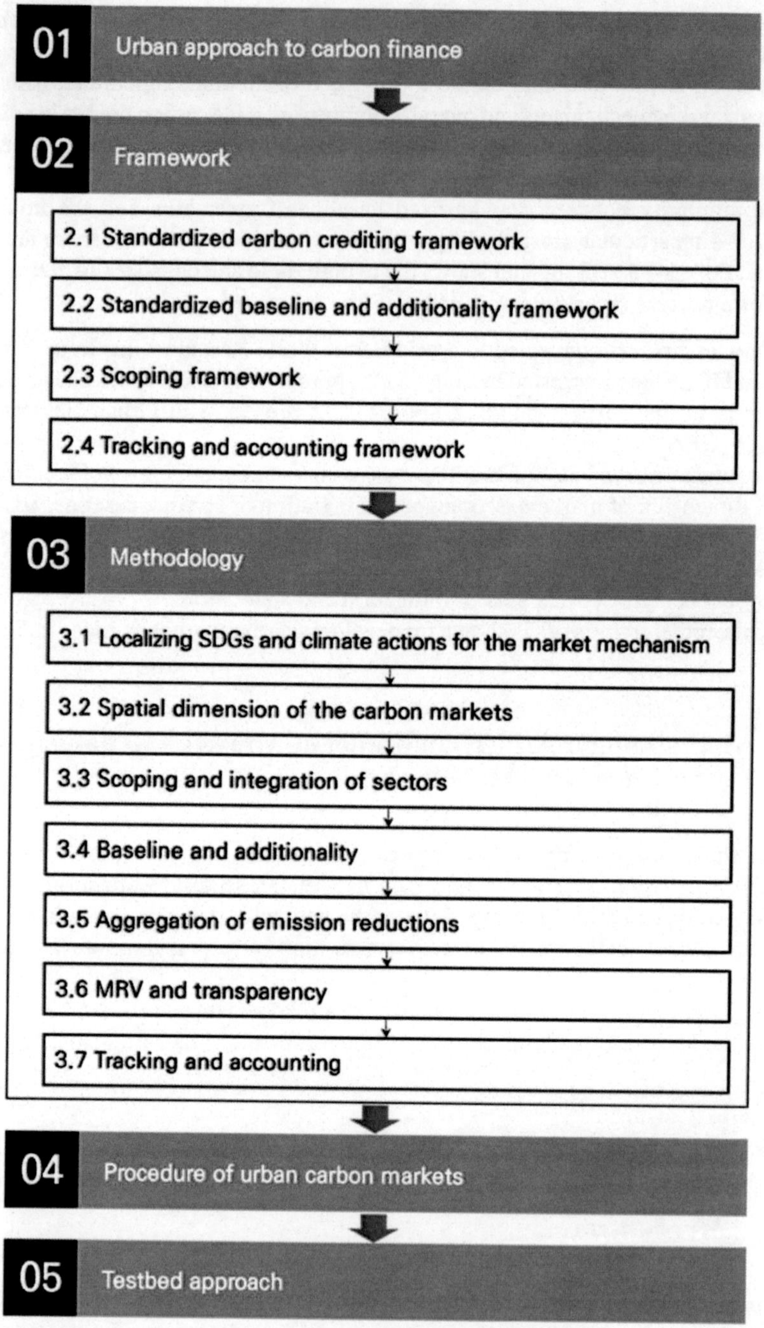

Fig. 7.1 A conceptual framework for developing urban carbon markets

The all-embracing sustainable development aims of post-2015-SDGs have first to be translated by local politicians and experts into their own languages and given tangible form. Another is that municipal planning commitments are made within a national statutory framework described in Chap. 1, and the laws have to be changed. Although local and regional governments can play a role within key international agreements such as the Post-2015 SDGs, Paris Climate Agreement, Fourth Industrial Revolution, and New Urban Agenda, cities are challenged with economic burden due to their financial vulnerability. The UNFCCC CDM movement has facilitated developing countries to take various climate actions. However, it is still rather piecemeal, fragmented and does not address specific city climate-related challenges in a holistic approach which could become one of the strategic areas of the current agreed international development agendas.

Smart technical developments in the era of the Fourth Industrial Revolution transform their potential into economic value domains for localizing the global agendas. A host of innovative projects and programs have been implemented to bridge connections between ICT and urban green developments. However, the urban market approach needs to include not only technical additionalities but also local policies and planning-related ones.

It is believed that cities can be empowered to catalyze a new comprehensive carbon market mechanism consistent with the principles of the current CDM and thereby support the implementation of the SDGs for reaching future sustainability of our cities and humankinds.

Under the SDG 7, city governments can contribute to energy efficiency directly by investing in energy-efficient buildings and green energy sources in public institutions (government offices, schools, etc.) and by introducing sustainability criteria into our procurement practices. In cities, local transport and urban land use planning policies as well as new smart digital technologies can have a significant impact on energy efficiency and carbon emissions. Furthermore, with regard to SDG 11, city governments have responsibilities for promoting the use of public transport in urban areas in order to improve road safety and reduce emissions. It is also city governments' job to provide citizens with safe, green public spaces, such as parks, squares, and gardens. In the context of rapid global urbanization, participatory urban planning is more important than ever if we are to prevent urban sprawl, tackle segregation, and reduce carbon emissions in cities. Sustainable solid waste management, with an emphasis on reuse, recycling, and upcycling, is also vital to the reduction of cities' environmental impact.

Concern should be expressed regarding the cost involved in translating the new development, in particular, climate smart and wise city development agenda, into reality. In this regard, the current CDM system should be considered as an interim economic tool to compensate climate change mitigation measures to be eventually replaced by a more streamlined procedure closely tied to integrated carbon-centered comprehensive (3Cs) urban planning (Kim 2014). It could make impact on our cities and thus can be our best interest.

7.3.2 Spatial Dimension of the Carbon Markets

Cities are recognized leaders in mitigating GHG emissions.

There is a need for an effectively coordinated carbon market mechanism that is necessary to integrate urban land use planning, transport, energy generation, building energy, solid waste, and wastewater and water efficiency with new technologies in a holistic manner. One of the attractive features of the urban methodology to the market mechanism is that it aims to develop and implement networked connections between projects rather than a project-based approach at the different levels of space (Fig. 7.2).

As described in Chap. 1, the spatial planning is often used in a generic sense to describe any land use or physical planning system and is centrally concerned with the coordination or integration of the spatial dimension of sectoral policies through a territorially based strategy. The strategy acts as a framework for the formulation and implementation of sectoral policy. One of the sectors is land use planning. In this sense, spatial planning seeks to identify the opportunities for synergy through the territorial strategy.

Good urban planning and spatial development can be thought of as proactive 'demand side' management for resource-efficient infrastructure systems. For instance, public transportation is financially viable only at certain threshold urban densities and is optimized under certain spatial layouts and land use patterns (World Bank 2010). Here, a matrix can be used to incorporate a list of GHG reduction activities by land-use types. A GHG reduction activity is identified at the interactions between an activity and a land-use type (Table 7.1).

Considering spatial dimension of the carbon markets, urban methodologies for CDM or Urban CDM hold much promise and other interesting opportunities as follows:

Firstly, in accordance with the principles and requirements of the Kyoto Protocol Mechanisms, urban methodologies could provide an organized approach to CDM for encouraging cities in developing countries to holistically contribute to emission reduction efforts, and helping cities in developed countries meet their emission targets at the city scale.

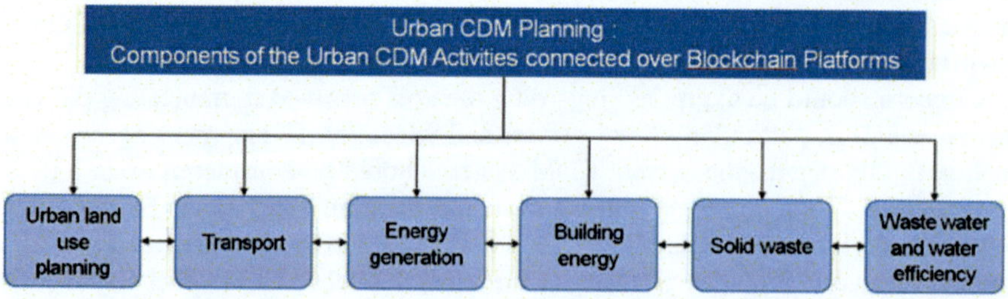

Fig. 7.2 Interconnections and integration of different sectors in a city for the urban methodology

Table 7.1 Interaction matrix between GHG reduction activities and land-use types: an illustration

Area		Residential	Commercial	Private buildings	Public buildings	Urban infra			Industry	Agriculture	Forest	Others
Sector						Water work	Sewage	Landfill				
Energy	Renewable (biomass)	Solar energy	Solar energy	Solar energy	Solar energy				Solar energy			
	Distribution	District heating and power system, digital data	Digital data	Digital data	Digital data	Digital data	Digital data	Digital data	Digital data	Digital data	Digital data	
	Energy saving (at home and at work city size)	Optimizing energy and resource use	Use of an area energy network for the efficient use of energy									
Solid Waste Management	Collection											
	Management (Heat Recovery)											
Sewage	Treatment											
	Reuse											
	Heat recovery											
Waterworks	Improvement of water supply					Smart water grid						
	Water saving											
Rainwater	Collection	• Rainwater penetration and storage rate(↑) • Rainwater evaporation rate(↑) • Rate of direct runoff of rainwater(↓)										
	Use											
	Rainwater garden					Blue-green infrastructure						

(continued)

Table 7.1 (continued)

Area		Residential	Commercial	Private buildings	Public buildings	Urban infra			Industry	Agriculture	Forest	Others
Sector						Water work	Sewage	Landfill				
Lighting	Replacement (including LED)											
	Newly set	Solar –LED lighting	Solar LED lighting									
New public transport		Electric and hydrogen cars	Smart parking						Transit-oriented development			
Land use		Compact city: land form, zoning reform				Risk-prone zones					NBS	
Spatial planning		High-density development(population density)			Intact public space							
Designing complex	Ventilation											
	Orientation											
	Landscaping											
	Passive design											
Wetland restoration	Marsh										Integra-ted system of natural infrastructure	
	Swamp											
Urban form	Horizontal city		Standards, code of practices									
	Vertical city											
Open space (ratio of open green space)											Carbon forest	
Buildings		Low energy retrofitting	Low energy retrofitting	Low-energy retrofitting	Low energy retrofitting							

(continued)

Table 7.1 (continued)

Area	Residential	Commercial	Private buildings	Public buildings	Urban infra			Industry	Agriculture	Forest	Others
Sector					Water work	Sewage	Landfill				
Low-carbon lifestyles	Socio-spatial integration										
Carbon-oriented urban regeneration	IoT smart homes	Infill development						Low-carbon smart zone			
Low-carbon Business Practices								Smart factory, carbon neutrality plan	Smart farming		
Use of natural climate control systems									Crop management	Carbon sink	
Smart grid system (use of an area energy network for the efficient use of energy)	Platform	Platform	Platform	Platform	Platform	Platform	Platform	Platform	Platform	Platform	
Composting							Recycling		Urban farming		

Secondly, nature-based solutions for climate emergency suggested in the 'UN 2019 Climate Action Summit' can be used for CDM. In cities, the restoration ecology is increasingly critical, as restoration science moves from built fabric to the urban 'natural environment' following the logic that degraded urban ecosystems may actually provide more economic opportunities than detriments.

Thirdly, emergence of carbon neutral eco-cities or climate smart and wise cities could provide the potential for CDM by using digital technologies. Climate smart and wise city approach to respond to climate change incorporates GHG reduction activities into low-carbon smart city development goals and planning framework. Climate smart and wise cities can be defined as ones that have digitalized connections of all sectors and functions of the city, as addressed in Chap. 1.

Fourthly, cities would have the flexibility to combine relevant technology options across different sections given their financial and development abilities, thus lowering transaction and administration costs that would otherwise have been required to set up individual carbon finance projects. Big data, deep machine learning, and AI in spatially connected cities are technologies for the market mechanism. They could be tools for CDM baseline and project scenario development through their prediction function. More and more advanced urban algorithms emerge every day to help us understand our city and predict the future in a holistic manner. Big data can be extensively used to support the urban methodologies. DLT/blockchain technology also can be used for GHG emission tracking and carbon accounting.

Fifthly and finally, in short, the proposed methodologies can be a mechanism which gives financial incentives and provides CERs as much as certified amount of cross-sectoral emission reductions in the field of urban land-use planning, renewable energy, building energy efficiency, transportation, waste and water at the adequate level of aggregation compared with baseline emissions at the city- scale. Urban carbon market planning directs each of the different sectors of a city and needs overarching considerations with their interconnections to maximize the efficiency and circularity of each of them (Table 7.2).

With the help of the GHG inventory and quantification of emission reductions from the implemented intervention, as shown in Tables 7.1 and 7.2, cities can plot their emission trajectory and control their cumulative impact on the local and global environment.

Many people acknowledge that realizing value from technological advancement is a complex and dynamic process that involves many spatial, economic, and social dimensions. Notwithstanding, we still have a limited understanding of how the potential of the technical developments of the Fourth Industrial Revolution can be harnessed to generate more economic value through localizing the Post-2015 Sustainable Development Goals (SDGs) and the Paris Climate Agreement in the urban space.

Table 7.2 Spatial dimension of urban sectoral interactions: potential for climate policy complementarity and tradeoffs: an illustration

Impact → (read horizontally)	Land-use zoning	Transportation	Natural resources	Building	Renewable energy	Waste and water
Land-use zoning: land-use zoning determines the density, height of buildings, and proportion of undeveloped land on each property		Segregation of land uses impacts travel distances and frequency; transit-oriented development zones encourage use of mass transportation	Zoning designates natural resource areas that may be set aside to reduce vulnerability to flooding or urban heat island effects	Zoning impacts placement and density of buildings, which in turn impacts building energy efficiency and vulnerability to flooding and urban heat effects	Zoning density can constrain on-site renewable energy production but can also increase efficiency of service delivery	Zoning density can determine the efficacy of delivery of waste, recycling and composting services; and the energy required for and efficacy of delivery of water services
Transportation: transportation policies determine the development and extension of road and mass transportation networks	Transportation infrastructure policies shape demand for land and acceptance of density increases		Transportation system impacts natural resource and preserved zones		Transportation policies can require renewable energy sources for mass transportation systems	
Natural resources: natural resource policies determine which areas are preserved from development and what uses are acceptable on them	Natural resource policies determine the limits of developed land-use zones	Natural resource policies affect the placement of road and mass transportation infrastructure	Natural resource policies can contribute not only to the carbon and biodiversity benefits, but also to the wellbeing and prosperity of people		Natural resources endowment makes certain renewable energies possible	

(continued)

Table 7.2 (continued)

Impact → (read horizontally)	Land-use zoning	Transportation	Natural resources	Building	Renewable energy	Waste and water
Building: building policies, including building codes, affect building materials, construction types, and other physical conditions	Building codes can increase acceptability of high-density zones by requiring design features to improve quality of high-density structures				Building codes can require the on-site generation of renewable energy	Building codes can require design and building materials that produce less construction waste and reduce water consumption in buildings
Renewable energy: renewable energy policies can increase on-site renewable energy production and share of energy produced by renewable sources				Renewable energy-connected smart grids can help citizens manage their own community and home energy schemes, and ultimately reduce their carbon footprint		Renewable energy production can involve high water consumption
Waste and water waste policies determine the means and extent of waste disposal. Water policies determine service extent, pricing, and water sources		Disposal of refuse in an optimal way can cut down wasted journey and unnecessary CO_2 emissions				Smart water grid can help to solve a number of problems related to water shortage, climate change and integrated management of water

7.3.3 Scoping and Integration of Sectors: A Standardization of Scoping Process

This section aims at categorizing and at identifying sectors and cross-sectors for measuring the dependence and impact, taking into consideration how such metrics can be integrated into sector cohesion.

There is a large number of sustainable projects in cities that generate a real positive impact on our planet and for bureaucratic reasons or lack of capital are placed outside the carbon market. Climate trade (http://35.180.169.143/) wishes to give these projects space on their platform, fostering a wide range of potential and true positive projects that will really make a difference in the goal of achieving global carbon neutrality.

The scope of new urban carbon market mechanisms should be broadened from sectoral perspective and play key role of national and local governments.

Growing awareness of the devastating impacts of climate change adds a new angle to urban infrastructure and resource management for saving the planet. Emergence of climate smart and resource-wise cities provides an opportunity for the introduction of innovative new market mechanisms from project to mitigation activities in decarbonizing cities and its benefits. An effectively coordinated carbon market mechanism is necessary to integrate land use and transport, energy generation, building energy, solid waste, and wastewater and water efficiency with new technologies in a holistic manner. Hegde (2018) has suggested a list of methodologies relevant to urban transport, urban land use planning, urban energy generation, urban building energy, urban solid waste, and urban wastewater and water efficiency (Table 7.3).

One of the attractive features of the Hegde list is that it includes urban land use planning in terms of low carbon and green land use and efficiency. The measures are transit-oriented development, infill/densification, and urban forestry and green space programs. There are currently no CDM methodologies available for land use planning, waste collection and transportation energy, and waste to energy.

The following list of items represents some very practical ideas and suggestions that can be useful in standardizing scoping process:

- Develop list of sectors/CPAs for any project.
- Select sectors/CPAs for consideration in the decision process.
- The sectors/CPAs can be selected depending on the availability of methodologies approved by UNFCCC CDM EB.
- Develop the interactions between the selected sectors.
- Choose optional combination of CPAs for the cross-sectoral integration with the greatest sum value.

In the scoping process, the optional combination of sectors/CPAs depends heavily on practical reality. Then, the combination can be expanded to cover more categories or sectors/CPAs for the implementation.

As an example, option 1 is to add a new CDM program of activities (CPA) to an existing CPA. On the other hand, option 2 is to combine two new CPAs and

Table 7.3 List of urban sector, measure, and CDM methodology

	Measure	CDM methodology
• Urban transport		
Mode shift	Build rapid transit system(bus, tramway, subway)	**Small-scale**: AMS-III.U, AMS-III.BM **Large-scale**: ACM0016, AM0031
	Operational upgrades for existing transit (Mapping, route optimization, ticketing upgrades)	
	Travel demand management (car sharing, congestion charging zones, parking restrictions and cycle hire program)	
	Bicycle infrastructure upgrades	
	Pedestrian infrastructure upgrades	
	Parking infrastructure management to disincentivize private car use	
Fuel switch and fuel efficiency	Public vehicle electrification	**Small-scale:** AMS-III.C, AMS-III.S, AMS-III.T, AMS-III.AA, AMS-III.AK, AMS-III-AP, AMS-III-AQ, AMS-III-AY, AMS-III-BC
	Taxi electrification	
	Electric vehicle infrastructure upgrades	
	Transit electrification	
	Bus electrification	
	Bus fuel switching	
	Private electric vehicle incentives (reduced fees, parking, tolls, taxes)	
	Low emission truck incentives (reduced fees, taxes)	
	Freight (delivery) system improvement (freight consolidation centers and improved logistic info, last mile bike delivery)	
• Land use planning		
Low carbon and green land use and efficiency	Transit-oriented development	
	Infill/densification	
	Urban forestry and green space programs	
• Energy generation		
District heating/cooling	Establish district heating/cooling	**Small-scale**: AMS-II.B **Large-scale**: AM0044, AM0058, AM0072, AM0117
	Efficiency improvements in district heating/cooling	
	Fuel switching in district heating/cooling	
Municipal renewable energy	Municipal renewable energy installation (solar, wind, biomass, etc.)	Heat for cooking, water and space (RE) **Small-scale**: AMS-I.I, AMS-I.J, AMS-I.K, AMS-III-AC Heat for cooking, water and space (EE) **Small-scale**: AMS-II.A, AMS-II.K
	Incentive program for residential and commercial building integrated renewable energy	
	Municipal clean energy procurement	

(continued)

Table 7.3 (continued)

• Urban transport		
	Measure	CDM methodology
• Building energy		
Private buildings	Establish mandatory building energy codes	Building efficiency/Renewable energy **Small-scale**: AMS-III.AE, AMS-I.F, AMS-I.J **Large-scale**: AM0091 Appliances **Small-scale**: AMS-II.C, AMS-II.E, AMS-II.J, AMS-III-M, AMS-II.N, AMS-II.O, AMS-II.Q, AMS-II.R, AMS-III-AV **Large-scale**: AM0046, AM0070, AM0091, AM0113, AM0060, AM0086
	Residential EE existing buildings retrofits (lighting, envelope, heating, cooling, water) incentives or support	
	Residential new buildings green building standards	
	Commercial EE existing building retrofit incentives or support	
	Commercial new buildings green building standards	
	Establish energy reporting, benchmarking program for residential and/or commercial buildings	
	EE appliances incentive program	
	Fuel switching	
	Centralized chiller energy efficiency	
	Rooftop solar PV	
	Rooftop solar water heating	
Municipal buildings	School building retrofits	Building efficiency/Renewable energy **Small-scale**: AMS-III.AE, AMS-I.F, AMS-I.J **Large-scale**: AM0091 Appliances **Small-scale**: AMS-II.C, AMS-II.E, AMS-II.J, AMS-III-M, AMS-II.N, AMS-II.O, AMS-II.Q, AMS-II.R, AMS-III-AV **Large-scale**: AM0046, AM0070, AM0091, AM0113, AM0060, AM0086
	Municipal rooftop solar PV	
	Municipal rooftop solar hot water	
	Municipal green building certification program	
	Municipal building EE retrofits (lighting, building envelope, heating, cooling, water)	
	Building energy monitoring	
	Fuel switching	
Municipal public lighting	LED street lighting replacement	**Small-scale**: AMS-II.L
• Solid waste		
Waste management (sorting, recycling, composting, treatment)	Residential/commercial /industrial recyclables collection	**Small-scale**: AMS-III-AJ, AMS-III-BA
	Residential/commercial/industrial food waste collection and compost	**Small-scale**: AMS-III.E, AMS-III.F, AMS-III.G, AMS-III.O, AMS-III.AF, AMS-III-AO, **Large-scale**: AM0053, AM0057, AM0069, AM0083, AM0093, AM0112, ACM001, ACM0022, ACM0024

(continued)

Table 7.3 (continued)

	Measure	CDM methodology
• Urban transport		
Waste collection and transportation energy	Waste transport vehicle efficiency/fuel switching	
	Waste transport vehicle routing/reduce km traveled	
	Waste treatment site operational efficiency	
Waste to energy	Install waste to energy plant	
• Wastewater and water efficient		
Wastewater	Wastewater treatment switching and optimization	**Small-scale**: AMS-III.H, AMS-II.I **Large-scale**: AM0080, ACM0014
	Biogas capture and use	
	Wastewater capture and reuse/recycling	
Water conveyance/delivery	Efficiency improvements (loss reduction, pumping)	**Small-scale**: AMS-III.AV **Large-scale**: AM0020, AM0086

more (Fig. 7.3). A LED project and a solar panel project are in the same category of activities, i.e., renewable energy, but a LED project and a geothermal heating project are in different categories (Fig. 7.4).

Figure 7.4 is designed to analyze possibilities of option 2 of Fig. 7.3, combining a new/expanded CPA (solar panel or LED lighting project) with a new CPA (geothermal heating project) as a pilot-testing project. It is important to note that once a PoA for LED or solar panel + geothermal is registered, the boundary can be expanded to cover new cities and countries. Letter of approval (LoA) from each of the countries where the country is expanded will be needed. Also, the PoA for more than one country can be registered if the LoA is available from all countries (UNFCCC, unpublished paper, 2013).

This process incorporates spatially networked connections between sectors for the alternatives under consideration for meeting a given need. This scoping process can be conducted in line with the procedures for urban methodologies set out in Sect. 7.4 of this chapter.

The innovative planning models described in Chap. 1 can offer more opportunities for new CPAs. Table 7.4 compares interconnected grid platform technology before and after the urban methodology.

Among other sectors, transport sector can be decarbonized for the net-zero energy transition and shared mobility. There are unlocking opportunities for transport decarbonization.

In terms of transportation open innovation from the angle of the urban methodology, new IoT-based technologies are allowing transport operators to make both short- and long-term changes to the city's transport systems to user's real needs, monitoring, and the constantly changing situation on the grounds, and using the data collected and insights produced to more effectively plan out the city's future transportation systems. As emerging tech creates new urban space innovation, we have to take a look at the transition to more sustainable, smarter, and resource wiser city. For example, the massive allocations of scarce urban land to parking should be avoided for the new world of safer and cleaner mobility. We must look at how intelligent

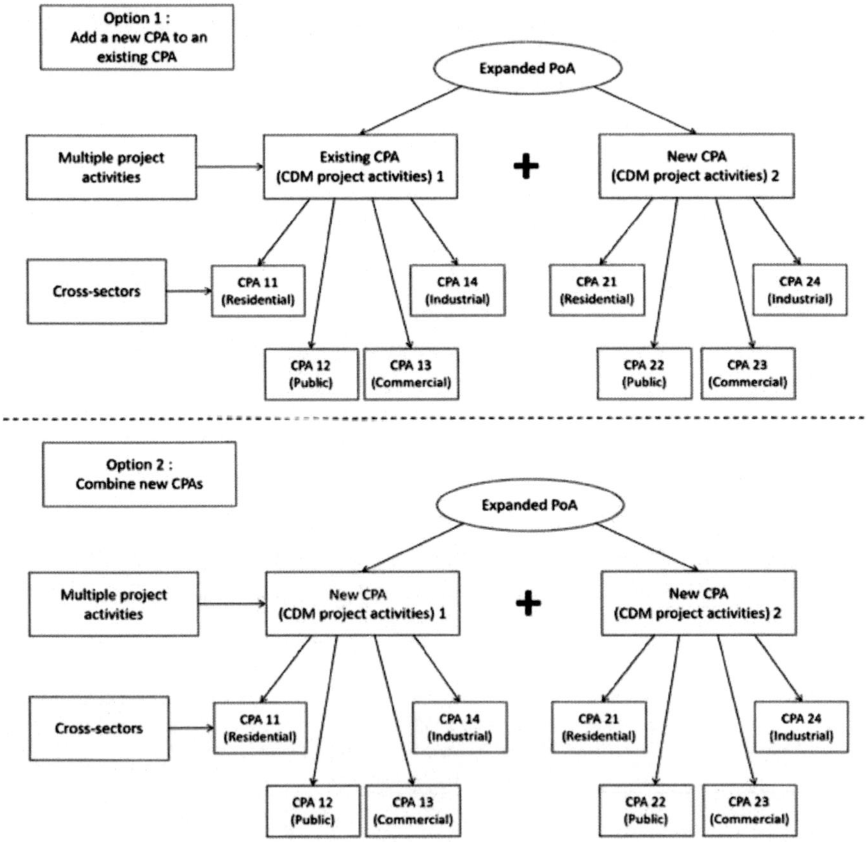

Fig. 7.3 Combination of different categories, sectors and CPAs

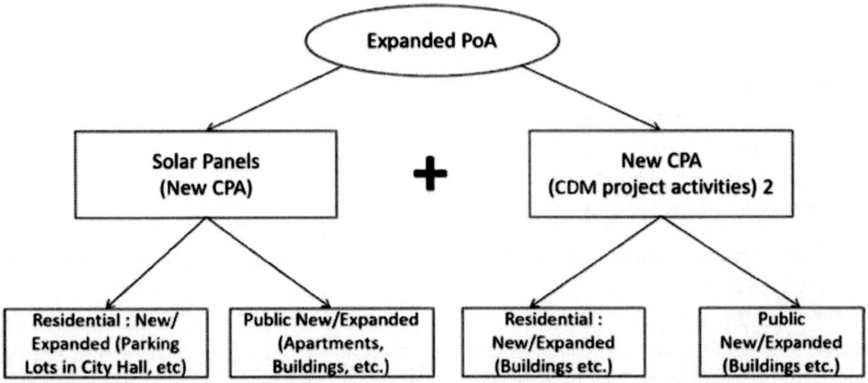

Fig. 7.4 Optional combination of two different categories of project activities

Table 7.4 Comparison in interconnected grid platform technology before and after the urban methodology

Before urban CDM	After urban CDM
Smart energy grid platform	• City-wide smart grid platform – Big data platform – AI smart city open platform – Smart digital sensor grid platform • Sectoral smart grid platform – Connected and autonomous vehicle(CAV) network platform – Conceptual urban 100% renewable energy grid platform – Green–blue grid platform – Integrated rainwater and drainage grid platform – Wetland connectivity grid platform • Community smart solar microgrid platform – Community smart solar microgrid platform without ESS – Community smart solar microgrid platform with ESS

Table 7.5 Comparison of transportation techniques before and after urban CDM

Current technologies for CDM	Future innovative techniques for urban CDM
Hydrogen car-based smart city (2050)	• Mobility innovation: eco-mobility network of car, rail, bike, pedestrian, scooter, etc. • Integrated transportation • Car sharing • Autonomous cars • Flying taxi (2029) • eBike (Electric bicycling) • EV and charging stations

vehicle technology, and iRoad will impact the design of future cities. For example, driverless cars will enable:

- Rethink of how our urban centers are designed and built.
- Relocation of parking lots and gas stations. They will be converted into parks and green spaces.
- Reduction of single passenger use.
- Reduction of curbside parking.
- Making everything electrical, point-to-point, pollution-free.
- Maintaining of freshness of data.

A comparison of transportation techniques before and after the urban methodology is shown in Table 7.5.

One of the attractive features of the proposed urban market mechanism in terms of scoping process is that it can be expanded or contracted from the list of methodologies (Table 7.3) to achieve the greatest degree of compliance with the umbrella considerations for the implementation of the urban market mechanism below.

• Project activities	• Technology selection
• Project boundary	• Accuracy
• Transparency	• Accountability
• Cross-effects	• Step-by-step approach: Scaled-up approach starting with testbeds at the community level

7.3.4 Baseline and Additionality

The baseline is the estimated or measured emissions from the electricity, fuel and/or water consumption of technology/technologies that would have been used in the absence of the project activity.

Non-mandatory standardized baselines can be developed for baseline setting and additionality demonstration for entire sectors and cross-sectors in cities, submitted by Designated National Authority (DNA) and approved by the CDM Executive Board (EB).

Closely related to the progress made on standardized baselines are decisions on simplifying additionality for micro-scale projects and on improvements on PoA procedures. Micro-scale activities including, e.g., efficient cookstoves, solar water heaters, lighting, in the areas of households, communities and small- and medium-sized enterprises are by EB62 Guidelines for demonstrating additionality of micro-scale project activities deemed to be automatically additional (positive list). These kinds of activities will typically be done within the framework of a PoA, and therefore, it is critical that the positive approach to additionality is also fully applicable to CDM program activities (CPAs) within PoAs. This is currently however not the case. The EB63 approved standard for application of multiple CDM methodologies for a program of activities permits different kinds of methodology combinations in the CPAs included in a PoA. This decision is critical to broader policy-driven activities including city-wide programs where it is not possible to define ex ante the combinations of measures (out of a range of pre-defined possibilities) that will finally be implemented under the PoA (World Bank 2010).

For the urban methodology proposed in this chapter, big data and AI technologies can be useful tools for baseline setting and additionality demonstration with some degree of transparency (Table 7.6).

An innovative baseline and additionality platform is described in '7.3.7' for enhancing the current EB62 Guidelines with sensors, data, and AI.

Table 7.6 Big data and AI technologies for the urban methodology in the era of AI

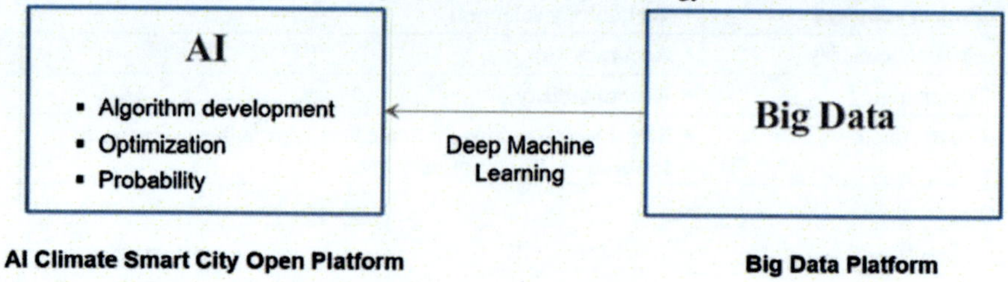

7.3.5 *Aggregation of Emission Reductions*

World Bank (2010) has suggested three methods to calculate and aggregate emission reductions for each sector:

- 'Measurement'-based approach that involves the aggregation of individual interventions using approved CDM methodologies;
- 'Estimation'-based approach that allows aggregation of activities using energy or GHG intensity-based methodologies. A city may use voluntary carbon market methods or sector-specific common practices to assess GHG reductions in a given sector; and
- Combined 'measurement' and 'estimation' approach that allows a city to choose between CDM methodologies and other accepted emission estimation methodologies (Fig. 7.5).

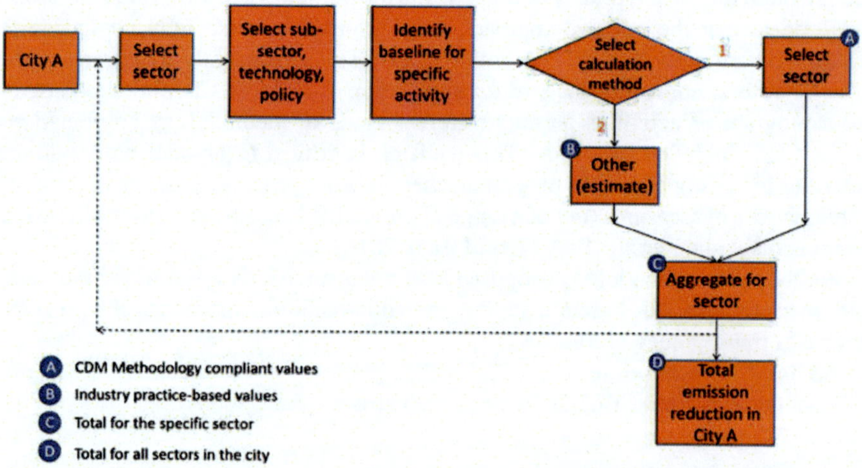

Fig. 7.5 Aggregation of emission reductions for the urban methodology. *Source* World Bank (2010) a city-wide approach to carbon finance, p 27

Table 7.7 Overall aggregation of GHG emission reductions by the proposed climate actions

Component project of activities (CPA)	Algebraic sum of emissions reductions
CPA1: Energy Generation	
CPA2: Building Energy Efficiency	
CPA3: Water Saving	
CPA4: Transportation (bicycling)	
Total emissions reductions	

Unit: $ktCO_2$

Spatial planning and cross-sector-wide approaches such as smart grid platforms which are across more than one sector need flexible calculation methodologies. Integrated platform approaches should be developed but are not elaborated in this section.

An algebraic sum of emission reductions by each project activity can be calculated and shown in the form of Table 7.7.

7.3.6 MRV and Transparency

There are the requirements for baseline and monitoring assessment required by CDM methodologies. While different sectors and technologies require different monitoring methods and procedures, the establishment of an overall framework, including measurement (on-the-ground monitoring), data storage, information routing, and reporting schedules would streamline the entire process of monitoring and reporting for verification and validation.

This process is described in the next section focusing on digital methodology (platform) for MRV with transparency.

7.3.7 IoT-Driven Climate Action Tracking and Accounting: A Blockchain-Based Carbon Market

More holistic approach which integrates all sectors into one common approach with IoT-driven methods will increase co-benefits of climate mitigation and adaptation. Cities can benefit from the new policy on blockchain-based urban market mechanisms as follows:

- Enabling cities and climate defenders to innovate by considering climate actions as essential to the urban planning and development.
- Keeping cities' climate actions continuously through financial gains by adopting and practicing the urban market mechanism.

- Supporting cities in increasing their access to carbon markets through the urban carbon market.
- Encouraging cities to raise citizens' awareness about climate change issues by sharing blockchain-based database on city climate status and future planning with regard to actions to be taken at the city and community level.
- Contributing to new climate economy for climate smart and wise cities.

This section aims at offering the hybrid of virtual and physical model of carbon finance from urban methodological point of view and at using a standard data platform project model for climate action tracing and carbon accounting.

Climate smart and wise cities, and the digital and carbon technology behind them hold much promise and other interesting opportunities for new carbon markets, including the urban methodology for carbon finance, services, and practices to the cities' stakeholders. The relationship between types of CDM and IoT digital technologies is shown in Fig. 7.6. The best available IoT digital technologies can be used for the cross-sectoral approach based on urban connectivity principles and market-ready smart tech in a holistic way.

Blockchain or distributed ledger technologies (DLT) have received wide attention in recent years as an innovative solution to provide a reliable, incorruptible, decentralized database that allows transactions between individuals who do not know and therefore do not trust each other. In recent years, blockchain technology has been tested in various international development projects, e.g., for financial services such as the transfer of remittances, peer-to-peer electricity trading, insurance, and the management of land titles and identities. However, currently, there is very little reliable information available about the impacts and lessons-learned from these applications (Burg J, Murphy C, Petroud JP 2018). The blockchain could be the missing link that brings consumers, businesses, and investors together on climate change.

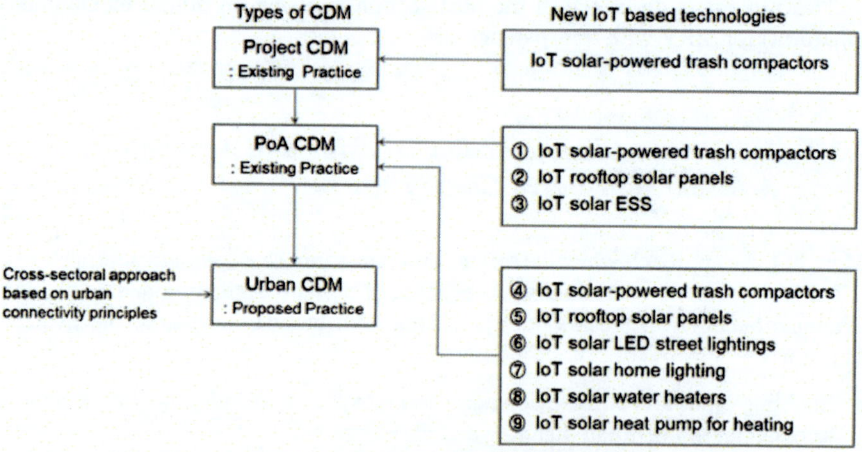

Fig. 7.6 Relationship between types of CDM and IoT digital technologies

Kim (2019) has designed four IoT/blockchain platform models for new market mechanisms (NMMs) and presented them at the UNFCCC COP 25. There are several variations of the Kim Blockchain model utilized for carbon financing for various types of projects.

Blockchain model A for carbon financing (Fig. 7.7) shows a GHG matrix for describing the relationship between typical measures and their potential ultimate impacts on urban development activities. The basic philosophy of the analysis is that a meaningful use of blockchain technology must ultimately be a tool for the determination of direct and indirect impacts on GHG emission reductions at the city scale.

Blockchain model B involves the three steps for tracking of emission reductions of each CPA, GHG reduction measures, and aggregation of each emission reduction by the measures (Fig. 7.8).

In blockchain model C (Fig. 7.9), each block illustrates different digital functional layers of climate markets which are connected to each other. Blockchain technology can synthesize and support the transaction of all types of GHG emission-related data (e.g., climate measures, additionality) in a shared visualized database.

Blockchain model D shows how ledger for urban methodology procedure can be distributed to stakeholders with the power of blockchain for the transparency tracking and control of urban carbon financing implementation (Fig. 7.10).

Urban CDM Project Elements which reduce GHG emissions		Urban development activities which impact climate					
		Land use planning	Transport	Energy generation	Building energy	Solid waste	Waste water and water efficiency
Measures	Transport (e.g. Transit oriented development)	Major GHG reduction	Major GHG reduction	Major GHG reduction	Minor GHG reduction	Its magnitude cannot be determined	Its magnitude cannot be determined
	Building Energy (e.g. Mandatory building energy code)	Major GHG reduction	Minor GHG reduction	Major GHG reduction	Major GHG reduction	Minor GHG reduction	Minor GHG reduction
	Land Use Planning (e.g. Infill/densification)	Major GHG reduction	Major GHG reduction	Major GHG reduction	Minor GHG reduction	Minor GHG reduction	Minor GHG reduction
	Energy Generation (e.g. District heating & cooling)	Major GHG reduction	Minor GHG reduction	Major GHG reduction	Major GHG reduction	Minor GHG reduction	Its magnitude cannot be determined
	Solid Waste (e.g. Waste transport vehicle routing reduces km traveled)	Major GHG reduction	Major GHG reduction	Major GHG reduction	Minor GHG reduction	Major GHG reduction	Its magnitude cannot be determined
	Waste water and water efficiency (e.g. Wastewater switching treatment and optimization)	Minor GHG reduction	Minor GHG reduction	Major GHG reduction	Major GHG reduction	Its magnitude cannot be determined	Major GHG reduction

Fig. 7.7 Blockchain platform model A for carbon financing

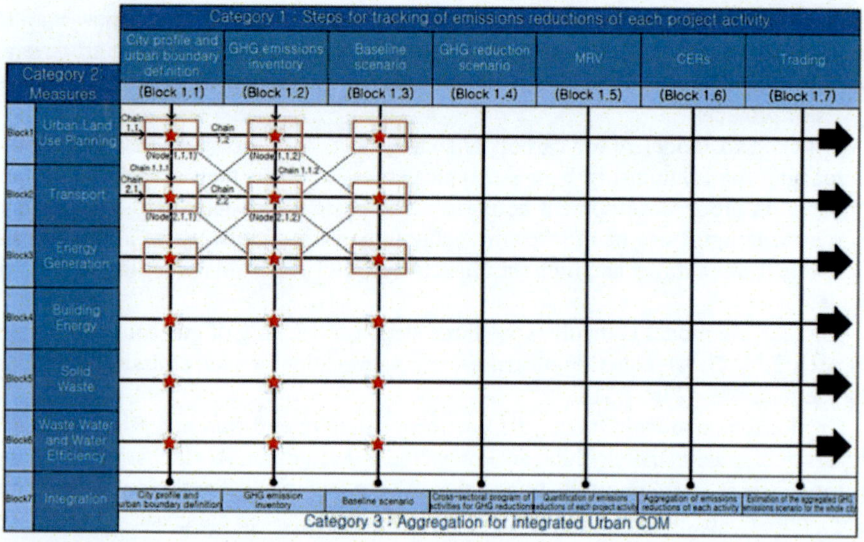

Fig. 7.8 Blockchain platform model B for carbon financing

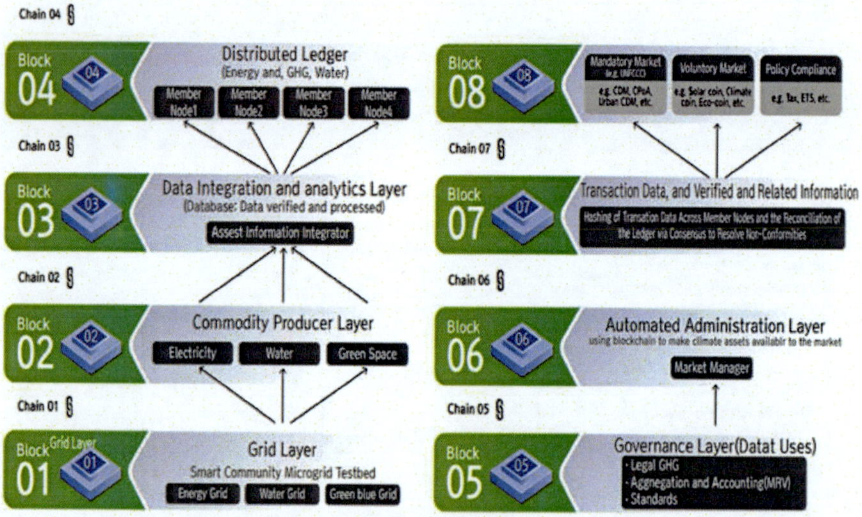

Fig. 7.9 Blockchain platform model C for carbon financing

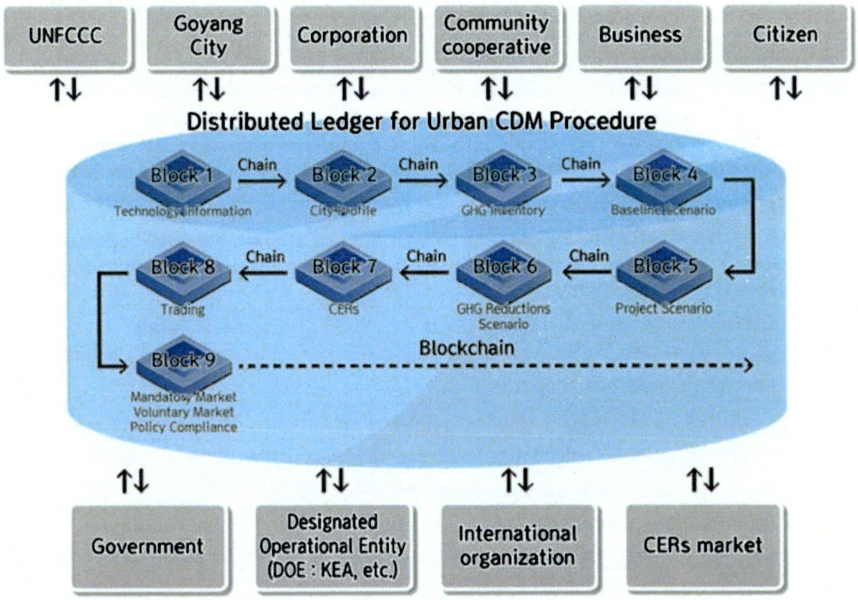

Fig. 7.10 Blockchain platform model D for carbon financing

7.4 Procedure of Urban Carbon Markets: A Step-By-Step Guide

An urban methodology to ensure environmental integrity suggested by Massamba (2019) considers ten steps that represent basic procedural requirements for urban carbon markets:

1. Identify the 20% type of mitigation activities that account for 80% of the potential of GHG emission reduction relevant to a city (e.g., clean cooling, upgrade of building performances, use of renewable electricity, scale-up use of EV, waste management).
2. Consider the sub-sectors where these GHG emission potentials lay.
3. Propose a model/approach to build for each of these sub-sectors an absolute emission pathway combining a dynamic benchmark emission intensity and a level of production output.
4. Request a description of each mitigation activity and a definition of how it will impact the baseline emission pathway.
5. Propose an approach for the monitoring of the mitigation activities undertaken in the subsector.
6. Propose an approach for the inventory of these in these sub-sectors.
7. Propose an approach for the monitoring of the actual key parameters of the models (GDP of the city, inhabitant, level of output of the targeted sectors).

Table 7.8 Steps for the blockchain-based integrated procedure for carbon financing

Step 1	Establish urban boundary (on-site and off-site)
Step 2	Identify sources of GHG emission and prepare GHG emission inventory report: Detailed steps and tools should be developed to complete a greenhouse gas inventory. The most important and time-consuming step in inventory development is the identification and quantification of emissions sources, as a lot of data is needed
Step 3	Formulate baseline scenario
Step 4	Identify the best available cross-sectoral program of activities for GHG reduction for the whole city: The program of activities for built form and urban infrastructure should be selected
Step 5	Deploy selected program of activities in the master plan at the city-scale
Step 6	Formulate planning policy and development mechanisms for implementation of the selected program of activities
Step 7	Quantify emission reductions of each program of activities: Measure or estimate
Step 8	Aggregate emission reductions of each program of activities for the whole city
Step 9	Develop a 'program of activities design document (PDD)'
Step 10	Estimate the aggregated GHG emission reduction scenario of planned program of activities
Step 11	Implement and monitor the activities
Step 12	Evaluate carbon effectiveness of existing CDM methodologies and newly proposed CDM methodologies: Economic feasibility analysis
Step 13	Validate or verify ER benefit from urban CDM

8. Propose an approach for the adjustment of the baseline emission pathway based on the monitored key parameters.
9. Measure the GHG emission reduction.

Kim (2018) has developed a blockchain-based integrated procedure for carbon financing and presented it at the side event, UNFCCC COP24 in Katovice, Poland. Table 7.8 outlines thirteen steps that urban carbon market mechanism developers should follow for carbon financing project development and operation.

The basic steps associated with project design document (PDD) for the proposed urban methodology for carbon financing are shown and compared with the PDD for conventional CDM in Table 7.9.

7.5 A City-Level Testbed Approach: Testing Standardized Baseline and Additionality Determination Under the Urban Carbon Market Mechanism

This proposal is an application for urban methodology which is a form of proposed new small-scale methodologies. It is a pilot project proposal at the city level toward an iconic carbon neutral eco-city which is a global dream. This proposal provides a

Table 7.9 Procedures for project design document (PDD)

Step and activities of conventional CDM		Urban Methodology	
Step	Activities	Step	Activities
Step 1: Using existing single project methodology or development of single project methodology	1. Project overview 2. Baseline methodology 3. Set the duration of project implementation and of CERs issuance 4. Monitoring methodology 5. Calculation of GHG emissions 6. Environmental effect evaluation 7. Comments from stakeholders	Step 1: development and registration of new methodology for urban CDM	Standardization of Urban financing methodology and registration to UNFCCC as a New project (This requires development or aggregation of methodologies available to use at a city-scale.)
		Step 2: planning	1. Formation of Project Team
			2. Capacity building/training for cities
		Step 3: project design	3. Identification of participating cities
			4. Establishment of baseline scenario
			5. Calculation of the emissions expected without urban methodologies
			6. Development of urban methodologies
			7. Formulation of policy actions and development plans at the city scale
			8. Preparation of urban methodology management plans
			9. Establishment of monitoring methodology to measure the actual level of emissions with PoAs
			10. Establishment of aggregation methodology
			11. Establishment of quantification and estimation methodology for cumulative volume of the emissions reduced with appropriate levels of aggregation
			12. Preparation of Project Design Document (PDD)
	*Note: Procedure for CDM project using existing methodology		*Note: Procedure for registration of new methodology

(continued)

Table 7.9 (continued)

Step and activities of conventional CDM		Urban Methodology	
Step	Activities	Step	Activities
Step 2: CDM national approval & validation		Step 4: national approval	13. Approval of PDD by designated national authority (DNA)
		Step 5: validation	14. Validation of designated operation entity (DOE)
			15. Request for registration
Step 3: CDM registration	2. Apply for registration of CDM project to CDM executive board (CDM EB) 3. Registration fee for CDM project varies according to projected average annual emission reductions	Step 6: registration	16. Registration of PDD by CDM Executive Board (CDM EB)
Step 4: Financial resources		Step 7: construction and operation	

(continued)

Table 7.9 (continued)

Step and activities of conventional CDM		Urban Methodology	
Step	Activities	Step	Activities
Step 5: Monitoring	4. MRV (Monitoring, Report, Verification) by CDM project implementing body or the third party (for certification) *Submit monitoring report to CDM Designated Operation Entity (DOE)	Step 8: Monitoring	17. Monitoring of GHG reduction activities
			18. Aggregation of monitoring results of PoAs
			19. Quantification or estimation of the cumulative volume of the emissions reduced, compared with the BAU baseline level of emissions
Step 6 : Verification and certification	5. The purpose of verification of CDM project is to identify achievements of actual GHG emissions reduction 6. CDM Designated Operation Entity (DOE) carries out evaluation of performance outcomes such as consistence between project design document and monitoring report, method used for monitoring, identification of GHG reduction amounts, project results, etc. 7. CDM DOE request for issuance of Certified Emission Reductions (CERs) in accordance with certified amount of GHG reductions 8. CDM DOE documents a certification report based on the submitted verification report	Step 9: Verification & Certification	21. Verification/certification of monitoring report by DOE & request for issuance
Step 7:	9. CDM EB receives MRV report and issues CERs unless objections are made in 15 days	Step 10: Issuance of CERs	22. Issuance of CERs by CDM EB
		Step 11: Publication	23. Publication of the report on Urban CDM Joint Program
		Step 12: Post PoA Management & Evaluation	24. Selling and trading of CERs
			25. Post-PoA management
			26. Periodic evaluation
			27. Final project Reporting

guide to planning blockchain-based testbeds based on the principles of carbon neutral cities through a process with twelve main stages described in Sect. 7.4.

Based on the testbed project, standardization should be extended to specific methodological approaches allowing to expand the CDM to new sectors and cross-sectors/types of activities and a testing field for design features of new mechanisms.

This testbed methodology aims to be a good practical showcase to demonstrate in a cost-effective way in line with a guide for '2 Degrees and SDGs Aligned with Urban Methodologies' where the emphasis of the umbrella term 'sustainability goals' is in place. The result of the methodology could be credited with new carbon market mechanisms in the urban context or depending upon how much it achieves the goals of the '2 Degrees and SDGs toward sustainable carbon neutral cities'. It may be a good test to investigate the operational potential of selected courses of climate actions at the city level. A testbed is a vision, and the methodology will require cost justification at every step before implementation and verification and before extensive application. The testbed could be an ideal place to explain more new thinking and to show how it links to urban methodology for carbon financing.

Table 7.10 shows component project activities selected for the proposed methodology, carbon emission reductions targets, and the availability of registered CDM methodology for each component project.

The steps for the testbed methodology are shown in Fig. 7.11. According to the steps, basic contents associated with the proposed testbed methodology are as follows:

Step 1: Title of the proposed small-scale methodology

Digital integrated methodology for energy generation, building energy efficiency, water savings, and transport measures in community renewal.

Step 2: Project description

The purpose of this project is to develop a testbed leading to city-wide carbon neutrality by 2050, building upon the Goyang Climate Action Plan by which Goyang City plans to achieve a 50% emission reductions by 2030. The city will propose to UNFCCC a revised form of new small-scale methodology which includes:

A. Induce or incentivize community developers to introduce higher energy efficiency standards as well as renewable energy standards;
B. Decrease home maintenance costs to homeowners through building energy saving through the introduction of net-zero energy building codes and energy efficiency technologies, including on/off community microgrid;
C. Diversify the sources of water heating through the usage of solar water heaters that involve less fossil-fuel electricity and LPG or natural gas consumption;
D. Extend the bike network;
E. Increase the use of rainwater through the integrated stormwater management system;

Table 7.10 Proposed urban methodology typology

Measure	Targets	Registered CDM methodology (Identification Number)	2 Degrees and SDGs contribution		
			Carbon reduction	Urban climate change resilience	Sustainability
Energy generation (Renewable energy)	• Baseline year 2018 • Target for 2023				
Building energy efficiency	• Baseline year 2018 • Target for 2023				
Water savings	• Baseline year 2018 • Target for 2023				
Transport (Bicycling)	• Baseline year 2018 Annual bicycle counts • Target for 2023				
Smart community on/off microgrid					
Aggregation and cross-effect analysis					

Note (1) The system may either be equipped with batteries (ESS) or interconnected to the local power grid

F. Create employment opportunities and generate income by introducing new policies and technology alternatives in the energy generation, building energy efficiency, water saving, and transportation (bicycling) sectors.

This approach could be a part of new carbon market platforms for a new digital generation.

Step 3: Issues—situation analysis

A situation analysis helps identify some of the key issues of the methodology to integrate relevant component project activities in order to address global and local climate change, and active citizen participation. It includes carbon issues, resilience issues, sustainability issues, and management issues.

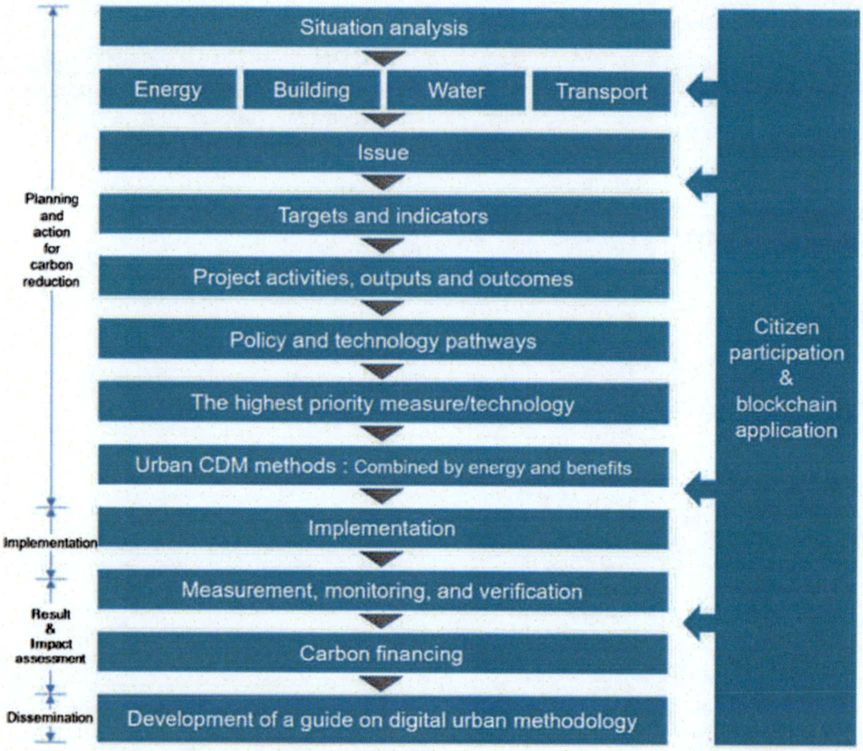

Fig. 7.11 Steps for the testbed methodology

Step 4: Project targets and indicators

One of the critical aspects of the methodology testing is to test relevant targets and indicators and/or metrics for tracking progress toward the testbed's climate goals which can be incorporated in the annual GHG inventory, baseline, and project activity scenarios. Table 7.11 shows examples of targets and indicators by project measures.

Step 5: Project activities, outputs, and outcomes

The methodology connects project activities to quantifiable outputs and ultimately desired outcomes of the proposed testbed. This enables reviewers to evaluate what the proposed project will achieve, if implemented. Detailed project activities, outputs, and outcomes are defined by a hierarchical system of 8 project activities, 17 project outputs, and 21 project outcomes.

Step 6: Policy planning options

We need to evaluate and act upon all of our policies and options across the energy, building, water, and transportation sectors, and forge a path to carbon neutrality. Our guiding principles for the carbon neutrality policies are:

Table 7.11 Project measures, targets, and indicators

Target	Indicator
Measure 1: energy generation (renewable energy)	
1.1 By 2023, increase substantially the share of renewable energy in the community energy mix	1.1.1 Renewable energy share in the total final energy consumption
1.2 Support and strengthen the participation of local communities in increasing the share of renewable energy	1.2.1 Number of houses, offices and industries participating in renewable energy shift program
	1.2.2 Investments in renewable energy-connected smart grid
Measure 2: Building energy efficiency	
2.1 By 2023, double the community rate of improvement in energy efficiency	2.1.1 Percentage of low-carbon domestic appliances
	2.1.2 Investments in building energy efficiency improvement (e.g., insulation)
	2.1.3 Number of rooftop gardens out of total number of houses
	2.1.4 Number of houses, offices, and industries participating in building energy efficiency improvement program
Measure 3: Water savings	
3.1 By 2023, increase substantially the use of water	3.1.1 Amount of water harvesting(captured water) and reuse
	3.1.2 Degree of integrated stormwater drainage system (0–100)
	3.1.3 Areas of rainwater parks
3.2 By 2023, increase substantially water-use efficiency	3.2.1 Number of efficient water-use activities and programs
	3.2.2 Change in water-use efficiency over time
3.3 By 2023, implement integrated water resources management at the community level, through city-wide cooperation as appropriate	3.3.1 Degree of integrated water resources management implementation (0–100)
3.4 By 2023, protect and restore water-related ecosystems, including mountains, forests, wetlands, rivers, aquifers, and lakes	3.4.1 Change in the extent of water-related ecosystems over time
	3.4.2 Number of trees planted for the water-related ecosystems
3.5 By 2023, support and strengthen the participation of local communities in improving water management	3.5.1 Proportion of local administrative units with established and operational policies and procedures for participation of local communities in water management
3.6 By 2023, increase treatment of urban runoff and capture	3.6.1 Amount of soaking in the first 24 mm of rainfall
	3.6.2 Amount of cleaning up the next 24 mm of rainfall

(continued)

Table 7.11 (continued)

Target	Indicator
	3.6.3 Amount of conveying rainfall greater than 48 mm in safe runoff routes
Measure 4: Transportation(Bicycling)	
4.1 By 2023, expand bikeways for sustainable transport systems	4.1.1 Percentage of bike riders out of total population and length of bikeways per capita

- Share Co-benefits: ensure that the benefits of climate actions extend to all community residents;
- Transparency and Accountability: The process to develop the policy should be inclusive and the information provided accessible. The policy should establish clear milestone and metrics that will be incorporated into the city's regular public reporting.
- Goal-oriented: Achieving carbon neutrality will require tough choices and bold action. The policy should provide help the community evaluate and make those choices (Washington DC 2018).

  Policy options could include:

- To integrate climate adaptation into methodologies to achieve carbon neutrality and to optimize climate co-benefits of climate mitigation and adaptation;
- To leverage feedback from stakeholder engagement;
- To propose a methodology for prioritizing based on evidence such as their emissions reduction potential (identified in the technical assessment) and social, environmental, and economic benefits;
- To identify high-priority actions that need to be accelerated in the near-term in order to advance on realistic trajectory to achieve carbon neutrality.

Step 7: Technology/Measures

There are many techniques that can be utilized to satisfy the features of a testbed program. Examples of technologies and equipments for each measure in the testbed are shown in Table 7.12.

Step 8: Testbedselection criteria and sustainable digital design of the selected community for energy, building, water, transport-integrated systems

This methodology can be tested for community redevelopments or renewal projects and new developments that employ climate actions in the areas of energy generation, building energy efficiency, water, and transport sectors to reduce greenhouse gas (GHG) emissions.

Table 7.12 Measure/technology and equipment

Measure	Technology		Equipment
Energy generation	Solar energy	• Solar energy technology • Heating and cooling technology • Solar LED lighting technology	• On-site solar LED light • Air conditioner for heating and cooling • Wind power turbine • Smart sensor networks and meters • Solar photovoltaic devices and batteries • Real-time controller programming units (controller network) • IoT • Charge devices or grid • Outdoor energy-generating exercise equipment (fitness bike and double-station leg press)
	Solar power	• Solar urban furniture and amenity technology	• Solar benches • Solar shade structures • Solar-powered trash compactors • Restroom occupancy sensors • Smart water fountains • Digital signs
		• Solar lighting technology	• Motion-activated sensors • LEDs and fiber optics as art • Off-grid light fixtures • Digital additions to LED fixtures • Lighting shields
Water Saving	Stormwater	• Pervious surface technology	• Wi-Fi hard-surface testing equipment – electronic sensors – lower-cost, portable equipment
		• Engineered soil technology	• Engineered soils which can remove pollutants and help water infiltrate better into the ground

(continued)

Table 7.12 (continued)

Measure	Technology		Equipment
		• Underground storage basins(or ponds) for retention or detention technology	
		• Filtration technology	
		• Above-ground rain gardens or retention pond technology	
Transportation		• Bicycling technology	• Automatic bicycle and pedestrian counters that track bicyclists and pedestrians • Smart digital bike racks • Battery-powered counters and located underground, on posts, or at street level – They use sensors, video cameras, or inflated tubs or metal strips(that detect bicycles or people passing over them) to obtain user counts
Building energy efficiency	• Insulation technology	• Carbon upcycled concrete	

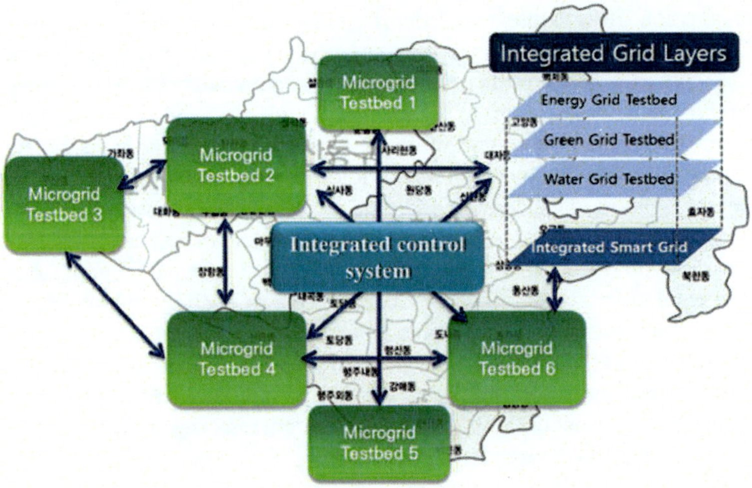

Fig. 7.12 Multi-layers of smart grids

The testbed could have the potential to establish a blockchain-based integrated community microgrid, a new approach to designing and operating an electricity-water-green space-transport nexus and providing indefinite, renewables-driven backup power to critical facilities. The approach could provide a foundation for developing an integrated holistic model of smart energy grid + smart water grid + smart green–blue grid + smart transport corridor grid for the nexus between energy, water, green space, and transport to be connected with sensors, IoT, DLT, and blockchain platform technology described in Chap. 1. Figure 7.12 shows illustrative applications of multi-layers of grids.

Step 9: Application of urban methods

This step is designed to test a technical approach and assessment to modeling policy and technology scenarios for achieving carbon neutrality by 2050 and a strategy for engaging stakeholders and technical experts in the development of those scenarios. This is another opportunity to emphasize the novelty of the approach, and how it can get applied to urban methodology for carbon financing addressed in this chapter.

7.6 Conclusions

In this chapter, we have elaborated on the specific view that new carbon market mechanisms would allow the climate smart and wise city planners to concentrate on the core question of urban methodologies for carbon financing such as spatial dimensions of the carbon markets. However, we have also stated that within particularly the CDM dossier, which concerns so many and such different methodologies available, the core question 'How to integrate different sectors and technologies for aggregation of GHG emission reductions?' on the basis of urban space is to be complemented with another one, 'which sector reduces how much?' and cross-effects of the integration.

Even though the cross-sector approach, methodologies, and procedures proposed draw upon the programmatic and multiple sector approach under CDM, this new methodology is not limited by CDM methodological requirements and is well beyond a simple aggregation of CDM methodologies for cities and communities.

In relation to the integration proposed, the methodology seems to require digital transformation platforms that are comprehensive in terms of sectoral smart grid platform project models described in Chap. 1. One of such methodological values is democratic accountability and transparency. In this article, we have suggested blockchain-based carbon markets which is IoT-driven climate action tracking and accounting. In this approach proposed, the democratic accountability seems transparently arranged by encouraging cities to raise citizens' awareness about carbon markets by sharing blockchain-based database.

With urban methodologies for carbon markets, we are however still in a learning phase regarding the possible application of the instrument and its related GHG emission reduction, and economic and other effects. In this vein, we have indicated the need for further assessment of the functioning of the urban methodology for carbon financing, focusing on a city-level testbed approach.

References

Burg J, Murphy C, Petraud JP (2018) Blockchain for international development: using a learning agenda to address knowledge gap. http://merltech.org/blockchain-for-international-development-using-a-learning-agenda-to-address-knowledge-gaps/. Accessed 20 June 2019

Kim KG (2012a) Outcomes of UEA gwangju summit & urban CDM, urban environmental accords members alliance, Gwangju City, p 36

Kim KG (2012b) Urban CDM: challenges and opportunities, the occasion of ICLEI global town hall, RIO+20 (presented on June 18, 2012)

Kim KG (2013) Urban CDM in the context of low-carbon green city planning, UNFCCC, Bonn, Germany (presented on January 23, 2013)

Kim KG (2018) Blockchain-based climate and urban CDM initiative, Goyang/CCC/UNFCCC, Side Event, UNFCCC COP 24, Kaovice, Polland, (presente on December 10, 2018)

Kim KG (2018) Low-carbon smart cities: tools for climate resilience planning. Springer, Heidelberg

Kim K G (2019) CDM in the building sectors, presented at COP25 side event: the role of the clean development mechanism (CDM) in decarbonizing cities and its co-benefits organized by UNFCCC, Madrid, Spain, 2–13 December 2019

UNEP, Gwangju City (2012) Cities and carbon finance, a feasibility study on an urban CDM. p 80

UNEP (2012) Framework elements for assessing urban environmental performance. p 44

Urban Environmental Accords (UEA) Secretariat (2014) The carbon banking system in Gwangju. World Bank, UEA, Gwangju, p 17

World Bank (2010) A city-wide approach to carbon finance. p 27

Chapter 8
How Can Global Carbon Markets Promote Low-Carbon Cities in Developing Countries? Lessons Learnt from the Clean Development Mechanism

Axel Michaelowa and Maike Sippel

Abstract Per-capita greenhouse gas emissions in cities like Bangkok or Shanghai have already reached emission levels of cities like London or Toronto. Large parts of the building stock and service infrastructure in cities in rapidly developing countries will be built in the coming decades—and may lock in high emissions pathways. A survey of projects under the Clean Development Mechanism (CDM) of the Kyoto Protocol shows that only about 1% of projects have been submitted by municipalities, mostly in the waste management and more recently in the transport sector. This is probably due to a lack of technical know-how, legal barriers, methodological challenges, long project cycles and limited "visibility" of projects for the electorate. A case study of city network ICLEI's experience with the CDM adds practical insights. We conclude that while the new market mechanisms under Article 6 may make it easier for municipalities to engage in international market mechanisms, new forms of cooperation between actors on multiple levels, potentially facilitated by ICLEI, are required to help to realize the urban potential in international market mechanisms.

Keywords CDM · Cities · Carbon markets · Transport · Waste · Buildings

8.1 Introduction

8.1.1 Global Climate Policy and CDM

As greenhouse gases are global pollutants, it does not matter where emissions are reduced. The Kyoto Protocol specified a greenhouse gas emission reduction target

A. Michaelowa (✉)
Institute of Political Science, University of Zurich, Affolternstrasse 56, 8050 Zurich, Switzerland
e-mail: axel.michaelowa@pw.uzh.ch; michaelowa@perspectives.cc

Perspectives Climate Research, Hugstetter Str. 7, 79106 Freiburg, Germany

M. Sippel
University of Applied Science, Brauneggerstr. 55, 78462 Konstanz, Germany
e-mail: maike.sippel@htwg-konstanz.de

K.-G. Kim and M. Thioye (eds.), *Planning Climate Smart and Wise Cities*,
The Urban Book Series, https://doi.org/10.1007/978-3-030-80165-6_8

of 5.2% below 1990 levels for 38 industrialized countries and countries in transition (so-called Annex B countries) for the first commitment period from 2008 to 2012. In order to achieve cost-effective emission reduction, the Kyoto Protocol introduced the flexible mechanisms International Emissions Trading, Joint Implementation and the Clean Development Mechanism (CDM).

The CDM allowed Annex B countries to meet their emission reduction targets by purchasing certified emission reductions (CERs) from greenhouse gas emission reduction projects in developing countries. CDM projects have to contribute to sustainable development in the host country, and they have to be additional, that means they would not have occurred without the funds generated by selling the CERs (Lecocq and Ambrosi 2007). However, there is a debate on the actual contribution of the CDM to sustainable development (Boyd et al. 2009; Holm Olsen 2007) and the interpretation of additionality has changed over time (Michaelowa 2009).

After a massive upswing between 2005 and 2012, the CDM languished since 2013 due to a decrease of CER prices by 95%. While the mechanism formally continues to exist, the inflow of new activities has slowed to a trickle. Since the negotiation and the entry into force of the Paris Agreement in 2015 and 2016, the focus has shifted from the CDM to the new market mechanisms of Article 6. However, these mechanisms are explicitly tasked to learn from the lessons of the Kyoto Mechanisms.

8.1.2 Climate Protection in Developing Country Cities

Besides international climate governance, cities play an important role in mitigating climate change (Betsill and Bulkeley 2007). Today, half of the world population live in cities. They generate greenhouse gas emissions from energy use, transport or waste management (Satterthwaite 2008). As cities in developing countries are starting to "catch up" economically, they have also been catching up in terms of greenhouse gas emissions: In Shanghai, annual per-capita CO_2 emissions had grown from 3.8t in 1985 to 16.7t in 2006 (Dhakal 2009). In terms of per-capita emissions, Shanghai together with Bangkok, Thailand (10.7t) or Cape Town, South Africa (11.6t) has already overtaken Geneva, Switzerland (7.8t), Prague, Czech Republic (9.4t) or London, UK (9.6t) (Kennedy et al. 2009). This is far beyond the global per-capita emissions threshold of about 2t which climate scientists are calling for. New investments in-built structure and infrastructure can either lock-in vast energy consumption or climate benefits for decades (Sovacool and Brown 2010).

This chapter focuses on local governments and whether and how the CDM was taken up by them, and what this means for local government engagement in the Article 6 mechanisms. Table 8.1 provides an overview of the different modes of governance, the influence a local authority can take via each mode, the relevance in terms of mitigation potential, and CDM/Article 6 activities a local authority may implement in its respective capacities.

Table 8.1 Modes of local climate governance and CDM/Article 6 projects

Role of local authority	Exemplary activities	Influence of local authority	Relevance (CO_2-emissions)	Examples of CDM project activities	Suited for CDM
Self-governing	Green fleets, municipal building energy management, purchasing green energy	++	–	"Improvement in energy consumption of a hotel" "Energy efficiency measures in 'Technopolis'"	O
Governing through enabling	Information campaigns, advice and grants for energy efficiency, loan schemes for renewable energies, education campaigns on green transport	O	+	"Visakhapatnam (India) OSRAM CFL distribution CDM project"	O
Governing by provision	Energy service providers/energy service companies, public transport service provider, waste management	Depending on ownership of operations	+	"Bandeirantes landfill gas-to-energy project" "BRT Bogotá, Colombia: TransMilenio Phase II to IV" and "Installation of low green house gases (GHG) emitting rolling stock cars in metro system" "Beijing 48 MW Guanting wind power project" "Beijing Taiyanggong CCGT Trigeneration Project" "BBMG cement WHR for 10.5 MW power generation project in beijing"	+

(continued)

Table 8.1 (continued)

Role of local authority	Exemplary activities	Influence of local authority	Relevance (CO_2-emissions)	Examples of CDM project activities	Suited for CDM
Governing by authority	City-planning to reduce transport generated by settlement patterns, supplementary regulations, e.g. for energy efficiency in the building sector	++	++	Not suitable as CDM project	–

Source Categories based on Bulkeley and Kern (2006), Sippel (2004), own evaluations

Under the Kyoto Protocol, developing country cities had limited interest in implementing climate protection activities (Bai 2008). Generally, they had no direct benefits from emission reductions, as their countries did not have emission targets under the Kyoto Protocol. In addition, Southern municipalities have very limited resources, and other more urgent policy issues to deal with.

Therefore, we ask (1) whether the CDM did motivate developing country cities to take climate action or help them to overcome existing barriers for local climate governance, and (2) which obstacles cities faced that wanted to engage in the CDM. From an analysis of local authorities' participation in CDM project activities registered so far, conclusions are drawn, e.g. which project types are attractive at the city-level, and why other project types are less common. A case study highlights the city network ICLEI's experiences with the CDM.

The chapter consists of three parts: a theoretical analysis of cities and the CDM, a survey of registered municipal CDM projects and a case study on ICLEI's experiences in promoting CDM projects in cities.

In order to get a theoretical perspective on cities and the CDM, motivation and barriers for municipalities to implement CDM projects were analysed. This analysis used a list of frequent motivation and barriers for municipal climate activities, as are reported in the literature on cities and climate change. Literature and case studies on municipal CDM projects were then analysed for the relevance of these motivation and barriers.

Secondly, a survey of municipal CDM projects engaged in during the "gold rush period" of the CDM was carried out (see Sect. 8.3). This survey was undertaken in November 2009 and updated in August 2011, based on the Point Carbon Project database, including all CDM projects submitted to the UNFCCC and presenting structured information on all critical project parameters (see Point Carbon 2011). The survey looked for CDM projects that include municipalities or municipal companies as project participants. As post-2011 projects remained few due to the CER price crash, the study period remains relevant up to today.

Thirdly, a case study analysed experiences that can be drawn from the city network ICLEI's experiences with the CDM. Texts and events that document ICLEI's activities around the CDM were used as a basis for this analysis. ICLEI experts contributed their experiences on concrete project examples. The CDM registry was further analysed for projects with the participation of ICLEI member cities.

In a concluding section, we discuss the relevance of municipalities in the context of the market mechanisms under Article 6 of the Paris Agreement.

8.2 Cities and the CDM—The Theory

8.2.1 The CDM

Any potential CDM project needs to be formally registered by the CDM Executive Board (CDM EB), the core international decision-making body who also issues the CERs. The resulting project cycle is shown in Fig. 8.1.

Given this project cycle which takes many months before registration is achieved and requires to hire specialist consultants as well as independent auditors, with costs that can easily exceed 100,000 € (see Michaelowa and Jotzo 2005), initial estimates of developing country potential to supply large amounts of CERs within short time were pessimistic (see, e.g., Black-Arbeláez 2003), with at best a few hundred projects materializing. However, the development of CDM projects exceeded these expectations (Rahman et al. 2010). Over 10,000 projects have been submitted to independent

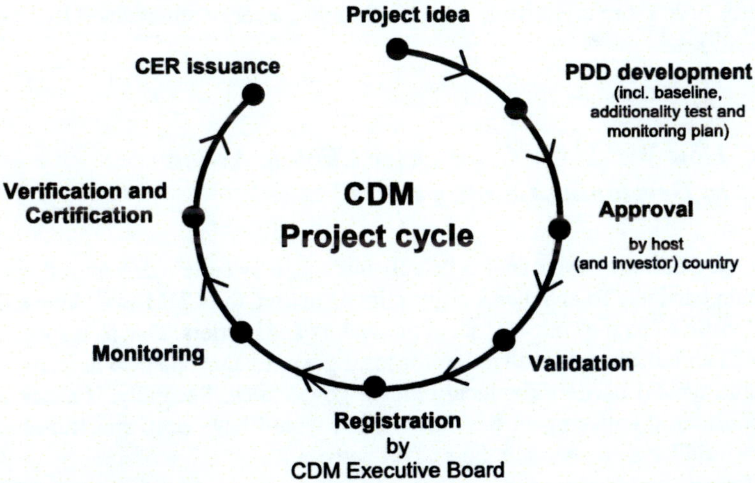

Fig. 8.1 CDM project cycle

auditors (Designated Operational Entities, DOEs) for validation and over 200 baseline methodologies have been approved for a wide range of technologies. The total volume of issued CERs reaches over 2 billion CERs. Overall, before the price crash starting in 2011 over 10 billion € had been committed or already spent on CER acquisition by governments and companies in industrialized countries.

8.2.2 Local Governments and the CDM

Some research has analysed local options for CDM projects, e.g. for Rio de Janeiro in Brazil (Schmidt Dubeux and La Rovere 2007) or Yinzhou District in Zhejiang Province, China (Zhao and Michaelowa 2006). Firstly, municipalities can develop CDM projects which reduce emissions that are produced by a local authority itself. A possible project type would be energy efficiency improvements in municipal buildings. Secondly, local governments can coordinate or facilitate emission reduction activities by local stakeholders. An exemplary project under the CDM could be the distribution of compact fluorescent lamps, as it reduces energy consumption and reduces energy bills for poor sections of the population. Thirdly, local governments may also act as service providers, e.g. managing waste from citizens or infrastructure to be used by citizens. Possible CDM projects in this field include landfill gas projects, renewable energy generation or energy efficiency improvement and public transport projects. Last but not least, local governments can to some degree regulate the behaviour of local stakeholders. However, regulatory activities are not eligible under the CDM. Challenges do also exist for demand-side energy efficiency projects under the CDM (Michaelowa et al. 2009). Direct regulation through building standards may have been more effective in realizing building energy efficiency (Ürge-Vorsatz et al. 2007).

8.2.3 How Would the CDM Drive Climate Action in Developing Country Cities?

Firstly, local climate governance is usually motivated by certain drivers and hindered by certain barriers. To enhance a city's climate action, the CDM would have had to bring "positive change" into this set of motivators and barriers. That is, it should have presented an (additional) motivation for local governments and also helped them to overcome typical barriers for local climate governance. Secondly, if cities would have considered engaging in the CDM, there would have been specific obstacles which would have constrained their CDM activities.

Motivators and barriers for local climate governance can fall into categories like economic, institutional or political/cultural. The effect of the CDM on these could have been ambiguous, as CDM project development by local governments could also have created new barriers for the rest of a city's climate action. For example, a CDM project would at the same time have helped a city to build expertise for climate action (both through additional finances available and experience gathered by staff during CDM project development), but also siphoned expertise away from other climate projects, as the CDM project management required skilled staff. The analysis includes both effects.

By selling CERs, a CDM project generated revenues. As local officials care about the revenues of their government, this would have been a motivator for cities to develop CDM projects. As project costs are an important barrier to local climate action, lower project costs may make the realization of projects more likely. Some evidence for these considerations is provided by Wilson (2007) for Nelson Mandela Bay Municipality in South Africa and by Qi et al. (2008), who studied CDM and local governments in China. In China, where the largest share of CDM projects was registered, many local governments engaged in CDM project development. This could be due to the important role of entrepreneurial local authorities in the transitions from a centrally planned to a market economy (see Gordon and Li 2012). This contrasts with most other countries, where CDM project development was essentially left to the private sector. Table 8.2 provides an overview of motivators and barriers for local climate governance, and how the CDM affects them.

In addition to economic aspects, a local authority's hope to realize co-benefits which can present political gains, e.g. by improved air quality, reduced traffic congestion, employment opportunities, improvement of waste and sewage management, improving social services and infrastructures may have been an important motivator (Clapp et al. 2010; Schmidt Dubeux and La Rovere 2007; Sharma 2007; Vennemo et al. 2006). The opportunity to improve the environment in the Laguna de Bay Basin, Philippines was a key motivation for local authorities in the area to develop a CDM project (Santos-Borja 2007). Wastewater, solid waste, public transport, buildings, metered services and street lighting may be sectors which hold particular potential for local co-benefits (Ritter 2009).

As the development of CDM projects was complex, it would absorb skilled local government staff, which then would have no longer been available for other local climate action. The effect could have been severe as many local authorities in developing countries suffer from a lack of human resources and expertise among their staff for climate action. However, the CDM could also have improved human capacity for local climate governance, as firstly, revenues from the CDM could have been used to finance additional staff or train existing staff, and secondly, staff would have acquired expertise during CDM project development, which it could then use for other projects, too (Santos-Borja 2007). However, staff that had experience with CDM has frequently been "poached" by the private sector due to salary offers that are much higher than the salaries a municipality was able to provide. This is a stronger risk than in other sectors due to the very specific knowledge that commands a substantial salary premium.

Three further barriers may have been affected by the CDM. Firstly, as CDM project development usually required the involvement of local companies, this would have improved a local authority's cooperation with the business sector and thereby have

Table 8.2 Motivators and barriers for local climate governance and CDM impact

		Motivator (M) or Barrier (B)	CDM impact *From "negative" (–) to "positive" (++)*[1]	How?
Economic	Revenues	M	++	Revenues from selling CERs
	Limited opportunities to generate/access funds	B	++	Revenues from selling CERs = additional funding source; Carbon credits may increase willingness to invest by private sector
	Costs	B	+	Project costs for municipality reduced by CER revenues
	Lack of financial and human resources	B	O	CDM involvement may draw resources from other climate action/CDM revenues may increase resources for local climate protection
Liveability	Improved air quality, reduced traffic congestion, reduced urban warming, increased energy security, social aspects	M	+	Possible as a contribution to sustainable development (esp. for transport and waste management projects, small energy projects)
Informational	Lack of expertise	B	O	CDM may draw expertise from other climate action/CDM may create expertise usable for other climate action
Institutional	Lack of cooperation with other stakeholders	B	+	CDM may require the involvement of local business, and thereby improve cooperation

(continued)

Table 8.2 (continued)

		Motivator (M) or Barrier (B)	CDM impact *From "negative" (–) to "positive" (++)*[1]	How?
Political/cultural	Short time-horizons of decision-makers	B	+	Long CDM project cycles may lock-in climate action (but may also prevent local government from commitment)
	Reputation	M	+	When local governments believe CDM does enhance reputation
Others	Limited availability and reliability of technologies	B	+	In case CDM projects involve technology transfer

Source Own evaluation, motivators based on Sippel and Jenssen (2010)
[1]Positive impact on barriers meaning: can help local governments to overcome barrier; positive impact for motivators meaning: can motivate local governments

positive effects on other local climate action, too. Secondly, as CDM project cycles were longer than election periods, CDM projects could have helped local authorities to lock in climate-friendly development pathways: Once a CDM project owned by a municipality had been started, and respective CER sales contracts signed for the whole CDM project duration, this would have required politicians to refrain from changes in the project. However, the requirement for long-term commitments may also have been a barrier to CDM project development in the first place, especially where more immediate needs for municipal action, e.g. on poverty reduction prevail (Wilson 2007). Thirdly, local governments may have believed they enhance their cities' reputation by engaging in the CDM (Santos-Borja 2007).

While some local governments may not have been aware of the CDM (ICLEI 2010; Wilson 2007), even the ones that did know it faced challenges in CDM project development. From a financial perspective, CDM projects required upfront investments such as the acquisition of energy-efficient equipment, especially in a situation where self-financing by project developers prevails (Michaelowa 2007). In addition, project development involved significant transaction costs for the elaboration of project documentation and payment of auditors (Michaelowa and Jotzo 2005). Municipalities would have refrained from involvement in the CDM (Santos-Borja 2007; Wilson 2007), if they lacked the required financial means (e.g. Clapp et al. 2010; Labriet et al. 2009), and their investment capacity was limited by state law. In the national context, capacity-building has proved to be necessary though not sufficient to mobilize CDM projects (Okubo and Michaelowa 2010). "Bureaucratic red tape" in the realization of projects, as well as "weak institutional capacity at city level" for the CDM, its integration into city priorities and the design of supporting policies have to be added as barriers (Santos-Borja 2007; Ritter 2009). This also included legal barriers such as the need to involve state agencies in foreign exchange transactions. Uncertainties about the CDM's future, or changing methodologies were institutional barriers on the global level. Furthermore, climate protection activities in the building or transport sector that would be important for the local level was faced by stringent rules (Labriet et al. 2009), and programmatic CDM, which can include demand-side activities in both sectors, came too late to make a difference (Jiang and Tovey 2010; Zegras 2007).

The complexity of the CDM procedure did not only lead to high transaction costs, as explained above, it also required project participants to acquire CDM-specific expertise and capacity (Michaelowa 2003). Local governments lacked the necessary staff, as well as the technical know-how needed for project development, and they had a slow learning curve for CDM rules (Clapp et al. 2010; Santos-Borja 2007).

The "incompatibility between [...] the nature of the political process (always a potential change of local government) and very long carbon project cycles" (Monroy 2009, p 4) has been reported to create uncertainty regarding local government staff assigned to a CDM project (Santos-Borja 2007). The problem was worse where political party interests led to divergent positions on a CDM project, as each party wanted to generate benefits to its members and supporters. From experience with landfill gas projects, Monroy (2009) concludes that local officials engaged less to make CDM projects a success than private sector employees.

8.3 Cities and the CDM—The Practical Side

Given the large upswing of the CDM in its early years, the absence of municipalities that had championed the CDM is striking. While the barriers discussed above explain part of the low participation, they cannot explain it all. There was no municipal government that actively marketed its role in implementing or supporting CDM projects. Few studies have focused on CDM projects implemented in cities (e.g. Labriet et al. 2009; Clapp et al. 2010). While obviously a substantial share of CDM projects was implemented on the territory of large cities, it seems that this is not due to any coordinated policy of the municipal government of those cities. Generally, CDM consultancies have scouted for project options and mobilized them, with the municipality normally acting more as a barrier than actively supporting the project (see, e.g., for landfill projects Terraza et al. 2007). An exception seemed to be China where municipalities have a broader entrepreneurial remit (see Gordon and Li 2012). According to Qi et al. (2008), Longnan City (Gansu Province) formed a coordination and leading group for CDM in March 2006. Its emphasis was on hydropower-related projects, of which Longnan had submitted two. The cities of Leshan (Sichuan), Nanyang (Henan) and Baoding (Hebei) formed governmental organizations for CDM development, with the latter signing a letter of intent for strategic collaboration regarding methane reduction from dairy farms in December 2007. But as no CDM

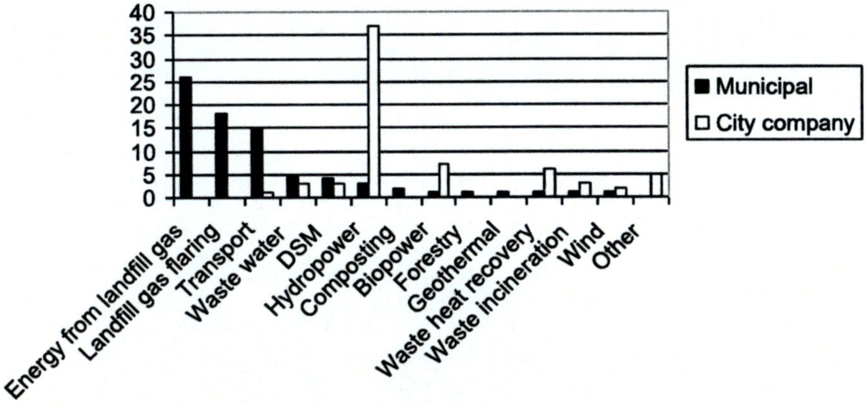

Fig. 8.2 Preferred project types. Data source: Point Carbon (2011)

project developed in these cities involves the municipal government it is impossible to check whether more projects have been added since.

Out of a database of 8038 CDM projects that had been submitted for validation by early August 2011, 81 projects (1.0%) have a municipality or a company formally labelled as municipal company as a project participant. Another 66 projects (0.8%), mostly from China, have a project participant whose name specifies "city". In this case, the name might indicate some link to the municipality, but could also just specify the geographic location. With regard to technology, waste management projects dominate for the municipalities, whereas renewable energy, especially hydro-dominates for "city" companies (see Fig. 8.2).

Regarding host countries, municipality-related projects have a high degree of geographical distribution, whereas "city" companies are concentrated in China (see Fig. 8.3).

8.3.1 Successful and Unsuccessful Projects

As discussed in the preceding sections, CDM projects in sectors managed by the municipality were particularly promising for municipalities. Traditionally, in many countries waste management, power generation and distribution for private households and public transport are organized by the municipality. Through land use regulation, municipalities have a strong influence on transport and buildings. We also look into water provision given the high energy intensity of water pumping systems.

Fig. 8.3 Host countries

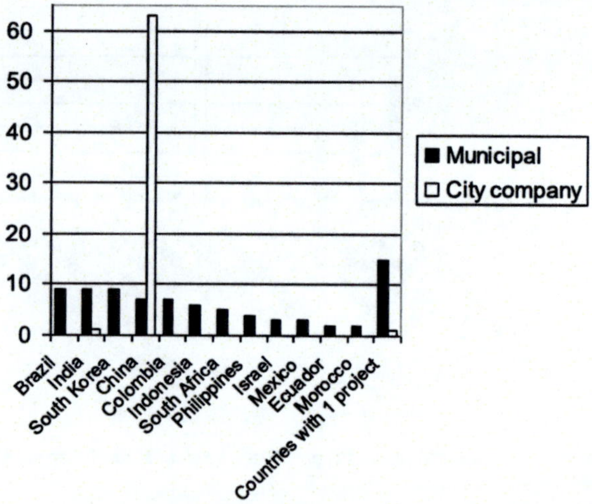

8.3.1.1 Waste

Waste-related projects dominated in municipal CDM, mainly regarding landfill gas collection. 78 MW of landfill gas were included. Landfill gas projects have been suffering from a low level of performance. This is mainly due to wrong estimates regarding the share of organic waste, unprofessional engineering on the landfill site regarding the management of leachate and a generic tendency of the model-based estimate of methane generation to overestimate methane generation. The six municipal landfill gas-to-energy projects with issuance have reached a performance of 5 to 65% of the forecasts made in the PDD. The four landfill flaring projects achieved just 5 to 37% of forecasts.

The city of Sao Paulo has maximized CER revenues from the two Bandeirantes and Sao Joao landfill projects by auctioning 1.5 million CERs through the Sao Paulo stock exchange in 2007 and 2008. This procedure prevented losses through brokerage fees and achieved a revenue of 26 million € at the time when CER prices reached their historical high (C40 Cities 2009). While issuance success of these landfills significantly deteriorated over time, they were initially having a high success rate, which enabled the city to benefit from the high prices for issued CERs in 2007–2008 (see Figs. 8.4 and 8.5).

A further sale of 0.53 million CERs occurred in mid-2012 at a price of 3.3 €, thus generating another 1.6 million € (Cruz and Paulino 2013, p. 128). Since then, no further sales seem to have been made. 50% of these revenues are formally to be allocated to the Environment and Sustainable Development Fund (FEMA), which is administered by the Municipal Department for the Environment. Allocation of these funds to projects seemingly did not benefit the population around the landfills, but other interest groups in the city (Cruz and Paulino 2013). Sao Paulo landfill CDM

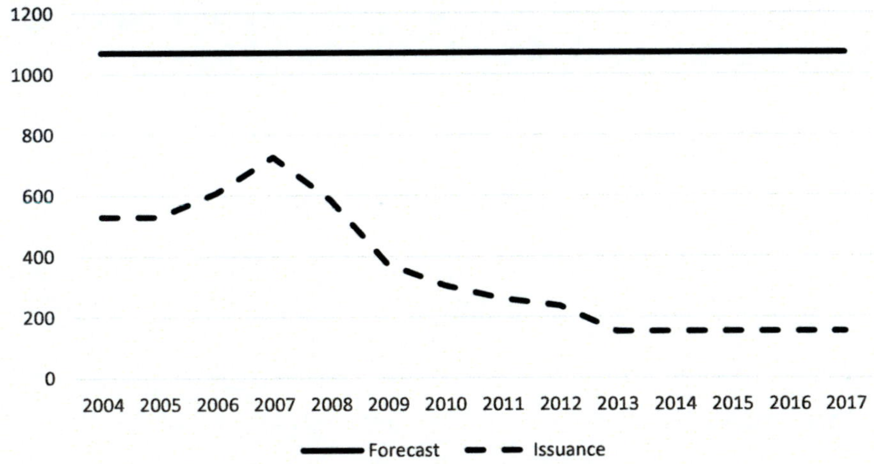

Fig. 8.4 Forecast and actually issued CER volume of Bandeirantes landfill project, Sao Paulo, Brazil (1000 CERs). Own figure, based on UNFCCC (2019a)

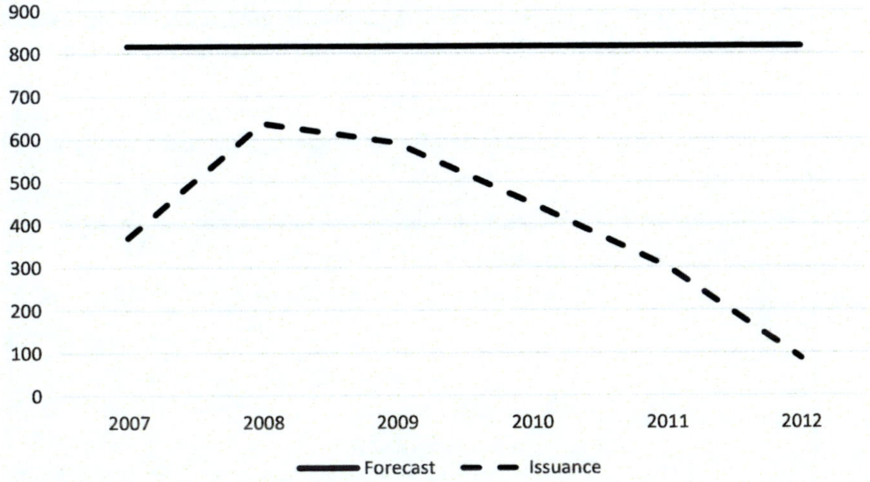

Fig. 8.5 Forecast and actually issued CER volume of Sao Joao landfill project, Sao Paulo, Brazil (1000 CERs). Own figure, based on UNFCCC (2019b)

projects can thus be characterized as "cashing in" during the CDM gold rush phase, but did not have a long-term effect.

Generally, private CDM project developers have complained about the slow decision-making and high degree of arbitrary changes in landfill gas project design and royalties to be paid to municipalities, particularly when city governments changed due to local elections. For example, project developers in Indonesia had to wait for several years before they could actually start their projects.

8.3.1.2 Municipal Power

Surprisingly, municipal power companies have not seriously ventured into the CDM. The only exceptions are Chinese "city" power companies that have invested mainly in hydropower, of which 639 MW showed up under the CDM. The main problem seems to be the relatively small size of municipal power plants and the lack of investment budgets for plant refurbishment.

8.3.1.3 Transport

Transport projects were rare under the CDM, but 14 bus lane transport projects in cities had been submitted by August 2011. Often, they were managed by a separate company that is not explicitly labelled as municipal company. Underperformance is often linked to construction delays and passenger overestimates (Grütter Consulting 2009). The first project with issuance, Transmilenio in Bogotá, achieved just 24% of the forecast in its first issuance. Performance improved subsequently, but did never reach the forecast (see Fig. 8.6).

Delhi Metro, which is probably the most successful transport infrastructure in India and has been universally acclaimed for its efficiency, has made strong use of

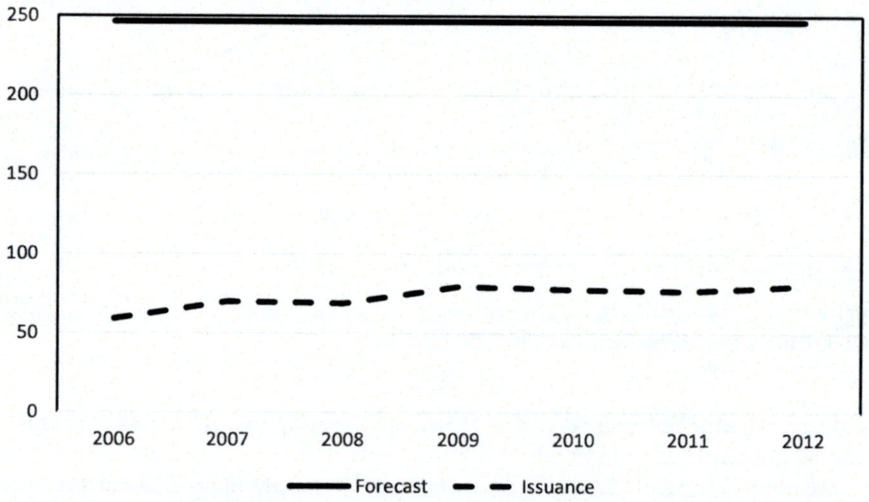

Fig. 8.6 Forecast and actually issued CER volume of transmilenio BRT project in Bogotá, Colombia (1000 CERs). Own figure, based on UNFCCC (2019)

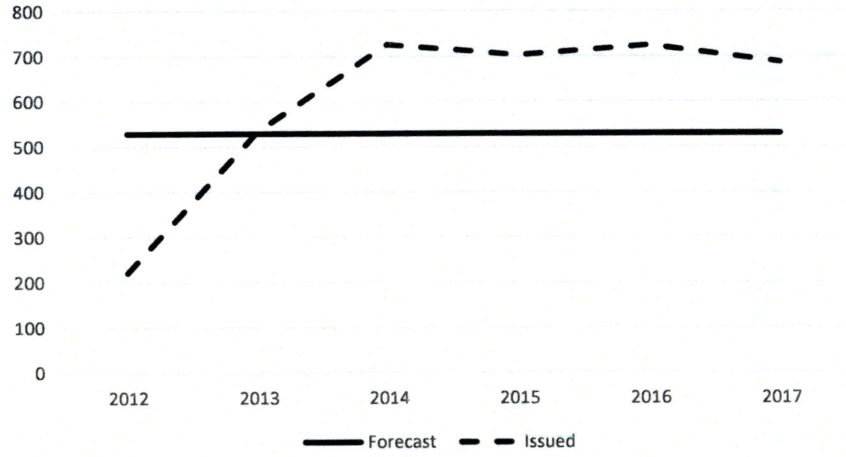

Fig. 8.7 Forecast and actually issued CER volume of metro modal shift project in Delhi, India (1000 CERs). Own figure, based on UNFCCC (2019c)

the CDM. Its first project was registered in 2007 and receives around 50,000 CERs per year for the regenerative braking systems of the metro carriages. After 2009, it has consistently achieved issuance above the forecast level of 43,500 CERs. CERs have been sold to a Japanese company at 4.5€/CER (India Power Trading Info 2011).

In mid-2011, the expansion of the metro system was registered and receives CERs for the shift of passengers from cars to the metro. This project has significantly overperformed due to the metro being used by much more passengers than expected (Fig. 8.7). The project is also special inasmuch it continues to issue CERs until today despite the low CER prices and tries to sell these on the free market (Delhi Metro Rail Corporation Ltd. 2019).

Delhi Metro generated significant revenues from CER sales between 2011 and 2014, but not since (see Fig. 8.8).

As Delhi Metro has made such positive experiences with the CDM in the late 2000s, it even registered a third CDM activity bundling 31 solar PV plants on metro installations with a total capacity of 14.9 MW in early 2017, forecasting 18,000 CERs per year (Shah 2017; UNFCCC 2019d). This project is part of a CDM PoA for various renewable energy projects in India managed by the Swiss CDM consultancy South Pole. The first monitoring report (UNFCCC 2018) shows a slight underperformance, with 85% of the forecast emission reduction.

Overall, it can be said that Delhi Metro has used the CDM in a very creative way, covering any aspect of its operation that leads to greenhouse gas reduction. All its CDM activities have been performing highly.

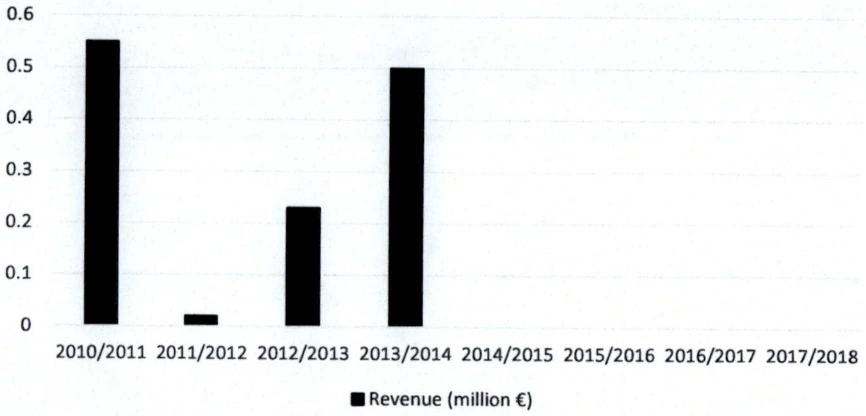

Fig. 8.8 Delhi metro revenues from CER sales (million €)

8.3.1.4 Buildings

The first municipality who developed a CDM project for energy-efficient buildings was Cape Town, with the Kuyasa project in the slum of Khayelitsha planning to retrofit 2300 houses with ceiling insulation, energy-efficient lamps and solar water heaters. The project which had been registered already in August 2005 was stalled for several years, as the upfront revenue from the forward sale of CERs to the British government at 15 € (Michaelowa 2005) only covered 30% of project costs, and the rest of the costs remained uncovered. In 2007, just ten pilot houses had been retrofitted. Eventually, the financing gap was closed through a Department of Environmental Affairs and Tourism poverty alleviation grant. By late 2009 more than 1200 houses had been retrofitted and in 2010, all 2300 houses were completed. The project provides 76 jobs. By late 2009 more than 1200 houses had been retrofitted and in 2010, all 2300 houses were completed. The project provided 76 jobs. While 9532 CERs were issued for the period 2005–2010 in July 2016, an earlier request for issuance of 10,527 CERs for the same period had been rejected by EB 85 in July 2015 due to changes in project design and monitoring inconsistencies. No issuance requests have been made since then (UNFCCC 2019e).

While there have been large-scale building energy efficiency projects in the CDM in the commercial sector, none of those involves a municipality. All but one other buildings-related projects are implemented by energy service companies or owners of large commercial buildings and do not involve local authorities.

8.3.1.5 Water Companies

In 2005, a baseline methodology for water pumping improvement was approved by the CDM EB, to which a project addressing several Municipal Water Utilities in Karnataka was attached. Nevertheless, it took four years for the project to be submitted for validation and it has never been registered.

8.3.2 *Insights on Problems with CDM Methodologies and Incentive Structures*

Why were municipalities unable to mobilize their substantial theoretical CDM potential? There are two key reasons besides the generic barriers discussed above. Firstly, the competencies required to write a PDD and accompany a project through the project cycle are not available in municipal administrations. Even if they were available, such skilled staff would be very much in demand and allocated to more urgent tasks. Therefore, specialized CDM consultants always have a competitive advantage compared to a municipality and can get project assignments. Even for landfill gas projects where a municipality should have a competitive edge due to its intimate knowledge of the situation on the ground, only 14% of projects had a municipality as project participant.

Secondly, municipal officials serve only for short periods. Thus, the incentive from CER revenues does not really reach them, as the long CDM project cycle means that CER generation will occur only years after the officials have left office. For the official, it is much more attractive to engage in a highly visible project which is "fashionable" with the voters. This is why Bogotá's mayor Moreno preferred the "glitzy" metro to the more mundane, but effective Transmilenio bus system (Gilbert 2008, p. 460).

8.4 ICLEI's CCP and the CDM

The International Council for Local Environmental Initiatives (ICLEI) is an international network of local governments working on sustainability issues. ICLEI coordinates the Cities for Climate Protection Campaign (CCP), which started in 1993. CCP's mission is "to reduce greenhouse gas emissions, improve air quality, and enhance urban sustainability" (ICLEI 2009a). By 2009, it had almost 1200 members (ICLEI 2011). More than half of them were located either in the United States or Australia and just 96 in non-Annex B countries. That means that the CDM was an option for only about 8% of CCP member cities (see Fig. 8.9).

By joining CCP, local governments commit to reduce greenhouse gas emissions in their cities, following the CCP milestone plan (ICLEI 2009a). Table 8.3 provides an overview of the milestones, and where the CDM could come in.

Another important aspect of CCP is that provides cities with a network of exchange and provides an overview of activities offered by the network to member cities, and whether they could be useful in supporting cities to engage in international carbon markets (Table 8.4).

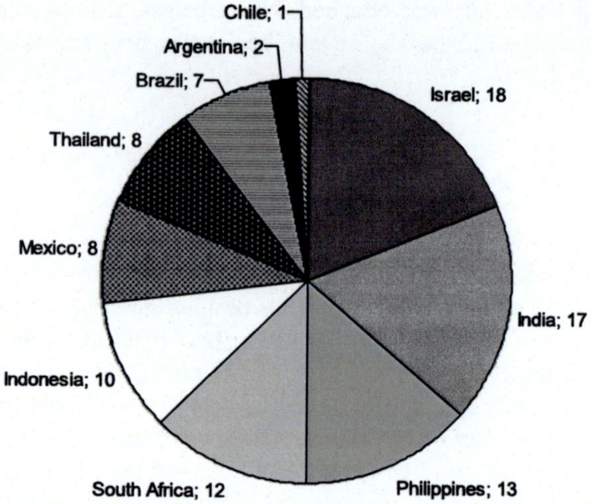

Fig. 8.9 CCP Membership Cities in non-Annex B countries. *Source Data from* ICLEI (2009a), *own graph*

Table 8.3 CCP milestone plan and CDM

Milestone	Required activity	CDM
1	Conduct a baseline emissions inventory and forecast	(X)
2	Adopt an emissions reduction target for the forecast year	
3	Develop a local action plan	X
4	Implement policies and measures	X
5	Monitor and verify results	

Source ICLEI (2009a), own evaluation

Table 8.4 CCP network of exchange and CDM

Activities	Description	CDM
Technical assistance	Inventory assistance, measures identification and quantification assistance	
Software tools	Designed to help with, i.e. inventories, quantification of emission reductions (e.g. HEAT)	
Information and policy assistance	Provision of case studies, fact sheets, sample resolutions, model ordinance language, links to key technical information	X
Training workshops		X

Source ICLEI (2009a), Lindseth (2004), own evaluation

By international advocacy, CCP tries to raise awareness for local governments' potential and needs on the national and international levels. To give an example, ICLEI led the Local Government Climate Roadmap to the UNFCCC Conference of the Parties (COP) in Copenhagen in December 2009. This included lobbying for a COP decision on cities/local authorities and climate protection, and city-relevant input to the negotiation process. However, ICLEI did never do active lobbying to make the CDM more city-friendly despite calling for CDM reform (ICLEI 2009b).

Nancy Skinner, founder and leader of CCP until 2004, explained in a fact sheet for CCP members that the "sale of emission reduction credits […] can provide capital to cover the costs of municipal projects" (Skinner n.d.). Upon entry into force of the Kyoto Protocol in 2005, Bob Price, Skinner's successor, told Indian CCP member cities that "the ratification of the Protocol means that the carbon reduction projects that you have been developing for many years will soon have real value as the international carbon trading market becomes a legal and practical reality and the Clean Development Mechanism (CDM) begins full operations" (Price 2004). Yunus Arikan, head of ICLEI's Bonn office, stated that the CDM is an instrument that can give local governments "access to both technology and finance" (Arikan 2009).

ICLEI's involvement in the CDM included supporting activities, e.g. for baseline analysis, in the development of Project Idea Notes (PINs) and Project Design Documents (PDDs), or in identifying project partners (Chaturvedula 2009). In a joint effort between ICLEI Japan and ICLEI South East Asia, and financed by the Japanese government, project development for two CDM projects in Indonesia was undertaken (Kishigami 2007, 2009a). The projects were organized in a city-twinning between Indonesian and Japanese cities; they include waste management and composting methods, and use of used cooking oil for municipal trucks. The results of the project are not available in English language (Kishigami 2009b), and the projects were never submitted to the UNFCCC. Another CDM project had been developed in 14 municipalities in Madhya Pradesh, India, including CCP member city Gwalior (Chaturvedula 2009). It is to improve energy efficiency in the street lighting system. As of late 2009, the project had been submitted for host country approval (Chaturvedula 2009). A CDM project in the solid waste sector was considered in the cities of Cochin, Raipur, Shimla and Varanasi in India (ICLEI South Asia 2009a). Shimla is a CCP member, and CCP assists GIZ in exploring CDM potential in municipal solid waste management in all four cities. So far, ICLEI South Asia has conducted preliminary baseline analysis and estimations for CDM projects in the four cities (Chaturvedula 2009). None of the four projects identified above has passed the whole way down from project development through to UNFCCC registration, which shows that they were eventually a failure.

At COP10 in 2004, ICLEI hosted a side event "CDM opportunities for Local Governments" and launched a guide including some information on "how city administrations may also get funding for climate protection through the CDM". In 2007, at COP13 in Bali, ICLEI Japan organized a side event "CDM for Local Governments Session" as part of the "Local Government Climate Sessions" (ICLEI 2007). ICLEI South Asia provided support to Indian local governments during the development of CDM projects. This includes technical support, like help with conducting baseline analysis, and developing Project Design Documents. It also included facilitation in finding project partners, e.g. by helping cities to "identify potential buyers of CERs through various networks" and "engaging technology suppliers, investors, utility companies and buyers on a common platform" (Chaturvedula 2009).

8.5 Discussion

On the one hand, cities in developing countries offer a substantial potential for greenhouse gas emissions reductions. On the other hand, municipalities only rarely engaged directly in the development of CDM projects, while private consultancies were able to get CDM projects in cities off the ground. Even the city initiative ICLEI was struggling to mobilize its members to engage in the CDM. While some of its members did CDM projects this happened mostly without ICLEI being aware of it.

Table 8.5 summarizes the identified barriers and presents ways to overcome them that may be explored in the design of the market mechanisms under Article 6. A key way forward would be an increased focus on the benefits other than carbon credit revenues that can be harnessed by carbon market activities. Enhancing visibility for the local politician and thus his chances to be re-elected can make a carbon market activity an asset instead of a liability which draws valuable resources in terms of staff and only brings benefits to the politician's successor.

It is clear that municipalities will only to a limited extent care for profit and thus always be sidelined by private companies solely motivated by profit. But the latter leave aside the more costly and difficult to mobilize "higher-hanging fruit". Therefore, the challenge will be how to combine private profit motives with the policymaker's aim to show to his electorate how he improves their lives. If these two motives can work in tandem, the future for carbon market activities in cities will be bright.

Under the market mechanisms of Article 6, a city's entire mitigation action plan and its implementation likely qualify. This would mean that the combined measures a local government takes to reduce greenhouse gas emissions would have to be bundled into one activity. This would probably include a variety of different technological solutions and also support programmes and regulatory activities. Such a mix was not feasible under the CDM, whose rules excluded regulatory activities.

Table 8.5 What prevented local governments to engage in the CDM?

Type of barrier	Barrier	Explanation	Possibilities to overcome barriers
Economic	Upfront costs of CDM projects/budgetary constraints of municipalities		Multilateral financing facility for municipalities, with payback in CERs after the first issuance; partnering with private actors
Institutional/legal	Limited authority of local governments	Restrictions by state level for local governments to invest/deal with foreign exchange	Devolution of state-local financial frameworks
Informational	Limited carbon market experience and technical know-how of local bodies		Partnerships between CDM consultants, local technical consultants and city governments, ICLEI CDM cell
Institutional	Uncertainty about CDM future post 2012		Timely reach international agreement
Institutional	Limited feasibility of important project types	Lack of programmatic approach, few approved methodologies for transport/buildings/etc	Top-down development of methodologies for urban sectors, including programmatic CDM
Institutional	Complexity of CDM procedure	Leads to high transaction costs, requires expertise	Partnerships between CDM consultants and city governments, ICLEI CDM cell, use of existing standard urban methodologies
Institutional	Length of project cycle	Incompatibility between nature of political process and project cycles	CDM EB to use part of its surplus to de-bottleneck the CDM process
Political/ cultural	Lack of ownership and political will	CDM saw as strange instrument parachuted from industrialized countries	Show in simple but drastic way how CDM can improve the daily life of citizens

(continued)

Table 8.5 (continued)

Type of barrier	Barrier	Explanation	Possibilities to overcome barriers
Political/ cultural	Changing priorities	With mayor electoral cycles or departure of key political supporter city priorities may shift	Ensure commitment by major local stakeholders and different political parties, also by contractual commitments (e.g. with private partners)

Sources Arikan (2009), Chaturvedula (2009), Clapp et al. (2010), Kishigami (2009a, b), Monroy (2009), Ritter (2009), Santos-Borja (2007), Wilson (2007); *last column: authors*

References

Arikan Y (2009) ICLEI Experience on climate change and ideas on access of local governments to carbon finance. Presentation at the Carbon Expo, Barcelona, 28 May 2009

Betsill M (2001) Mitigating climate change in US cities: opportunities and obstacles. Local Environ 6(4):393–406

Black-Arbeláez T (2003) The state of development of national CDM offices in central and South America. Barriers to the deployment of CDM potential. The Andean center for economics in environment, Bogota. www.ucd.ie/envinst/envstud/CATEP%20Webpage/Powerpoint/paris%20pres/thomasblack.pdf Accessed 26 Jul 2011

Bulkeley H, Kern K (2006) Local government and the governing of climate change in Germany and the UK. Urban Stud 43:2237–2259

Betsill M, Bulkeley H (2007) Looking back and thinking ahead: a decade of cities and climate change research. Local Environ 12(5):447–456

Bai X (2008) Integrating global environmental concerns into urban management: the scale and readiness arguments. J Ind Ecol 11(2):15–29

Boyd E, Hultman N, Roberts JT, Corbera E, Cole J, Bozmoski A, Ebeling J, Tippman R, Mann P, Brown K, Liverman DM (2009) Reforming the CDM for sustainable development: lessons learned and policy futures. Environ Sci Policy 12(7):820–831

C40 Cities (2009) Sao Paulo, Brazil. Sao Joao and Bandeirantes landfills. CC40 Cities. http://www.c40cities.org/docs/casestudies/waste/sao-paulo-landfill.pdf. Accessed 23 Nov 2009

Chaturvedula S (2009) ICLEI South Asia personal communication, 13 Nov 2009

Clapp C, Leseur A, Sartor O, Briner G, Corfee-Morlot J (2010) Cities and carbon market finance: taking stock of cities' experience with clean development mechanism (CDM) and joint implementation (JI). OECD environmental working paper no. 29. Available via OECD Publishing: https://www.oecd.org/env/cc/46501427.pdf. Accessed 3 Dec 2020

Cruz S, Paulino S (2013) Local use of resources from clean development mechanism projects in landfill sites in the city of Sao Paulo. Ambient Soc 16:117–137

Dhakal S (2009) Urban energy use and carbon emissions from cities in China and policy implications. Energy Policy 37(11):4208–4219

Delhi Metro Rail Corporation Ltd (2019) Notice calling quotations. Selling CERs, GS-VER, GS-CER. http://www.delhimetrorail.com/otherdocuments/922/Quotation-for-Sale-of-CERs-and-GS-VERs-18-July-2019-Final-19072019.pdf. Accessed 20 Sep 2019

Gilbert A (2008) Bus rapid transit: is transmilenio a miracle cure? Transp Rev 28:439–467

Grütter Consulting (2009) Monitoring report CDM project 0672: BRT Bogotá, Colombia: trans-Milenio Phase II-IV Monitoring Period 1.1.2008–31.12.2008. http://cdm.unfccc.int/filestorage/

F/J/H/FJHZMG35I72D9WUY6L4QXPATB1VSEK/Monitoring%20Report%203.pdf?t=V1R8bHhnMG90fDDcqg1dutKbEyhxdyr_JOXw. Accessed 7 Jan 2012

Guiza D (2009) Metro de Bogotá: is this really what the city needs right now? http://www.bluehaired.com/2009/09/metro-de-bogota-is-this-really-what-the-city-needs-right-now/. Accessed 23 Nov 2009

Gordon R, Li W (2012) Provincial and local governments in China: fiscal institutions and government behavior. In: Fan J, Morck R (eds) Capitalizing China. national bureau of economic research and university of Chicago Press, pp 337–369

Holm Olsen K (2007) The clean development mechanism's contribution to sustainable development: a review of the literature. Clim Change 84:59–73

ICLEI (2007) CDM for local government session. (draft 26 Nov 2007)

ICLEI (2009a) Cities for climate protection (plus subsites). http://www.iclei.org/index.php?id=800. Accessed 22 Oct 2009

ICLEI (2009b) Local government climate roadmap: status report. Jun 2009

ICLEI South Asia (2009a) ICLEI South Asia assisting GTZ in evaluating CDM Potential. http://www.iclei.org/index.php?id=10015. Accessed 4 Sep 2009

ICLEI South Asia (2009b) Energy efficiency-CDM project in 14 cities of Madhy Pradesh. http://www.iclei.org/index.php?id=6920. Accessed 4 Sep 2009

ICLEI (2010) Cities in a post-2012 climate policy framework. Climate financing for city development? Views from local governments, experts and business. Bonn, Germany

ICLEI (2011) Cities for climate protection (CCP) campaign (plus subsites). http://www.iclei.org/index.php?id=10829. Accessed 18 May 2011

India Power Trading (2011) Delhi metro sells 82,000 CERs at € 4.5/CER. Times of India says. http://indiapowertrading.info/delhi-metro-sells-82000-cers-at-e-4-5cer-times-of-india-says/. Accessed 20 Sep 2019

Kishigami M (2007) Alliance of local governments for the development of CDM projects. Presentation at the ICLEI parallel event to the UN climate change conference 2007 in Bali, CDM for local governments session, 10 Dec 2007

Kennedy C, Steinberger J, Gasson B, Hansen Y, Hillman T, Havranek M, Pataki D, Phdungsilp A, Ramswami A, Villalba Mendez G (2009) Greenhouse gas emissions from global cities. Environ Sci Technol 43(19):7297–7302

Kishigami M (2009a) Strategy of ICLEI towards sustainable urban development and challenge of the climate change. Presentation at CITYNET Yokohama Congress, 8 Sep 2009

Kishigami M (2009b) Personal communication, email 17 Nov 2009

Lindseth G (2004) The cities for climate protection campaign (CCPC) and the framing of local climate policy. Local Environ 9(4):325–336

Lecocq F, Ambrosi P (2007) The clean development mechanism: history, status, and prospects. Rev Environ Econ Policy 1(1):134–151

Labriet M, Caldes N, Izquierdo L (2009) A review on urban air quality, global climate change and CDM issues in the transportation sector. Int J Glob Warm 1(3):144–159

Michaelowa A (2003) CDM host country institution building. Mitig Adapt Strateg Glob Chang 8:201–220

Michaelowa A (2005) CDM: current status and possibilities for reform. HWWI research paper 3, Hamburg. https://www.hwwi.org/uploads/tx_wilpubdb/HWWI_Research_Paper_3.pdf. Accessed 3 Dec 2020

Michaelowa A, Jotzo F (2005) Transaction costs, institutional rigidities and the size of the clean development mechanism. Energy Policy 33(4):511–523

Michaelowa A (2007) Unilateral CDM—can developing countries finance generation of greenhouse gas emission credits on their own? Int Environ Agreements: Politics, Law Econ 7:17–34

Michaelowa A (2009) Interpreting the additionality of CDM projects: changes in additionality definitions and regulatory practices over time. In: Freestone D, Streck C (eds) Legal aspects of carbon trading. Oxford University Press, Oxford, pp 248–271

Michaelowa A, Hayashi D, Marr M (2009) Challenges for energy efficiency improvement under the CDM—the case of energy-efficient lighting. Energy Effic 2(4):353–367

Monroy M (2009) Creating urban carbon access: developer perspective. Presentation at Carbon Expo, Barcelona 28 May 2009

Mr JP, Tovey K (2010) Overcoming barriers to implementation of carbon reduction strategies in large commercial buildings in China. Build Environ 45(4):856–864

Okubo Y, Michaelowa A (2010) Effectiveness of subsidies for the clean development mechanism: past experiences with capacity building in Africa and LDCs. Clim Dev 2:1–20. https://doi.org/10.3763/cdev.2010.0032

Price B (2004) Message from CCP program director. In: ICLEI CCP India monitor, 4/2, Nov 2004

Point Carbon (2011) Carbon project manager. Database of CDM and JI projects owned by Point Carbon. http://www.pointcarbon.com/trading/cpm. Accessed 6 Aug 2009

Qi Y, Ma L, Zhang H, Li H (2008) Translating a global issue into local priority: China's local government response to climate change. J Environ Dev 17:379–400

von Ritter K. von (2009) Capacity development—one element in cities agenda for a future climate regime. Presentation at the cities, climate change and finance symposium, Barcelona, 26 May 2009

Rahman S, Dinar A, Larson D (2010) Diffusion of Kyoto's clean development mechanism. Technol Forecast Soc Change 77:1391–1400

Skinner N (no date) Carbon offset opportunities. ICLEI CCP Fact sheet

Sippel M (2004) Global climate policy and corresponding activities on a city-level. HWWA discussion paper 280, Hamburg

Santos-Borja A (2007) CDM for local governments: an innovative approach for lake basin management. Presentation at the ICLEI parallel event to UNFCCC COP 13 in Bali, CDM for local governments session, 10 Dec 2007

Schmidt Dubeux C, La Rovere E (2007) Local perspectives in the control of greenhouse gas emissions—the case of Rio de Janeiro. Cities 24(5):353–364

Sharma P (2007) Clean development mechanism (CDM) opportunities for MP local governments. Presentation at the ICLEI parallel event to UNFCCC COP 13 in Bali, CDM for local governments session, 10 Dec 2007

Satterthwaite D (2008) Cities' contribution to global warming: notes on the allocation of greenhouse gas emissions. Environ Urban 20:539–549. https://doi.org/10.1177/0956247808096127

Sippel M, Jenssen T (2010) What explains cities' climate policy-making? A review of drivers and barriers. J Environ Manag Tour 1(1):39–56

Sovacool B, Brown M (2010) Twelve metropolitan carbon footprints: a preliminary comparative global assessment. Energy Policy 38:4856–4869

Sippel M (2011) German cities, their climate mitigation activities and the potential of city-partnerships. J Resour Energy Dev 8:37–48

Shah N (2017) Delhi Metro. UNFCCC registers DMRC's solar power initiative as CDM project (March 28, 2017). https://www.metrorailnews.in/delhi-metro-unfccc-registers-dmrcs-solar-power-initiative-cdm-project/. Accessed 20 Sep 2019

Terraza H, Willumsen H, Guimaraes H (2007) Landfill gas capture. Design versus actual performance and the future for CDM projects, http://siteresources.worldbank.org/INTLACREGTOPURBDEV/Resources/840343-1178120035287/EditedLFGWorkshopReportAugust14.pdf. Accessed 7 Jan 2012

UNFCCC (2009) Decisions adopted by the conference of the parties serving as the meeting of the parties to the kyoto protocol. FCCC/KP/CMP/2009/21/Add.1

UNFCCC (2011a) CDM: CDM statistics. http://cdm.unfccc.int/Statistics/index.html. Accessed 18 May 2011

UNFCCC (2011b) CDM: project activities. http://cdm.unfccc.int/Projects/index.html. Accessed 2 May 2011

UNFCCC (2018) Monitoring report for 'renewable energy PoA in India. https://cdm.unfccc.int/filestorage/Y/C/0/YC0VBZNG2XASUJT94H8OMIF1R6W7LQ/DMRC%20Monitoring%20Report_v1.1.pdf?t=NGl8cHk0dXV1fDDi9_0RtzVNp5Gw6N8OMNWJ. Accessed 20 Sep 2019

UNFCCC (2019a) Project 0672: BRT Bogotá, Colombia: TransMilenio phase II to IV. https://cdm.unfccc.int/Projects/DB/DNV-CUK1159192623.07/view?cp=1. Accessed 20 Sep 2019

UNFCCC (2019b) Project 0164: bandeirantes landfill gas to energy project. https://cdm.unfccc.int/Projects/DB/DNV-CUK1134130255.56/view?cp=1. Accessed 20 Sep 2019

UNFCCC (2019c) Project 0373: São João landfill gas to energy project. https://cdm.unfccc.int/Projects/DB/DNV-CUK1145141778.29/view. Accessed 20 Sep 2019

UNFCCC (2019d) Project 4463: metro Delhi, India. https://cdm.unfccc.int/Projects/DB/SQS1297089762.41/view. Accessed 20 Sep 2019

UNFCCC (2019e) CPA 6161-P1-0002-CP1: DMRC solar PV PRoject. https://cdm.unfccc.int/ProgrammeOfActivities/cpa_db/A9WKHVXZJ7FT1OEDC8RMIYU2BNG56Q/view. Accessed 20 Sep 2019

UNFCCC (2019f) Project 0079: Kuyasa low-cost urban housing energy upgrade project, Khayelitsha (Cape Town; South Africa). https://cdm.unfccc.int/Projects/DB/DNV-CUK1121165382.34/view. Accessed 20 Sep 2019

Ürge-Vorsatz D, Koeppel S, Mirasgedis S (2007) Appraisal of policy instruments for reducing buildings' CO_2 emissions. Build Res Inf 35(4):458–477

Vennemo H, Aunan K, Fang J, Holtedahl P, Hu T, Seip H (2006) Domestic environmental benefits of China's energy-related CDM potential. Clim Change 75(1–2):215–239

Wilson C (2007) Barriers and drivers to the implementation of the clean development mechanism within the Nelson Mandela bay municipality: a case study. Dissertation, Rhodes University

Zegras PC (2007) As if Kyoto mattered: the clean development mechanism and transportation. Energy Policy 35(10):5136–5150

Zhao X, Michaelowa A (2006) CDM potential for rural transition in China case study: options in Yinzhou district, Zhejiang province. Energy Policy 34:1867–1882

Chapter 9
Climate Change Mitigation Action Enablers for Climate Smart Cities

Massamba Thioye

Abstract Cities governments can play an important role as climate actors. They are in a unique position that entitles them to conduct a broad range of possible climate actions. They can:

- Facilitate emission avoidance by fostering change of consumption behavior of their citizens leading to displacement of carbon intensive products
- Enable GHG emission reduction by the citizens in their jurisdiction;
- Reduce the emission related to the infrastructures they build and operate.

The first category of climate action is the most transformative. Cities, through city action plan, are well-equipped to mobilize this potential by mainstreaming disruptive climate action into city development action.

This chapter describes key categories of mature climate change mitigation solutions related to finance and technology that can be readily deployed to support climate actions by cities.

Green bond is identified as an important enabler for climate action by city. However, for that instrument to be effective, it must be reformed to address identified barriers. A reformed green bond framework to support climate action by cities is described and its benefits discussed.

Digital technologies can facilitate both the development and the deployment of these climate mitigation actions by reducing cost and by streamlining processes.

Keywords GHG emission reduction · GHG emission avoidance · GHG emission reduction enabler · Climate solutions · Transformative climate action · Financing instrument · Reformed green bond framework · Digital technology · Core needs

Disclaimer The content of this chapter has no relationship whatsoever with the UNFCCC process or the work of the secretariat. Its findings, interpretations and conclusions do not reflect the views of the secretariat. The secretariat does not vouch

---

M. Thioye (✉)
UN Climate Change Secretariat, Bonn, Germany
e-mail: mthioye@unfccc.int

Haus Carstanjen, Martin-Luther-King-Strasse 8, 53175 Bonn, Germany

K.-G. Kim and M. Thioye (eds.), *Planning Climate Smart and Wise Cities*,
The Urban Book Series, https://doi.org/10.1007/978-3-030-80165-6_9

for the accuracy, currency or completeness of the data or information contained in the article and excludes all liability for any loss or damage arising (including through negligence) in connection with and as a result of any reliance on the data and information contained in the article and/or its use. Questions concerning the analysis and views expressed in the article shall be addressed to the author.

9.1 Introduction

Global warming is one of the biggest challenges ever faced by human social, political and economic systems. Its impacts are now 'playing out in real time,' with the extreme weather we have experienced these last years. Their severity could increase exponentially in the absence of bold responses. The stake is the survival of humankind. These disruptive issues raised by climate change require, to address them and hopefully turn them into opportunities, a combination of solutions leveraging disruptive responses from all level, including at the city level.

For cities to respond to that threat, considerable low carbon and climate resilient investments, including in cities hard (e.g., buildings) as well as soft (e.g., IT services) infrastructures are required to achieve the SDGs and the 1.5 °C goal. Indeed, according to an article published in nature climate change (Felix Creutzig et al. 2016) and the KPMG report on sustainability addressing cities infrastructure (Nick C et al. 2012), urban infrastructure has become one of the most pressing challenges facing the world today. This has been echoed by the World Bank (The World Bank 2020). The urban infrastructures in cities' action plans make them important for climate mitigation action: their implementation can lock cities in for generation with specific infrastructures and land use patterns. They are also important for adaptation: they can enhance the cities' adaptive capacity to climate and disaster risk.

To be aligned with the climate goals, cities need to go beyond pursuing incremental changes. They are expected to (i) deploy disruptive leadership and reinvent entirely new way of satisfying citizens' needs through transformative changes as well as (ii) secure the well-being of and a sustainable living for both their actual and future generations of citizens. This requires sustainably addressing (i) challenges related to the core needs of citizens as well as (ii) resilience to the global challenges. Disruptive leadership in that context means moving away from short-termism and bounded perspective, zooming in and out to see the bigger picture but also the interactions across issues with different time horizon and stimulating integrated approaches as holistic response to the short (e.g., COVID-19), medium and long-term global challenges (e.g., climate change). It also means collaboration between cities and their enabling and enabled entities. More than anytime in human history, cities need responsible leaders, able to create global public goods and achieve long-term benefits. Incremental and sector-based changes alone are not enough to achieve the triple bottom lines objectives for short, medium and long-term global challenges. Need and system-based transformative solutions are required, leveraging a culture of innovation that addresses technology, policy making, financial instruments and business model. Proaction to contribute in preventing medium to long-term global challenges become as important as reaction to short-term city level issues.

With such disruptive leaders, cities can become climate 'smarter' and 'wiser'. Cities' government as asset owners but above all as key enabler of sustainability actions, can deploy innovative climate policies and financial instruments, enabling their citizens to access more sustainable products and services, for the satisfaction of their needs. This will trigger a virtuous circle as heading toward sustainability through the deployment of these climate solutions, will enable cities to build talent, stimulate the creation of new businesses with disruptive business models and inspire to investors disruptive financial products, which in return will accelerate cities' movement toward sustainability.

9.2 Climate Action by Cities and Related Enabling Instruments

9.2.1 Features of the Enabling Instrument for Climate Action by Cities

As such, there is not one optimal pathway for the sustainable development of cities, but rather a range of possible pathways. In addition, none of these pathways is a one-size-fits-all solution. It rather depends on contextual factors and priorities. However, there are common aspects. The transformational change expected from cities requires their action plans to enable the massive deployment of existing mature climate solutions but also the development of new disruptive ones. The climate solutions such as innovative climate leadership, policies, technologies, financial instruments and business models are to be integrated. These combined climate solutions should provide to citizens' but also to their key enablers (financiers, private sector suppliers of good and services, technology providers):

- Direction for action
- Incentive to act
- Power to act
- Capability to act
- Opportunity to cooperate on action.

Direction for Action

The policies and regulations should be part of a clear vision articulated through a short, medium and long-term strategy, enough stable to give to all the actors predictability about the future evolution of the policy and regulatory framework.

It will include for example (i) a climate solutions road map covering both deployment of existing climate solutions and vertical transfer of future disruptive ones, (ii) the related investment plan as well as (iii) the enabling climate and investment policies. Indeed, finance has been a significant barrier to the deployment of climate solutions in many cities and very often required policy intervention to provide greater certainty to investors. It remains a major challenge in many developing countries. Targets and timetables are also needed.

Incentive to Act

Enabling actions by city government should target, for climate solutions relevant to cities, the reduction of their:

- Capital cost
- Financing cost and
- Operation cost.

The capital cost of cities' climate solutions is reduced by promoting investment in R&D. For example, the development of third-generation thin film cells based on earth-abundant materials and with higher conversion efficiency to replace silicon-based solar photovoltaic energy would lower its cost and enhance its performances.

The financing cost of cities' climate solutions is reduced by de-risking the related projects and providing them access to low-cost financing. Climate solutions very often needs public finance support, particularly in early stage project development, in the form of grants, subsidies and guarantees to facilitate its development where it is too risky for private sector to invest. The scarce public finance should less and less be used for self-financing or direct lending where the private sector would anyhow have invested. This would crowd out the private sector. It should rather be used to crowd in the private sector by helping meet its risk/return expectations where it would otherwise not invest. The IFC Amundi launch in March 2018 of the world's largest targeted green bond fund focused on emerging markets is an example of initiative that can help central governments give to cities, access to low-cost financing. It facilitates the issuance of green bond by other financial institutions in emerging market through active investment in their green bond issuance. The capital market is in a phase of exceptional liquidity. Policy makers, through the International Financial Institutions could multiply fundraising campaigns from a larger network of institutional investors worldwide to scale up and replicate this initiative, with the view to giving to their cities access to low-cost financing.

Policies encouraging competitive markets will relatively reduce the operating cost of climate solutions if they level the playing field by (i) removing subsidies to carbon intensive value-chains, (ii) and internalizing their negative externalities as well as the positive externalities of climate solutions.

Power to Act

Policies and regulation should leverage innovative technologies to empower all potential climate actors, including individual citizens. For example, individual consumer should be able to choose the source of energy they would like their consumption to come from. They should also be empowered to be prosumer able to discharge their surplus of electricity in a dynamic distributed electricity ecosystem that leverages Distributed Ledger Technologies functionalities, ubiquitous connectivity, intelligent sensors and devices, information and operations technology, and data-driven machine-learning functionality across the grid value chain.

Opportunity to Cooperate on Actions

City governments are shaping up as key actors accelerating cooperation on actions. They are more and more putting in place municipal projects, initiatives, policies and funds to help finance the green cooperatives. These cooperative cities are partnering with their citizens to foster greater co-design and delivery of local services. This can facilitate the implementation of climate actions by cities, enabling change of consumption behaviors leading to GHG emission reductions by the citizens as well as displacement of carbon intensive products and value-chains. It allows also to reach citizens that are most in need, in an efficient manner. Citizens get what they need at the right time and place.

9.2.2 Climate Action by Cities and Related Enabling Instrument

There are 5 main types of climate actions relevant to cities.

Avoiding emissions through replacement of existing carbon intensive products or value-chains

The first type of possible climate actions replaces existing carbon intensive value-chains intended for the satisfaction of given citizens' core needs with alternative 1.5 °C aligned value-chains satisfying those same core needs. The disruption of the value-chain arises mainly from city's action plan and the outcome is scope 4 avoided emissions. Example of such types of climate action includes enabling smart and wise cities, compact, complete, connected and coordinated to expand the 3C model (Floater et al. 2017), where homes, jobs, schools, community services, parks and recreation facilities are easily accessible. This will encourage active transportation and efficient use of infrastructure. It also includes the use of material of lower carbon footprint for the construction of cities' infrastructures as well as green procurement by the city government. While there are some readily available climate solutions in that category, a lot of effort is required for the development of new innovative ones, able to affect the way citizens' core needs of shelter, nutrition/health, access and lifestyles are satisfied, disrupting existing carbon intensive value-chains for a transformative response to climate change. The outcome will enable cities to radically expand the space of climate action. Incubators and accelerators, technology innovation hubs, policy innovation hubs, finance innovation hubs, business model innovation hubs, venture capital funds seeded with public finance at a first loss position accessible to startups will be critical to fully mobilize the potential of this category of climate actions. Technology push policies supporting climate solutions developer will also be required to complement the existing technology pull policies focusing on users of climate solutions. Green bonds can finance the scaling up of climate solutions that have reached a certain level of maturity.

Enabling GHG emission reduction through enhanced efficiency of citizens' consumption of goods and services

The second type of climate action enhance the efficiency of citizens' consumption of goods and services. The impact is scope 3 emission reductions arising for example from the city investing in infrastructures and equipments that enable enhanced circularity of the economy, extending the time products stay in a high value state circulating in value-chains and facilitating the reduction of waste. The infrastructures made available by cities track products in the economies and enable the recovery of as much value as possible from them, even after the end of their economic lifetime. Example of such climate action include developing integrated waste management systems to improve recycling (e.g., constructing waste recycling center or making available equipment such as glass container), with waste-to-energy facilities to recover energy from waste that cannot be recycled and proper disposal of residual waste that cannot be combusted. Another type of climate action in that category is the promotion of sharing economy. Examples of such climate actions include having dedicated space in public parking for car sharing. These climate actions require predominantly the deployment of existing mature climate solutions but also the development of new ones. On policy solutions, laws, taxes, policies and deregulations will be relevant. On financing, the following instruments can be relevant:

- Self-financing
- Debt financing
- Need satisfaction company (NSC)
- Citizen cooperative
- Crowd funding
- Venture capital fund for development.

Enabling GHG emission reduction through the decrease of the carbon footprint of products and services consumed by the citizens

The third type of climate actions decrease the carbon footprint of products and services consumed by the citizens. The impact is also scope 3 emission reductions arising from the use by citizens of the cities' infrastructure such as electric vehicle charging station for e-mobility, information and awareness campaigns on consumption of sustainable goods and services, capacity building and access to low-cost finance for energy efficiency improvement and enhanced deployment of renewable energies. This also includes promoting the production, distribution and use of renewable energy, including by enabling the development of energy storage facilities where groups of households can send their surplus of renewable electricity and from where they can get renewable electricity. Another example is the city building new low-carbon transport systems, upgrading existing transport systems and enhancing their interoperability to promote sustainable local mobility. These transport systems include Bus Rapid Transit, high-quality rail, river and sea transport. The same enabling instruments as the one identified for climate action enhancing the efficiency of citizens' consumption of goods and services apply also here.

Enabling GHG removal by the citizens

The fourth type of climate actions is the promotion of urban trees. It contributes in enhancing GHG emission removal but also have many others impact on mitigation. Indeed, it can contribute to a compact, complete, connected and coordinated city by increasing the local food and nutrition security. Trees can also protect building walls against direct solar radiation and help cool indoor air, reducing the energy required for air conditioning by 30% and for heating in winter by more than 20%.

Reducing the GHG emissions or enhancing the GHG removal of cities infrastructures

The last category of climate actions reduces the emissions related to the operation of cities' infrastructures. It leads to scope 1 and 2 emission reductions. These include reducing emissions related to the operation of public buildings, social housings, district heating and cooling systems, urban transport (transport fleets, freight logistics in urban areas, and modal change and shift), street lighting, water and waste water treatment, solid waste treatment, purchase of green energy. Green bonds, although not widely used could be the most appropriate enabling instrument, if some barriers are addressed. It will give the city government access to low-cost financing. This category also includes the emission removal arising from cities green infrastructures.

9.3 City Action Plan and Need-Based Solution-Oriented Approach to Climate Action

9.3.1 Cities Are Product Consumption Centers

Manufacturing plants have left cities to move to the non-city areas. The service industry is currently the main driver of urban employment. This means cities are more consumer than producer of high-carbon footprint products (energy, cement, iron & steel). They have externalized the emissions related to most of their consumption. Therefore, the sector-based problem-oriented approach to climate action aiming at reducing the GHG emission of big polluter is not very relevant to cities. The need-based solution-oriented approach mainstreaming climate action into SDG action is much more appropriate for cities. The city government, with the support of its enabling entities, can build a framework conducive to a change of pattern in the way its citizens are satisfying their core needs.

9.3.2 Importance of City Action Plan on Climate Change Mitigation

For cities, the potential of scope 4 avoided GHG emissions is much more important than the scope 1, 2 and 3 emission reduction potential. It is transformative, while the latter is incremental. Cities are in a unique position to mobilize this untapped potential through city action plan. This is what positions them as key climate actors. Through action plan, cities can use backcasting normative scenario, to translate their citizens' core needs into enabling needs and options for infrastructures, supporting their low emission and climate resilient development. Then cities can use forecasting exploratory scenarios as well as system analysis to identify how the different options for infrastructures serving a given core need will impact the way other core needs are satisfied. The system analysis will also identify how the options for infrastructures will impact the planet's needs. This will ensure an integrated approach to address holistically the interconnected pathways for the satisfactions of the citizens' core needs.

9.3.3 Need-Based Solution-Oriented Approach to Climate Action

Climate actions are currently conducted mainly to decarbonize the world's economy. The sectors where production is taking place and GHG emissions is occurring, are the field of deployment of these climate actions and they act on the problem: the GHG emissions. Indeed, effort is put (i) on the reduction of the GHG emissions of polluters by changing their production processes, their technologies, their raw materials or (ii) on the sequestration of their emissions. This approach to climate action is sector-based and problem-oriented. Under that framework, climate action is mainly undertaken as a response to actual regulations or an anticipation of future regulations. GHG emitters, in silo, uses incremental improvement to reduce the carbon footprint of their products and services at the lowest cost.

This scope of climate action is too narrow and not enough transformative to lead to climate solutions that can, in aggregate, achieve the 1.5 °C goal, as it does not cover the disruptive solutions leading to deep change in the way the world satisfies its needs. It also put into action only traditional climate actors such as corporates and national policy makers. It will not deliver the bold response needed to avoid the exponential increase in the severity of the climate change impact. That is the first gap climate action by cities could fill. The incremental and manufacturer-focused climate action from corporate which is sector-based and problem-oriented must be complemented with a need-based and solution-oriented approach to climate action from cities, that is disruptive, end-user focused and with a much broader and impactful domain of application.

Where the current incremental and manufacturer-based framework for climate action tries to reduce the GHG emission of the polluter, mainly through the development and deployment of new climate technologies and production processes, the need-based and solution-oriented approach of cities could explore alternative solutions for a sustainable satisfaction of the core need of their citizens. The former approach can accommodate that companies keep the same business model and products/services, simply changing the inputs, technologies and/or processes they use. The later however, will disrupt their markets and require companies in carbon intensive supply chains to shift their business model and/or change their products or services. Figure 9.1 indicates the focus of the sector-based and bottom-up approach of traditional climate actors as well as the focus of the need-based and top-down approach of cities. Cities contribute in further expanding the space for climate change mitigation actions.

For example, where national governments promote the GHG emission reduction of the cement and iron and steel sectors under the current incremental manufacturer-focused climate action framework, and supports the development of climate technologies to abate their emissions, under the need-based solution-oriented framework, cities can effectively complement their actions by promoting climate actions that enable the satisfaction of the basic human need related to shelter at the lowest level of emission. They can facilitate the development and deployment of climate solutions from startups, architects, construction companies, universities and research centers, aiming at introducing alternative enabling needs (e.g., digital technology making redundant part of the building floor surface needed for offices where possible), alternative products (e.g., where there is no negative impact on biodiversity, use to replace cement and concrete of sustainable wood for construction, recycled at the end-lifetime of the building) to satisfy the core need related to shelter at the

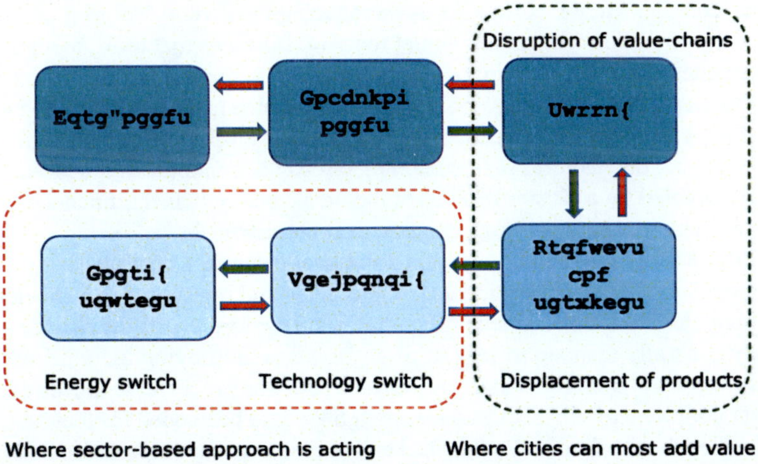

Fig. 9.1 Space where cities have the potential to add more value to climate change mitigation actions

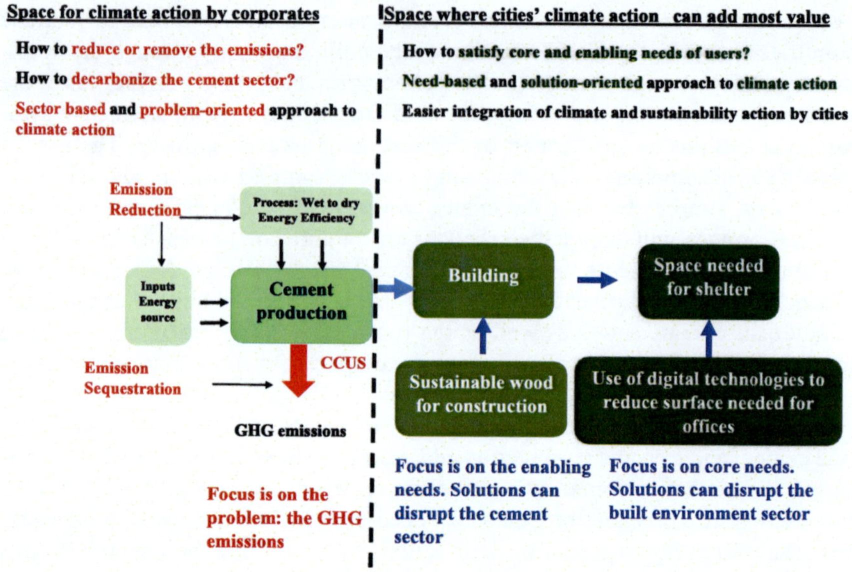

Fig. 9.2 Example of the value-chain to satisfy the core need of shelter; space for climate mitigation action by corporates and space where cities can most add value

lowest GHG emissions. Figure 9.2 shows how cities and companies can each best contribute to decreasing the carbon footprint related to the satisfaction of the core need of shelter.

Cities can take the lead in expanding the space for climate actions by mobilizing the potential related to need-based solution-oriented type of climate actions. They will then complement the current focus on reducing the carbon footprint of existing products by corporate with an additional focus on solutions that look for alternative products, alternative value-chains to satisfy the world's core needs. Climate action by cities will then add the low-emission development agenda to the pure decarbonization agenda. As it responds to the question how we can develop while being aligned with the 1.5 °C goal, this requires mainstreaming climate action into SDG action.

Cities enabled by a cluster of entities including policy makers, financiers, technology providers and individual citizens, can contribute in delivering integrated 1.5 °C low-energy demand compatible solutions, through a combination of innovative policies, financial instruments, technologies and consumption behaviors. In the current climate action framework, polluters are the key targets for climate action and they generally respond to regulation in isolation, improving incrementally the carbon footprint of their products, services or production processes at the lowest cost. They are not prone to self-disruption, although it could be needed for their survival, for example when they must 'disrupt or be disrupted'. On that aspect also, cities can demonstrate leadership in climate action. They can facilitate the creation of climate solution clusters, with networks including climate solution users, climate solution developers and climate solution enablers, tasked to leverage digital technologies and identify new low-carbon market opportunities cities can support (i.e., through testing environment such as sandboxes). The climate solution cluster can include corporates, city government policy makers, financiers, technology providers and citizens. It will undertake integrated climate action that includes the development and deployment of effective climate policies, technologies, financial instruments, business models and leadership, for solutions that sustainably contribute to the satisfaction of core needs of the citizens and are aligned with the climate goals.

Cities are in a unique position to expand the climate action space and:

1. Establish additional roles for new climate actors such as urban planners, architects, construction companies: sources of emissions such as corporates are no more the only key target for climate actions, but also cities as solutions providers;
2. Reframe the challenge: From improving existing systems sector by sector, facility-by-facility to developing new transformative system solutions that identify the best low carbon, people-centered, sustainable way to meet needs in society;
3. Move to clusters: Instead of looking at individual stakeholders as climate actors (i.e., investors, regulators, project developers), focus on clusters and especially the enablers of climate action.

The expansion of the climate action space will enable a broad range of new transformative and impactful types of climate action and at unlocking an important untapped potential of GHG emission reduction.

9.4 Climate Solutions Related to Finance

Cities play an important role in building infrastructure. Urban economic growth needs to be supported by services including housing, transportation, water and sanitation. As part of their action plan, cities need to develop financing roadmaps for the realization of new green infrastructures and the upgrade of existing ones, considering their specificities. However, in all cases, these financing roadmaps are expected to limit the pressure on public budget. Indeed, cities are facing challenges to sufficiently finance the fast-expanding demand for urban services from swiftly growing populations. This failure to fully deliver the urban services has more negative impact on vulnerable citizens. To implement their action plan, it is difficult for cities to rely only on self-financing. Where this is the case, budgetary constraints can be a barrier to adequately deliver the urban services expected by the citizens. Indeed, cities' green infrastructure generally have a long lifetime and high-upfront cost. Therefore, debt should almost always be an important part of the financial mix, where the municipal borrowing regulatory framework allow cities to borrow. Apart from the regulatory constraints, the exclusive use of self-financing is most of the time related to limited awareness of all the available options for mobilizing low-cost financing from the public and the private sector that could complement self-financing. In countries with established national-level financial maturity (Floater et al. 2017), there are often national and international financing programmes relevant to cities in a form of grant, loan, guarantee or equity. To use them in the most effective way, they should be blended with private finance. Indeed, the considerable amount of financial resources required for cities' infrastructures cannot be provided by the public sector alone. Commercial banks have been lenders to cities. But, in many countries, they have relatively high-interest rate and short to medium-term loan maturity date. Finance from the capital market, including from institutional investors, should be considered. It can fill the gap. Indeed, finances are largely available. Global capital markets are in a phase of exceptional liquidity with the real interest rate on risk-free assets hovering around zero. However, raising debt from the capital market to finance cities' projects is uncommon, particularly in developing countries. The private sector is not enough involved in investment for city infrastructure projects, perceiving them as being too complex and diverse, not common and financially risky, with very limited risk mitigation tools, individually too small, hence presenting liquidity risk and not warranting the costs of developing the necessary due diligence. This section discusses some of the financing options available to cities, with a focus on conditions for cities to have a better access to the capital market.

9.4.1 *Self-financing*

Cities' self-financing is a model whereby they invest only using their own revenues they have generated. The self-financing of cities' plan has some economic and social advantages. They can learn how to design and implement suitable policies, regulatory and institutional arrangements and manage them to finance their own development, rather than borrowing money (Turnbull 2002). Indeed, it requires from cities greater fiscal accountability, transparency, and democratic public engagement, to be able to generate the necessary local revenues, that have the advantage of being more stable, reliable, and consistent compared to other forms of finance (Nelson 2020).

Another benefit of this financing model is the absence of financial cost and therefore the lower burden it put on taxpayers. However, while this financing model is very attractive, it is not always feasible as it requires enough own resources from local taxes for the upfront investment as well as enough revenues generated by users and service fees which are material only exceptionally.

9.4.2 *Citizen Cooperative Financing*

Citizen cooperatives rely on their members in the local community or municipality to invest in projects which benefits them after implementation, when in operation. The members of the cooperative can be individual citizens, households, non-commercial organizations, public entities and SME's which have purchased a cooperative share.

All these members can benefit from the profits. Citizen cooperatives can be involved in a variety of projects covering entire cities. They get easier access to the services delivered by the projects they are involved in. In the energy sector, the cooperative can disrupt the centralized market dominated by large utilities and facilitate the emergence of decentralized market with many active energy citizens. Cities can adopt financing model whereby citizens' cooperatives co-own and co-manage cities infrastructure. This can contribute in reducing the pressure on cities' budget.

9.4.3 Crowd Funding

Crowd funding raises money from many people, typically via the internet, each contributing a relatively small amount. In return for their contributions, the funders can get tangibles or intangibles benefits, depending on the nature of the campaign. The campaign generally takes place on crowd funding internet-based platforms.

Cities can use all the three categories of campaigns to mobilize financial resources:

- Donation-based campaign whereby contributors receive nothing in return for their contributions.
- Reward-based campaign whereby donors receive goods or services in exchange for and proportional to their contributions.
- Lending-based campaign which are peer-to-peer lending whereby debt providers are repaid at a maturity date with an interest rate in exchange for financing a project of the city.

Unlike banks and other traditional financial intermediaries, crowd funding platforms do not borrow or lend money. They will simply connect the fundraisers and backers, providing information and advice about the activities that will be financed with money fundraised.

9.4.4 Need Satisfaction Company (NSC)

Energy efficiency improvement is an important climate change mitigation action. By reducing the energy-expenditure required to produce a certain amount of good or satisfy a given service, it reduces the associated GHG emission if the energy is fossil fuel based. Massive energy efficiency potential exists in cities, in the built environment and the transportation sector. Energy efficiency improvements are among the climate mitigation actions with negative marginal abatement cost.

It may be interesting to broaden the concept of energy efficiency from 'energy efficiency in production' to 'efficiency in satisfying needs'. Figure 9.3 below shows the difference between energy efficiency in production and efficiency in satisfying the core needs.

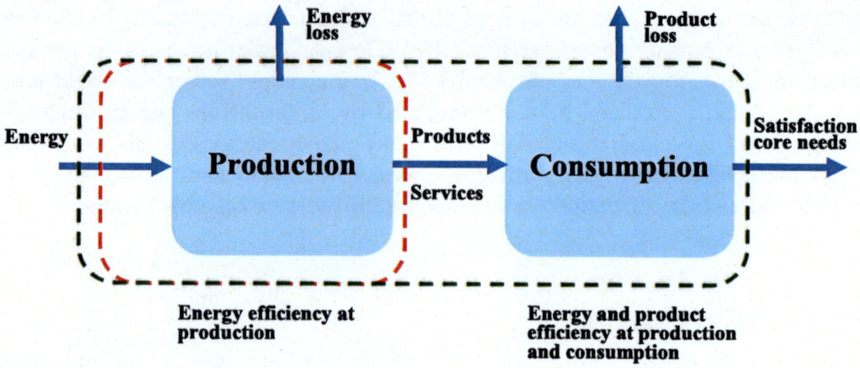

Fig. 9.3 From energy efficiency in production to efficiency in satisfying the core needs

While the energy efficiency focus only on how the production can take place with the lowest amount of energy, the efficiency in satisfying the core needs addresses in addition:

- Inefficiency in making products or services accessible to the final consumer.
- Inefficiency of the products or services in enabling the satisfaction of the core need.
- Amount of products losses.

Where municipalities lack financial resources, expertise and skills to enhance efficiency in satisfying the core needs of the citizens, they could promote the development of need satisfaction companies (NSCs). They could be companies with the required technical skill and financial profile that can be interested in investing in projects increasing the efficiency in the satisfaction of core needs of citizen in exchange for a share in the profit created by this investment. Energy service companies (ESCO) would be a specific case of these NSCs.

Need satisfaction companies could leverage digital technologies and support smart and wise cities in increasing the circularity of core need satisfaction value chain through value-added usage of by-products (recover, recycle), extended lifetime of products or their parts (repurpose, remanufacture, refurbish, repair, re-use), promotion of smarter purchase and use (reduce, rethink, refuse). For example, in the nutrition/health value chain, cities can contribute in avoiding the loss of part of the 1.6-Billion-ton of food lost or wasted every year, with a total value of $1.2 trillion, by supporting the development of NSCs. As for ESCOs, the NSCs will share the gains, in the form of products which loss has been avoided, with the relevant stakeholders.

9.4.5 *Bank Debt Financing*

An important challenge for cities willing to implement green infrastructures is the high-upfront investment needed. This is especially the case for small cities, with constrained financial resources. Bank debt is a financing instrument available to cities in established national-level financial maturity. It has the advantage of giving them access to capital they lack for self-financing. However, the cost of financing can impact the profitability of the cities' green infrastructure project. Debt financing is therefore only suitable where the cost of financing is not too high to make the viability of a green project get below the cut-off viability level. Revenues from municipal taxes and revenues generated by users and service fees pay back the debt. Cities can obtain a bank loan from available public lending programs, National Development Banks or Multilateral Development Bank or a commercial loan from a commercial bank.

9.4.5.1 National Development Banks and Multilateral Development Banks Funds

The development banks have an important role to play in supporting sustainable development. Most of the National Development Bank and the Multilateral Development Banks for instance have prioritized sustainable finance and made a pledge to green their portfolio. As examples, six multilateral development banks made a joint statement at Paris in 2015 to substantially enhance their climate financing (eib 2015). The World Bank has pledged $200 billion for climate action between 2021 and 2025 (The World Bank 2018). The development banks can contribute in creating an enabling environment for blended finance, by de-risking green projects to crowd in the private sector, although they are still more provider of direct low-interest loans to both public and private sector. Development banks are also deploying other financial instruments including dedicated credit lines through local financial intermediaries. However, development banks can be constrained by their mandates to directly finance cities. This prompted the preparation by a group of stakeholders including C40 cities, of a paper scoping a dedicated green cities development bank (Alexander et al. 2019).

9.4.5.2 Commercial Banks

Cities do not always have access to financing from development banks even if, based on their national regulation, they are able to borrow. In such case, they can turn to the commercial banks that offer loan at market interest rate and with shorter maturity date. The interest rate that will be applied to them will depend on their creditworthiness.

9.4.6 Green Bond

Banks have been one major supplier of loans to local authorities. However, commercial bank's loans tend to have higher interest rate compared to the financing cost available from the capital market. Furthermore, it is anticipated that bank loans will be more expensive due to the new financial regulations under Basel III.

Traditional sources of financing cities' infrastructures, such as self-financing and bank financing have proved to be insufficient to address the needs for sustainable development of cities. Green bond issuance can contribute in filling the gap. However, so far only a few big cities have issued green bonds to mobilize resources from the capital markets. Indeed, there are challenges hindering the use of such financial instruments.

9.4.6.1 Benefits Cities Can Draw from Green Bonds Issuance

Bonds in general are especially suitable for issuers with guaranteed revenue streams, implementing projects with long-term investment horizons and high-capital costs. Indeed, the cost of financing is lower than the cost of commercial bank debt financing and the maturity date can be adapted to the specific needs of a city. Compared to the bank debts, bonds can have longer term to maturity, e.g., up to 20 years or even more. The green attribute of the bonds enables the diversification of the source of financing and give cities access to non-traditional group of financiers for cities, including institutional investors such as pension funds, insurance companies, sovereign wealth funds, asset management companies and hedge funds that have sustainability objectives. It can also enhance the reputation of the city, building climate-friendly infrastructures.

9.4.6.2 Challenges Faced by Cities for the Issuance of Green Bond

Despite these benefits cities can reap by issuing green bond to finance their sustainable development, there has been only very little use of this financial instrument for the financing of green city action plan even where this complies with regulations, and just by large cities. The followings are identified as challenges:

- Issuance by cities do not always meet the risk/return expectations of the capital market, while for green bonds, the creditworthiness and the greenness of the issuers are both critical. The perception of risk is exacerbated as cities are not frequent borrower in the financial market.
- Issuing green bonds requires expertise and skills that are not always available in the city government.
- Issuing green bonds enhance the transaction cost of the issuance and does not necessarily lead to premium price for the issuer.

- The issue size of the green bond issuance needs to be enough large (minimum issue volume) to attract the financial markets and to cover the transaction cost related to the bond issue, such as advertisements and promotions, rating, verification of the use of proceed, certification of the green attribute of the bond.
- Cities governments need to display an overarching vision of investments across their project boundaries (KfW 2019) as well as an integrated approach to their implementation. (Floater et al. 2017) also bring out the need for coordinated governance, which form with compact urban growth and connected infrastructure the three C of the 3C model of urban development they have proposed. Cities' plan can facilitate that.
- Both the issuer and the buyer are exposed to the reputational risk of being accused of greenwashing.

To address some of these challenges, the entity issuing the green bond could be a Local Government Green Funding Agency (LGGFA) instead of an individual city. To address the remaining challenges, a new green bond framework is proposed.

9.4.6.3 Local Government Green Funding Agency as Green Bond Issuer

Green cities can create a LGGFA and use it as a vehicle to access the capital market for the financing of their action plans. The LGGFA acts as a cooperative bank owned by the cities. It consolidates their financial needs derived from their investment roadmaps and borrows from the capital market on their behalf through the issuance of a green bond, to finance their actions plans. The cities owning the LGGFA are the guarantor of last resort. For the LGGFA to operate effectively, the following are required:

- Cities are individually responsible for their solvability
- There is a strong cooperation among the cities
- Cities apply strict budgetary rules and good governance
- Cities members of the LGGFA have green action plans aligned with the climate goals.

The LGGFA addresses some of the challenges identified 8.6.4.2. with the following benefits:

- Cities reduce their financing costs and diversify the source of their financial resources
- The LGGFA can build technical and institutional capacities it would be more challenging to build for the individual cities
- The issue size of the LGGFA will be more attractive to financiers and will absorb the transaction cost
- Green bond backed by several cities, guarantor of last resort, diversify risk and can be tailored to meet the risk/return expectations of institutional investors.

9.4.6.4 A New Green Bond Framework that Address the Challenges Identified Above

Under this section, an issuer can be a city or a LGGFA.

(1) Green bonds to support impactful climate actions by cities

For green bonds to support climate actions by cities aligned with the climate goals, it should be scalable and effective in enabling additional greenhouse gas (GHG) emission reductions, arising from the implementation of plans for a green compact, complete, connected and coordinated city. Indeed, the use of green bonds to enable the development and implementation of green city action plans will be effective only if it can foster GHG emission reductions additional to what would happen if green bond was not issued. Otherwise, the green attribute of the bond would be useless, not adding value to the issuance of a vanilla bond. It is also useless to develop an instrument that can be effective in ensuring enhanced GHG mitigation outcomes but is not embraced by the targeted users, because it is not sufficiently attractive. Such an instrument would be relevant but not effective. The current green bond framework does not enable additional GHG mitigation and is not sufficiently conducive to the scaling up of the green bond market. It cannot address the barrier related to the reputational risk related to the accusation of greenwashing. It also does not justify why a bond buyer should pay a premium price. Finally, it introduces additional transaction costs, with no additional climate value created but just the green labeling of activities that would also take place even in the absence of a green label.

(2) Current definition of green bonds

In most of its current definitions, green bonds are differentiated from regular bonds solely by the demonstration and verification of exclusive use of the proceeds to finance or refinance 'green' projects, assets or business activities (European Commission 2016; EU Technical Expert Group on Sustainable Finance 2019; Ehlers et al. 2017; The World Bank 2015; UNDP 2016). While broadly accepted, this definition could be the key barrier limiting the capability of green bonds to serve the climate goals. In the course of preparing this book, only the Economist (2017) set out an alternative definition, explaining that 'simply using the proceeds for environmentally friendly investments is not enough' to make bonds green. However, that article does not provide guidance on what else is needed to make bonds green.

(3) Use of proceed of the green bond

While under the current green bond framework, for project bonds and asset-based security bonds the debt recourse is a project or group of projects, for UoP green bonds the recourse is the issuer even if the proceeds of the green bonds are earmarked for specific green projects. This means that the debt is provided to the city/LGGFA issuing the green bond based on the evaluation of its credit quality and not on the financial viability of specific infrastructure financed with the proceeds of the green bond. Furthermore, the different projects to build a compact, complete, connected and coordinated city are all interlinked, which requires a holistic approach for their evaluation. Therefore, the green attribute of the bond should not be evaluated solely based on the green attribute of specific infrastructure financed by the proceeds. This criterion needs to be complemented with additional criteria relevant to all the other projects in the action plan of the city issuing a green bond or the plans of the city members of the LGGFA issuing the green bond. The action plan of the issuer, taken holistically and including the projects not financed with the proceeds of the green bond, should contribute to the achievement of the climate goals. Otherwise, a city with an action plan which is highly carbon intensive, would be able to segregate the cleaner infrastructures from the action plan and issue a green bond for those projects. The positive contribution to the climate goals of the cleaner projects would be recognized, while the negative impact of the highly carbon intensive projects of the action plan would simply be ignored. The issuer of the bond would appear as a green city with a green action plan even though it is not.

It is nevertheless important to stress that the proposed green bond framework is forward-looking and addresses only activities contained in the city's action plan that are yet to be financed with the proceed of a green bond. The proposed framework does not consider infrastructures already built based on previous plans. For example, a city would like to change its 'misaligned' urbanization model and in future, develop in a way that is aligned with the climate goals. That city would be locked in with carbon intensive infrastructures but can still issue a green bond, while it continues to operate these carbon intensive infrastructures, resulting from its past misaligned urbanization model it plans to change. What is important is that its future activities in its current city action plan taken holistically including investment in new infrastructures, upgrade of existing infrastructures as well as early asset closures, reflected in its city action plan, are aligned with the climate goals.

With the focus on the city's action plan, one of the requirements for a bond to be green is that the action plan of the issuing city taken holistically should be green. It is not enough that the proceeds finance some specific green projects. However, the issuing city's past investments are not required to be green.

(4) Additionality of bonds labeled 'green'

a. Additionality requirements

The bond market is largely dominated by use of proceeds (UoP) bonds that best meet the risk/return expectation of investors: more than 80% of the bonds declared 'green' by their issuers are UoP bonds. Yet there are more and more voices challenging the suitability of UoP bonds, considered 'green' according to the above definition, in contributing to the climate goals. However, these skeptical voices have not always identified and correctly formulated the environmental integrity issue undermining the current green bond framework. Some arguments point out the fact that investors do not pay a premium and buy green and UoP vanilla bonds at the same price in the primary market. Paying a premium compared with vanilla bonds is neither a necessary condition for a bond to be green nor a sufficient condition. Others insist on the fact that bonds are primarily refinancing instruments, question whether they can trigger the decision to invest in green activities and challenge their effectiveness as a climate instrument. Given that the maturity date of debts from commercial banks generally occurs between 5 and 10 years after issuance, the perspective of possible refinancing can reasonably be considered as an important criterion for those commercial banks when financing a long-term project or a city implementing long-term projects. Indeed, in most developing countries, the availability of long-term green debt from the capital market for the refinancing of short-term green debt from local commercial banks can be a key enabler for green investment by cities. Finally, other arguments against the current green bond framework concern the absence of a causal relationship between the financing conditions of the green bond and the implementation of the green activities. However, to assess the effectiveness of a green bond, there is no need to establish such a causal relationship or to assess the extent to which the cost and availability of the proceeds has contributed to the investment decision-making process. It is also not necessary to assess whether the benefit provided by a green bond buyer, compared with buying a regular bond, is substantial enough to influence the cities' investment decision. These questions could, however, be relevant for determining the shade of green of the bond.

To fully serve the climate goals, green bonds issued by cities should be sufficiently attractive to mainstream financiers and contribute to scaling up activities that enable the development and implementation of a green city action plan that would otherwise not be implemented. "Otherwise" means that in the absence of the green attribute of its city action plan, the issuer would not be able to access the financing conditions offered by the proceeds of its green bond, and without those financing conditions, the interconnected activities which give to the city action plan its green attribute would not be viable. The financing conditions offered by the proceeds of the green bonds should therefore be more beneficial for the issuer than the prevailing baseline financing conditions. The benefit is measured not against the financing conditions that the buyer would offer for a regular vanilla bond but against the financing conditions that the city/LGGFA issuer would have access to in the absence of the green attribute of its city action plan. It is therefore not a necessary condition that the buyer pays a premium. Only where the financing conditions of the market from which the buyer obtains the UoP vanilla bonds are the same as the prevailing baseline financing conditions accessible to the issuer, is the benefit measured by comparing the financing conditions of the UoP green bond with those of a UoP vanilla bond, which means that the buyer has to pay a premium.

The buyer paying a premium, or more generally the city/LGGFA issuing the bond, drawing an economic benefit from the issuance of a UoP bond compared with the prevailing baseline financing conditions, is also not a sufficient condition for the bond to be green, even if the city's action plan is green. In addition to the requirement of being green, the activities in the city action plan should present a viability gap that the economic benefit contributes to filling. The proceeds of the green bond finance activities aimed at developing and implementing a city action plan that would otherwise not be implemented. Here, it is important to stress that filling the financial viability gap of green activities that would otherwise not be implemented is the condition, not financing green activities. The difference between them depends on whether the city action plan of the issuer would have been financed anyway with the prevailing baseline financing conditions or not.

A city action plan with a financial viability gap that the proceeds of the bond should finance for the bond to be considered green could be defined as a city action plan that is still not implementable (e.g., below the cut-off viability mark) under the available baseline financing conditions after all available sources of improvement of its viability have been exhausted, with the exception of accessing lower external financing cost. Possible sources to improve the viability include:

- Efforts from citizens accepting to pay higher taxes;
- Central government providing grants;
- Technology development reducing the cost or enhancing the efficiency of the climate and development solutions enabling the successful implementation of the city action plan;
- Citizens paying a premium for the users and service fees.

Green infrastructures aligned with the climate goals but with a financial viability gap under the prevailing baseline financing conditions will only be included in a city action plan if additional revenues or cost reductions that can contribute to filling that gap are available. The proceeds of a bond that do not provide a lower financing cost to the issuer or any other form of economic advantage compared with the prevailing baseline financing conditions will not contribute to filling that financial viability gap. Indeed, the proceeds of such bonds can only be used to finance activities that are viable under the prevailing baseline financing conditions. Instead of increasing the share of finance used to develop and implement green city action plan, such bonds will simply replace financing that would be available in their absence. Therefore, those bonds will not support the climate goals.

There is no need to establish a causal relationship between the issuance of the green bond and the viability of the city action plan it finances. It is enough to substantiate that the green bond has contributed to filling the financial viability gap. The benefit provided by the green bond's financing conditions may not by itself be sufficiently substantial to make the viability of the city action plan cross the cut-off mark, but it can contribute to doing so where it is blended with other benefits such as the payment of a premium by citizens for the users and service fees or grant from central government, as well as efforts from the suppliers of climate solutions to decrease their cost and increase their efficiency.

b. Why additionality of green bonds is necessary

Figure 9.4 illustrates the extent to which, depending on its definition, green bonds as a financial instrument can impact the ratio 'investment in climate solutions/total investment'.

Case A is consistent with the current definition of green bonds that does not require the filling of a viability gap. The development of the green bond market will, to a large extent, only promote the green labeling of financing activities directed to viable city action plans that would be implemented in any case, including in the absence of a green bond framework. Within the share of investment-in-climate solutions (the area

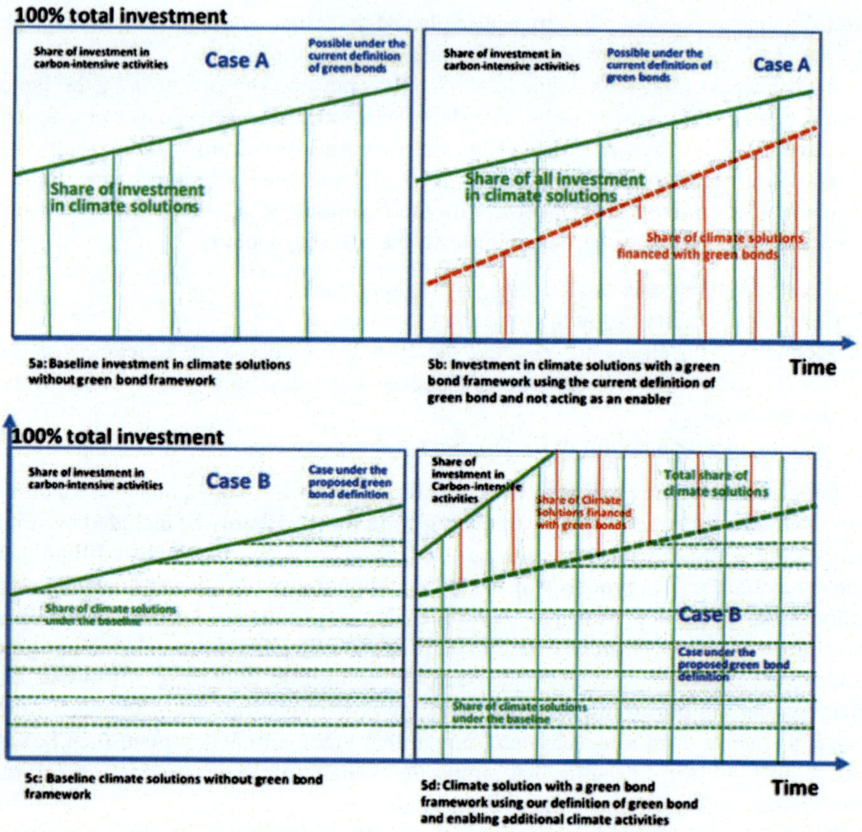

Fig. 9.4 Total share of investment in climate solutions part-financed through green bonds

below the solid green line) that would be financed in any case with or without green bonds, the current green bond framework can provide an incentive for increasing the green labeling of the bonds (the area below the red dotted line can increase). However, it does not enable the implementation of additional climate solutions (it does not lift the solid green line). Although a green bond framework applying the current definition of green bonds can be appealing for all stakeholders, it is not contributing to the achievement of the climate goals. Indeed, it allows:

(1) issuers to claim that they are implementing climate solutions;
(2) investors to claim that they are financing climate solutions; and
(3) validators and verifiers to further develop their business (the rise of the red dotted line).

However, the green continuous line representing the ratio 'investment in climate solutions/total investment' is not affected by the existence of a green bond framework if it does not require the filling of a viability gap (the ratio is the same under case

5a relating to a baseline without a green bond framework and case 5b relating to a baseline with a green bond framework).

Case B illustrates the implications of additionality in the green bond definition. As the green bond must fill a viability gap and finance climate solutions that would otherwise not be financed, the green bond as a financial instrument lifts the curve representing the ratio: 'volume of investment in climate solutions/volume of total investment.' The share of climate solutions financed with green bonds (the solid green line) will be on top of the share of climate solutions viable under the baseline financing conditions (the dotted green line). Under these conditions, it is clear that green bonds as financial instruments are contributing to the climate goals.

(5) Scaling up green bonds

Making genuine green bonds attractive for mainstream financiers enables to scale up the implementation of city action plans that would otherwise face a financial viability gap. That condition must be met for any green bond framework to foster the effective participation of the financial sector, beyond angel and climate-conscious investors, for the successful implementation of the Paris Agreement and the 2030 Agenda for Sustainable Development. This will make mainstream financial flows become an enabler for scaled-up and impactful implementation of green city action plans and is therefore an important part of the climate solution toolbox.

(6) New framework for green bonds

Compared with the prevailing baseline financing conditions, the proceeds of a green bond are expected to provide the issuing city/LGGFA with some economic benefits that contribute to filling the financial viability gap of a green city action plan that would otherwise not be viable. In addition, if the purpose of labeling a bond as 'green' is so that alongside the trading of the bond there is some form of recognition of the buyer's contribution to the achievement of the climate goals, simply buying a bond that finances or refinances 'green' projects, assets or city action plans, even if those activities support the climate goals, is not enough for the bond to be labeled 'green'. If the 'green' label associated with a bond means that the owner of the bond can claim impactful climate action contributing to the climate goals, the green attribute associated with the implementation of the financed infrastructures, in the form of mitigation outcome security (MOS), should be transferred from the issuer to the buyer of the bond, alongside the bond, in exchange for some benefits for the issuer. In the same way that a bond is an instrument of financial indebtedness of the bond issuer to the holders of the bond, the MOS is an instrument of carbon unit indebtedness of the bond issuer to the holders of the MOS. The bond issuer owes the holders of the MOS equivalent mitigation outcome units and is obliged to pay them at the maturity date, which can be the date when the real mitigation outcome units arising from the implementation of the city action plan are expected to be issued. This is comparable with the way in which the green electricity market operates, with the only difference being that renewable electricity certificates are issued ex-post after the consumption of renewable electricity, while the MOS are securities issued ex ante before the generation of the mitigation outcome units. The fact that the electricity

Fig. 9.5 Two households receiving conventional and green electricity from the same independent power producer

generated by an independent power producer (IPP) is green is not an enough criterion for the same electricity consumed by a household receiving electricity from that IPP to be considered green.

In Fig. 9.5, the same IPP provides electricity from the same wind power plant to two households, A and B. The electricity generated is green. Household A pays a premium to buy the electricity generated but also its green attribute, represented in this case by the renewable electricity certificates. Household B is interested in buying conventional electricity, not green electricity, and therefore, does not pay a premium or get the green attribute in the form of renewable electricity certificates. The certificates corresponding to the electricity bought by household B are not transferred but kept by the IPP. Although both households are fed with the same generated green electricity, household A is consuming green electricity, while household B is consuming conventional electricity. The transfer of the electricity's green attribute against the payment of a premium is the only difference that makes the electricity consumed by household A green and the electricity consumed by household B not green. Renewable electricity producers can sell either green or conventional electricity to final consumers depending on whether a renewable electricity certificate has been transferred from the renewable electricity producer to the final consumer.

Likewise, a bond issued by a city implementing green city action plan that supports the climate goals can be considered either a green bond or simply a plain vanilla bond depending on whether the green attribute of the activities in the city action plan (in this case materialized by the MOS) is transferred.

In the same way that the final electricity consumer pays a premium in order to obtain the renewable energy certificate which enables it to claim consumption of green electricity, the bond buyer should provide economic benefits compared with the prevailing financing conditions in order to receive the MOS and be able to claim that it possesses a green bond. The bond buyer will have the flexibility to buy a green bond with a desired shade of green or a vanilla bond. From the same bond issuance, some bond buyers can buy dark green bonds or light green bonds, while others buy vanilla bonds depending on whether they pay a premium or provide any other additional economic benefits, so that in return some MOS are appended to their bond. The shade of green will depend on the amount of MOS appended to the bond. The mitigation outcome units that will be used to pay for the appended MOS should be entirely, or at least largely, sourced from the city's action plan that was targeted to be financed with the proceeds of the bond at the point of issuance of the green bond. For example, mitigation outcome units cannot be bought from the carbon market and simply be combined with a regular bond to make a green bond without any mitigation outcomes coming from the implementation of the city action plan financed with the proceeds of the bond. The MOS is expected to represent the green attribute of the financed city action plan. However, mitigation outcome units can be acquired by an issuer from the carbon market and be used to further darken the shade of green of an already green bond, including to compensate for any deficit of mitigation outcome units that the issuer committed to delivering to the green bond buyer. The key difference here is that the mitigation outcome units acquired from the carbon market are used to pay only partly for the MOS of a bond that is already green because its proceeds have financed activities that have generated mitigation outcome units (but not as much as initially expected) and not to green a regular vanilla bond financing activities in a city action plan that are, taken holistically, not generating mitigation outcome units.

Figure 9.6 below shows a comparison between the green electricity market based on the trading of conventional electricity and renewable electricity certificates and the green bond market based on the trading of vanilla bonds and MOS.

The green bond framework proposed in this paper departs from the current green bond framework in that the later does not:

(1) Require providing the issuer with additional benefits compared with the prevailing financing conditions;
(2) Require the use of these additional benefits to fill the viability gap of city action plans aligned with the climate goals;
(3) Require the formal transfer of the green attribute of the activities in the city action plan (the mitigation outcomes associated with the implementation of the financed city action plan) from the bond issuer to the bond buyer as part of the green bond transaction;
(4) Address climate performance gaps in the activities of the issuer that are not financed with the bond's proceeds.

The lack of formal transfer of the green attribute in the form of MOS from the issuer to the buyer in the current green bond framework could explain why bond buyers are not eager to pay a premium. It also makes double counting and double claiming possible. From this analysis, it can be derived that the necessary and sufficient conditions for a bond issued by a city to serve the climate goals and therefore be labeled green are fivefold:

(1) The proceeds of the UoP bond are exclusively used to finance or refinance city action plans generating mitigation outcomes;
(2) The proceeds of the UoP bond provide an economic benefit to the city or LGGFA issuer compared with its prevailing baseline financing conditions;

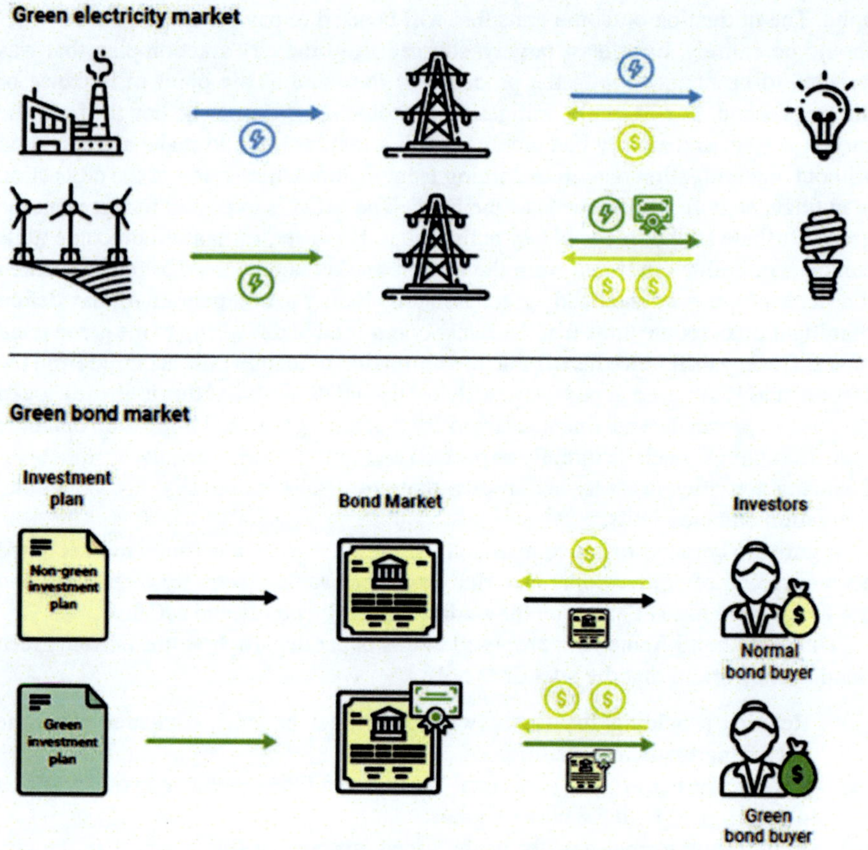

Fig. 9.6 Comparison between the green bond and the green electricity market

(3) The above economic benefit contributes in filling the financial viability gap of city action plans that would otherwise not be developed and implemented and the related mitigation outcomes would not be generated;

(4) In exchange for the economic benefit, all or part of the green attributes of the financed or refinanced activities in the city action plan, in the form of MOS, are transferred, appended to the bond, from the issuer to the buyer; with a robust carbon market, the bond buyer will make gain by selling the MOS;

(5) If the city action plan of a city, member of the LGGFA, has generated less mitigation outcome units than the MOS it produced, this city needs to fill the gap of climate performance by acquiring mitigation outcome units from other cities members of the LGGFA, if they have surplus of mitigation outcome units. Otherwise, the city with the deficit of mitigation outcome units should acquire them from the market. If the crediting of mitigation outcome units is not done at the city action plan level but activity by activity, then before

transferring the mitigation outcome units to pay for the MOS, any climate performance gap from other activities of the city's action plan should be offset using the mitigation outcome units generated by the green activities. Only the leftover mitigation outcome units will be used to pay for the MOS. If no mitigation outcome unit is left to pay for the MOS, the city is no more eligible to continue to claim the issuance of a green bond and it should acquire mitigation outcome units from the other cities, members of LGGFA, or from the market to compensate for the mitigation outcomes it did not generate.

a. Redefining green bonds

In the primary market, a green bond is characterized as follows:

- A bond acquired and to which are appended, MOS; future repayment of the MOS is made using mitigation outcome units that are at least partly generated by the city action plans its proceeds finance. These city action plans generate the mitigation outcomes by developing or deploying climate solutions mainstreamed in the city development actions. In exchange for the MOS, the bond buyer pays a premium or provides to the issuer another type of economic benefit compared with its baseline source of financing;
- The future mitigation outcome units intended for the repayment of the MOS and that will be generated by the activities in the city action plans financed by the proceeds of the bond, where credited at the activity level, should be used firstly to compensate any climate performance gaps in the city's activities. After these offsets, mitigation outcome units should be left for repayment of the MOS to the bond buyer. Otherwise, the city is no more eligible to continue to claim the issuance of a green bond and it should acquire mitigation outcome units from the other cities, members of LGGFA, or from the market, for the bond to be confirmed green;
- In return for the economic benefit arising from the issuance of green bonds, the cities fill the viability gap of their action plans which generate mitigation outcomes. It is to be highlighted that there is no need to track the proceeds of the green bonds to confirm what they are used for, in the implementation of the city action plan. It is enough to establish that the issuer has implemented a city action plan with a viability gap, which has generated mitigation outcome units and that after the offset of any climate performance gaps in the city action plan, there are mitigation outcome units left over, to transfer to the bond buyer for the payment of the MOS. What the proceeds were used for does not matter;

In the secondary market, a green bond is a bond acquired with, and to which are appended, all or part of the MOS that were transferred from the bond issuer to the bond buyer in the primary green bond market. The shade of green of the bond is measured by the amount of MOS appended to it per unit of financial resources.

It is important to further stress that while mitigation outcome units generated by activities different from those financed by a green bond can be acquired from the carbon market and used for the repayment of the MOS that were appended to a green bond to fill a gap in the mitigation outcome units generated by the activities financed with the proceeds, they cannot be appended to a normal bond to make it become green. This avoids a situation whereby a bond whose proceeds are not used to finance activities generating mitigation outcomes can become green simply because some mitigation outcomes generated by green activities and bought separately are attached to it.

While the current definition of a green bond was based only on the financing or refinancing of city action plans that generate mitigation outcomes, the above proposed definition requires the buyer to provide economic benefits compared with the prevailing financing conditions that enable the generation of mitigation outcomes that would otherwise not be generated, and the mitigation outcome units are transferred from the green bond issuer to the green bond buyer. Figure 9.7 provides a

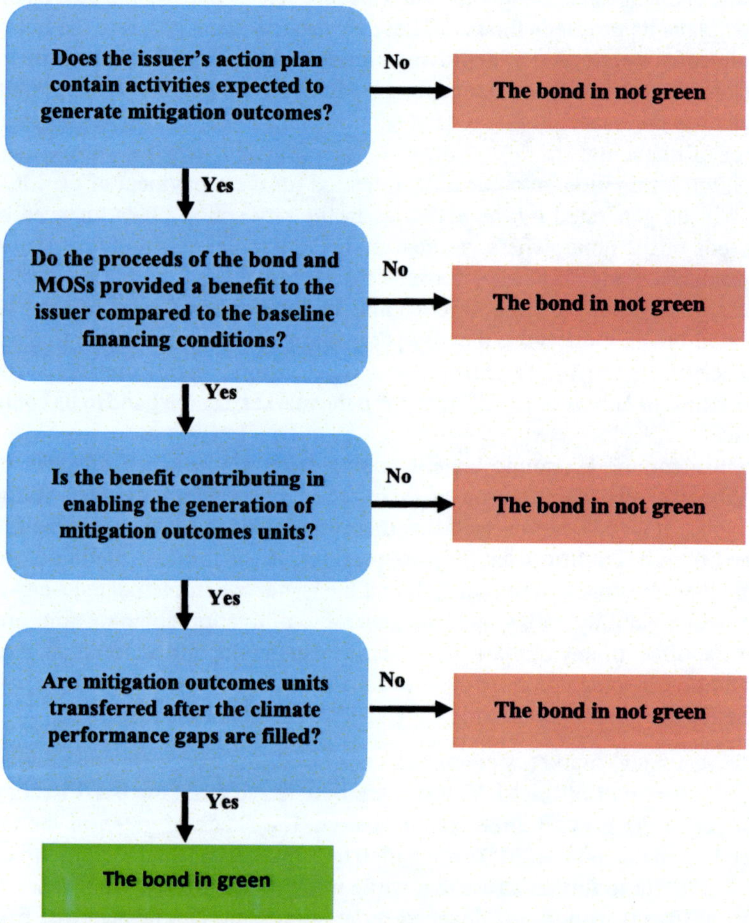

Fig. 9.7 Decision tree to assess whether a bond is green

decision tree to assess whether a bond is green.

b. Use of proceeds

There is no need to demonstrate that the proceeds of the bond have been used to finance infrastructures or activities generating mitigation outcomes. The issuer transfers to the financiers, alongside the bonds, mitigation outcome units that represent the green attribute of the green bond. The issuer should simply establish that:

- Mitigation outcome units generated by action plans financed with the proceeds have been used to offset any climate performance gaps related to the issuer's projects financed after the issuance of the bond.
- The leftover mitigation outcome units are transferred to pay for the MOS.
- .Compliance of issued green bonds with the proposed green bond framework.

The Swedish Energy Agency-Nigeria virtual pilot green bond appears to be the first issued green bond that offers an option that can comply with the proposed definition of green bonds. It is a sovereign green bond issued in Nigeria to mobilize low-cost financing from the capital market and enable Nigeria to exceed a mini-grid capacity target based on the use of renewable electricity that has been derived from its nationally determined contribution. International investors can opt to receive the mitigation outcomes in return for a reduced interest rate. This means that the interest to be paid by this sovereign green bond will be lower if the mitigation outcomes are transferred to the investors. The proceeds of the green bond will be used as concessional loans. This will contribute to improving the limited profitability of mini-grid investments and make them meet the risk/return expectation of potential project developers.

This sovereign green bond issued by Nigeria will therefore contribute to filling the competitivity gap of the mitigation projects it finances and be among the first green bonds in compliance with the proposed definition of green bonds. The anticipated incentive for the financier is the promise of mitigation outcome units that the project will generate. However, it will be an incentive only if the financier can make a capital gain from it. Hence, the importance of designing and implementing a framework that can transform the financier's acquisition of these mitigation outcome units into a capital gain.

(7) Conditions for the operationalization of the proposed green bond framework: integration of the green bond and carbon markets and related benefits

Carbon finance from the traditional carbon market, generated through results-based financing mechanisms, is not appropriate to fill the viability gap of green projects with no access to low-cost financing because:

- It is not an ex ante enabler that can make a green city action plan with a viability gap cross the cut-off mark of viability but rather an ex-post hypothetical reward and therefore does not always facilitate compliance with the risk/return expectations of the capital market;

- The limited access to the international capital market means that only financial resources with a high-interest rate will be available in many developing countries for green projects, resulting in an even more important viability gap to fill.

Instead of aiming at filling the viability gap of a city action plan or more generally a green projects in an investment plan of an issuer with no intend to leverage low-cost finance from the capital market, climate finance—including carbon finance—could be used where possible to make these green projects meet the risk/return expectations of the international capital market.

a. Integrating the green bond and carbon markets

One condition for a financier to be able to claim that it holds a green bond is that the MOS (future repayment of which is made using mitigation outcome units generated at least partly by the issuer's activities) must be appended to the bond. The bond buyer pays a premium or provides an alternative economic benefit equivalent to the cost of the appended MOS. The shade of green of the bond will be determined by the ratio MOS/price of the bond. The MOS are determined based on an ex ante calculation, while the related mitigation activities are only at the planning stage in the issuer's action plan.

A liquid market for the MOSs, driven by demands from governments, corporates, cities, other financiers and individuals, will be needed to attract financiers in an integrated green bond and carbon market as illustrated in Fig. 9.8 below. With a sustainable high demand for mitigation outcome units, mainstream financiers will be strongly incentivized to buy genuine green bonds that have MOS appended to them, which means investing in city action plans with activities that promise to generate mitigation outcomes. They will accept to pay a premium for a green bond if they are able to make an additional capital gain from it through the sale of the MOS in the carbon market. The more the demand is robust, leading to attractive prices for the

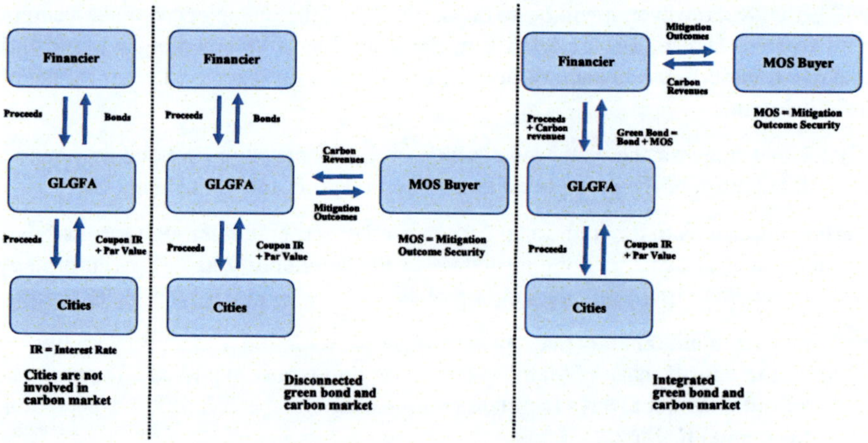

Fig. 9.8 Disconnected and integrated green bond and carbon market

MOS, the more investors will accept to pay the premium that will lower the financing cost of implementing city action plans. The green bond and carbon markets will then be integrated. The development of the carbon market leading to higher demand for mitigation outcomes will pull forward the green bond market that will provide the supply of mitigation outcome units. In addition, the development of the green bond market will foster the scaling up of activities supporting the climate goals.

To strengthen this incentive, governments and private sector stakeholders should (1) have ambitious mitigation objectives aligned with the climate goals; and (2) acquire mitigation outcome units to compensate any gaps in meeting those mitigation objectives.

For cities not involved in crediting mechanism, the city action plan is implemented without carbon finance. If there is no viability gap to be filled with a climate or carbon finance, then the activities in the city action plan are not additional and the bond is not green.

For cities involved in carbon markets disconnected from the green bond market, if the city action plan would not be implemented without carbon finance, then it is additional. However, the green attribute of the activities is not transferred to the bond buyer, therefore the bond is not green.

For cities involved in carbon markets connected with the green bond market, if the city action plan would not be implemented without carbon revenues from the green bond market, then the city action plan is additional, and the bond is green.

b. Key benefits of an integrated green bond and carbon market

The integration of the green bond and carbon markets as proposed above will have several benefits for the climate.

(a) Enhanced transparency, objectivity and environmental integrity

The proposed green bond framework will increase the transparency of the green bond market by providing an objective definition of green that does not provide room for misinterpretation and that is easy to assess. This will protect investors and cities from the accusation of 'green washing.' It will also address the issue of multiple claims in the current green bond framework under which both the issuer and the buyer can claim the green attribute of the activities.

(b) No need to track the use of proceeds

In the current green bond framework, significant efforts need to be deployed for the tracking of the UoP. However, this is not effective as it does not address the environmental integrity issue arising from the issuance of a green bond by a city which city action plan is dominated by carbon intensive infrastructures. Indeed, even if the city can establish that the proceeds of the bond are financing ring-fenced green projects in its carbon intensive city action plan, it should not be able to claim that its bond is green. The proposed framework addresses this issue while requiring no tracking of the UoP. Mitigation outcome units generated by the city action plan should be left for transfer after the offset of any climate underperformance of the activities in the city action plan. This (i) requires the investment plan to be green overall and

(ii) establishes that the proceeds are used to finance the implementation of a green investment plan.

(iii) Optimal use of public climate finance to enable impactful climate action

Ensuring the optimal use of public climate finance to enable impactful climate action is an important argument supporting the integration of the green bond and carbon markets.

To achieve the objectives of the Paris Agreement, the use of scarce public climate finance must be transformative, effective and efficient. The 'power to transform' attribute requires public climate finance to be used in a way that enables and catalyzes the actions expected from all stakeholders (subnational stakeholders, central government policymakers, technology providers and financiers) for the implementation of impactful climate solutions. It can have as a metric tons of carbon dioxide equivalent (t CO2 eq) reduced as a result of the public climate finance. The attribute of effectiveness requires the public climate finance to mobilize, for a climate solution, private finance that would otherwise not be available and in the absence of which the climate solution would not be possible. The related metric could be the volume of private finance leveraged by the public climate finance. The attribute of efficiency requires the maximum impact for a given amount of public climate finance. The related metric could be tCO_2 eq per unit of public climate finance.

(iv) Advantages for green financiers

Financiers will have the choice to either cancel the mitigation outcome to demonstrate a climate contribution or trade the mitigation outcome. The MOS can be an important source of capital gain for mainstream financiers. In addition, green bond buyers are protected against the incrimination of 'green washing'. As there will be no possibility of 'green washing', green bond buyers will be able to fully enjoy reputational advantages compared with financiers investing in activities that are misaligned with the climate goals.

(e) Governments buying from financier mitigation outcomes generated from developing countries' cities can claim the mobilization of private climate finance for the benefit of those developing countries

In addition to increasing the effectiveness and efficiency of climate finance, the proposed green bond framework could provide some other interesting benefits to governments buying MOS appended to green bonds issued from developing countries' cities. Indeed, buying those MOS would contribute to enabling green activities in developing countries and provide them with access to low-cost financing.

If mobilization of private finance for activities serving the climate goals is a core component of the climate finance, there is a need for a sound, objective and transparent approach to its accounting. There are many voices recognizing the limitations in the current accounting system for mobilized private finance. Under the proposed green bond framework, governments buying mitigation outcomes appended to green bonds issued from developing countries would directly contribute to the mobilization of their proceeds which, if considered as private climate finance, would be easy to quantify. Developed countries buying the MOS associated with the issuance of green bonds in developing countries would be able to claim the transfer to those developing countries of part of the green bond proceeds, proportional to the amount of mitigation outcomes they have bought. This could be an important incentive for developed countries to buy mitigation outcomes appended to green bonds from financiers rather than buying them directly from corporates or cities. Financiers can therefore expect a liquid market for the MOS transferred to them as the green attribute of the green bonds they have bought. In that sense, the development of the carbon market will drive the development of the green bond market. In addition, by enabling activities generating mitigation outcomes, the development of the green bond market will also support the development of the carbon market. The participation of financiers in the carbon market will also enhance its liquidity, facilitate price discovery and increase economic efficiency.

(f) Operationalizing the disclosure of exposure to climate risks and opportunities

Another important benefit of integrating the green bond and carbon markets is that it can support the operationalization of the disclosure of climate-related risks and opportunities.

Investors' financial assets can be exposed to high climate risks or opportunities, including physical climate risk originating from extreme rapid onset events, such as intense typhoons or wildfires, or from slow onset events such as sea level rise or ocean acidification. They can also be exposed to transition risk arising from policies and regulations implemented to combat climate change. Finally, the risk can originate from technological innovations developed as a response to climate change threatening the business models of a financier's investee organization. These risks can have financial implications for a financier's portfolio. It is therefore important to operationalize the regulators' recommendations on disclosure of climate-related risks and opportunities not only for corporates but also for financiers.

Countries buying MOS appended to green bonds could set as a criterion for a financier selling mitigation outcomes to participate in their bidding processes, the disclosure by that financier of the level of exposure of its financial assets to climate-related risks and opportunities.

This could be an effective incentive instrument to enhance financiers' willingness to (1) request their investee organization to disclose their climate performance; and (2) use these climate performances to measure the climate performance of their own portfolio.

(g) Turning carbon finance into an ex ante enabler to access the international capital market

In the current carbon market, carbon finance can be just an ex-post reward for cities that are able to mobilize finance and address the challenge related to the high upfront cost of their green projects. To date, countries with low capability to access low-cost finance have been able to implement only a limited number of the infrastructure projects they need for their development and have therefore prepared very few clean development mechanism projects under the Kyoto Protocol. This led to an imbalance in the regional distribution of clean development mechanism projects. The need to achieve the Sustainable Development Goals coupled with the scale and urgency of climate action requiring prompt action in all regions does not provide any room for the recurrence of a situation where less than 20% of Parties not included in Annex I to the Convention accounted for more than 80% of all clean development mechanism projects.

In an integrated green bond and carbon market, carbon finance will rather act as an ex ante enabler, an instrument that will facilitate access to the international capital market, of green city action plans in low capability countries, particularly when coupled with global de-risking instruments. This will boost the green bond market, diversify the supply side and help shift the USD trillions from mainstream investment toward green investments, including in developing countries. Carbon finance will no longer reward only cities that have already the capability to access finance. It will incentivize investors to buy genuine green bonds, the proceeds of which will finance additional green projects, and pay a premium for the MOS appended to them.

(h) Benefit compared to the EU Taxonomy combined with the EU Green Bond Standard

The combination of the EU Taxonomy with the EU Green Bond Standard improves significantly the current green bond framework. Indeed, if under the EU Taxonomy, only activities aligned with the 1.5 °C goal, are eligible, this is an important step forward. However, it is not enough. Because that framework does not include an instrument to fill any viability gap of green activities and financiers will not invest in green activities just for moral reason, it will promote only the green activities without any viability gap meaning those that would have been implemented anyhow even in the absence of the EU taxonomy and the EU Green Bond Standard. Therefore, this will not increase the volume of mainstream finance redirected toward green activities. It will simply increase among the financing activities that would have anyhow occurred, the share that will be labeled green bond.

The EU Taxonomy has also the advantage of being able to consider an entire city action plan with several infrastructures in it. It can address the issue of cherry picking whereby a city could bring out the green activities in its city action plan and be silent about the brown ones. However, the aggregation done across the portfolio cannot consider the shade of green or the shade of brown. For example, it cannot distinguish between (i) a combination 100 MW wind plus 100 MW oil and (ii) a combination 100 MW wind plus 100 MW coal. This issue is addressed by the green bond framework proposed in this chapter of the book.

The framework the book proposes is strictly forward-looking. It does not consider the existing operating infrastructure of a city. It focuses on the upcoming activities in its city action plan. Indeed, this is what is relevant to a debt provider as this is what its money will be used for.

(i) Empowering financiers to play a greater role in addressing climate change

The carbon market is an important instrument to make green investment economically competitive. Providing investors with an opportunity to directly influence its development alongside policymakers will empower them to contribute to achieving the climate goals.

Bond buyers will provide the financial resources needed to build the green infrastructures for the 2030 Agenda for Sustainable Development and the Paris Agreement. If they are accessible to cities at the condition that they issue bonds that are green with MOS appended to them, as a criterion, the cities' action plans will turn green. They should therefore be incentivized and empowered to play a role in the carbon market, supplying to governments the mitigation outcomes needed for compliance. This will result in financiers that are very influential stakeholders, being more engaged in requesting policymakers to make more ambitious commitments for an enhanced demand of MOS.

(j) Transfer of a tangible product the buyer must pay for as a condition for the bond to be green

The proposed definition of green bonds inherently addresses the issue of benefit for the issuer by requiring, as a condition for the bond to be green, the purchase by the bond buyer of MOS appended to the bond by the issuer.

(k) Flexibility in selling, purchasing, not purchasing or purchasing part of the mitigation outcome arising from the financed activities

The bond and the green attribute of the bond, in the form of mitigation outcome, can be sold separately. In the primary market, cities, issuer of the bond, can decide on its shade of green by appending more or less MOS to the bond. The buyer can also decide to buy bonds with or without their green attribute, as well as on their shade of green.

Bond owners can sell the bond in the secondary market, with fewer green attributes.

To make capital gain from the green bond, the bond owners will not rely solely on demand from climate-conscious investors in the secondary bond market. They can also sell separately the bond in the secondary market as a vanilla bond and the green attribute of the bond in the form of MOS in the carbon market.

9.5 Climate Solution Related to Technology

The world is experiencing an unprecedented period of technological upheaval: Internet of Things (IoT), Artificial Intelligence (AI), Distributed Ledger Technologies (DLTs), cloud-based quantum computing, 3D printing, 5G wireless, virtual and augmented reality, and others. While each of these technologies present exciting opportunities and significant challenges for sustainable development, their combination can be truly transformational. For example, combining IoT, AI, and DLT is critical to the full management of the data value chain and to radically streamline and enhance existing processes, create entirely new business models, and develop innovative products and services for a new generation of consumers. They enable the data discovery (collection, preparation and organization), the data distribution and integration as well as the data exploitation.

Indeed, these new technologies can be important enablers of climate and SDG actions by cities, thus serving the people and the planet. More specifically, they can be used to address the four issues identified below for a more effective climate policy making.

9.5.1 Challenge Related to MRV of Climate Action by Cities

9.5.1.1 The Tracking of MOS Can Be Challenging

An important feature of the proposed reformed green bond framework is the production and transfer of tradable mitigation outcome securities from the issuer of the bond to the buyer of the bond. These mitigation outcome securities differ from existing carbon units because they get their value from two sources: the GHG emission reductions they represent and the amount of finance they have shifted toward activities aligned with the climate goals.

Therefore, the framework should be designed so that one mitigation outcome security (representing the promise product), with a given capital intensity (amount of capital per ton of CO_2 abated) is paid with one mitigation outcome (the delivered product) representing the reduction of one ton of CO_2, achieved through a mitigation activity with at least the same capital intensity. As the mitigation outcome security is a promise to deliver a specific product, the description of the product should be immutably recorded on the mitigation outcome security. The information related to the actual amount of finance mobilized to undertake a mitigation activity should also be tied up to the mitigation outcomes that is generated by a mitigation activity. This information in the form of capital intensity of the climate action needs to be immutably linked to the mitigation outcome, in addition to a serial number and other information to trace the origin of the mitigation outcome.

The transparent reconciliation of the mitigation outcome security with a mitigation outcome used to pay for it requires trust on the production and transfer without alteration of the data they are associated with. The use of the emerging digital technologies including Internet of Thing, digital methodologies, smart contract and Distributed Ledger Technologies allows to weed out the use of a trustworthy intermediary to build trust on the data production as well as a third-party verification that can be burdensome, costly and time consuming.

9.5.1.2 Costly and Time-Consuming Processes Are Required to Trust the Production of Data

In climate policy making, transparent MRV of the impact of climate action is important. It allows for example a city to claim the effect related to its action without exposing itself to the accusation of green washing. It allows also the policy maker to recognize and where needed incentivize GHG emission reductions while being confident that they comply with the requirements set in its standards. The transparent determination of the impact of climate action also requires confidence on the quality of the data relevant to the determination of the impact. Therefore, a third-party verification that can be burdensome, costly and time consuming is required to build trust on the data production. With the use of IoT, DLT and AI, a large quantity of data can be automatically collected, securely transferred and accurately processed.

9.5.1.3 Costly and Time-Consuming Processes Are Required for Decision Making Related to the Climate Performance of an Activity

A city seeking recognition of the impact of its climate action is required to process the related data and use the outcome to establish its claim and demonstrate that it complies with the requirements set by the regulator. Because of the complexity of the requirements, demonstrating compliance can be very challenging. The regulator in turn, will analyze the claim, process again the data that have been verified before taking the decision to approve or not the claim. The decision is expected to be objective and consistent with its previous decisions.

Using digital technologies, the regulator can transform the requirements in its regulations into a smart contract. Instead of making a claim, the city will simply upload the data relevant to its activities. The compliance of its activities with all the relevant requirements will be checked automatically and if they are all met, the impact of its climate action will be calculated and automatically recognized. In some cases, digital sensor can be used for an automated acquisition of the data, with no human intervention.

9.5.1.4 Processes Required to Track Carbon Credit Attached to Bonds and Avoid Their Multiple Use is Costly and Time-Consuming

The mitigation outcomes securities are tradable assets. They are expected to be exchanged through traceable transactions that do not provide any room for double use. The trading of the mitigation outcomes securities represents value-transfer from the seller to the buyer. The use of digital technologies can ensure the transfer of the mitigation outcomes securities without the intermediation of a third party and without double use.

9.5.1.5 Ex Ante Attribution Based on Infrastructures in the City Action Plan

Cities are to a large extend enabler of climate action by their citizens. This represents a scope 3 climate contribution. Compared to scope 1 and 2 climate contributions, the measurement and attribution of scope 3 climate contributions, presents some additional challenges DLT can help address.

For the existing crediting mechanisms focused mainly in ex-post rewarding scope 1 and 2 GHG emission reduction of an entity, it makes sense to undertake the measurement of the impact of a climate action on the existing physical assets of that entity. The recognition and the reward will be result-based and done ex-post upon the implementation of the physical assets and based on their climate performance when operating. As it is an ex-post reward for result achieved and not an ex ante enabler to make possible the expected GHG emission reductions, it creates a signal that can enhance the willingness to act of the recipient but not its capability to respond to the signal. While ex-post reward of climate action is reactive and backward-looking, an ex ante enabler of climate action would be proactive and forward-looking.

A framework aiming at fostering scope 3 GHG emissions reductions cannot be based on a reactive and backward-looking mechanism. The ex-post attribution of scope 3 GHG emissions to a city as a result of the implementation of an action plan resulting in carbon intensive activities by its citizens, would look like a punitive measure in the sense that in most cases there is no more possibility of economically acceptable recourse, to avoid the emissions arising from the implementation of the action plan. Instead, if the scope 3 GHG emissions are attributed to the city on the basis of its action plan, it will be encouraged to change the plan and strive to improve its climate performances. This is a preventive measure to the development and implementation of carbon intensive city action plan. It is also preferable in the context of a framework enabling scope 3 GHG emission reductions, to provide to the city and its financiers, ex ante proactive and forward-looking enabling incentives rather than an ex-post reactive and backward-looking reward, available only after the implementation of the city action plan, the operation of its infrastructures and the achievement of the climate performances expected from them. Another important feature of the proposed green bond framework is that it put the accountability where it should be. Under the current green bond framework, financier evaluating the creditworthiness of a city, considering its possible future revenues from carbon credits, bear the risk if these revenues do not materialize. Under the proposed reformed green bond framework, because the MOS are commitment to deliver mitigation outcome units, the risk is borne by the green bond issuer.

Now, recognizing climate contribution based on information in a plan raises some new challenges. What if the built infrastructures are not the ones that were described in the city action plan? What if they are the same, but are operating with a climate performance below what was expected from the city action plan? Future correction of the attributed climate contribution will be required based on the gap between the performance of the implemented city action plan and the performance of the registered city action plan, based on which the climate performances were credited.

However, while a physically implemented infrastructure cannot be altered and is transparent, a simple information in a plan can be altered. Then the challenge is how to ensure that the information that was in the registered city action plan and that has been the basis for measurement and attribution of climate contribution will not be altered? This information is required for the correction to be trustworthy. Here again, the immutability function of DLT can help address the issue.

9.5.1.6 Measurement of Avoided Emissions or Scope 4 GHG Emissions

The disruption of carbon intensive markets and supply chains and the replacement of their products with less carbon intensive products satisfying the same need is one of the possible outcomes of climate actions by cities. To measure the climate contribution related to the consumption of the replacing products, their carbon footprint is compared with the baseline carbon footprint of products satisfying the need. This requires the tracking of embodied GHG emissions in the replacing products across their supply chains (Thioye 2021). DLT is needed to address this challenge.

9.5.1.7 Measurement of Impact of Action Enhancing the Efficiency in Satisfying Needs

Increasing the circularity of core need satisfaction value chain through (i) value-added usage of by-products (recover, recycle) and (ii) extended lifetime of products or their parts (repurpose, remanufacture, refurbish, repair, re-use) is an important climate action in cities. However, measuring its impact is challenging. It requires allocating carbon footprint to the by-products and tracking them where they are recovered or recycled, to ensure they displace products with higher carbon footprint. Otherwise, the allocated carbon footprint has to be re-allocated to the main product. It also requires the tracking of the lifetimes of products or their parts where they are repurposed, remanufactured, refurbished, repaired or re-used. Here again, DLT can help address this challenge.

9.5.2 Use Cases of Digital Technologies to Support MRV of Climate Action by Cities

The combination of IoT, DLTs and AI can serve the MRV of impact of climate action by cities. Indeed, it allows to address the issues identified under 8.5.1.

9.5.2.1 The Role of IoT

The IoT will enable to build trust around the production of the data upstream of the data value chain without the use of a third-party verifier. Where possible, signal from a sensor of a measuring instrument will generate data and send it to a DLT without human intervention. This strongly reduces the possibility of data manipulation at the production stage and enhance trust.

9.5.2.2 The Role of DLTs Combined with AI

The DLT will record the data with a time stamp and this information will be cryptographically sealed on an append only data, which makes it immutable. It is then distributed to all the nodes of the DLT. Digital methodologies related to the DLT will transform these data into climate contributions directly recorded and distributed to all the nodes. The immutability of the record and its distributed nature ensure that it cannot be altered. This functionality is critical for the tracking of carbon embodied in products along their value-chains. Where carbon footprint is allocated to a by-product, it enables its tracking without no possibility of corrupting the carbon footprint value. It also enables the tracking of the lifetime of products and their parts and where they are repurposed, remanufactured, refurbished, repaired or re-used, the measurement of their extended lifetime. If the information from the measuring instruments is confidential, the nodes will only receive the climate contribution calculated based on the confidential information.

There is no need for the cities to demonstrate and calculate anything and no need for the climate regulator to assess anything. The AI of the digital infrastructure will ensure objective and consistent decision making based on an automated assessment. Combined with IoT and cloud computing, DLTs and AI facilitates cheaper data-transfer, larger data storage, faster processing and enhanced accuracy. Without simplifying the requirements of the methodologies (such simplification can lead to inaccuracy in the determination of the climate contributions), the claim of climate contribution and the assessment of the claim will be much more cost effective and streamlined up to being close to real-time.

Where carbon units are issued and traded, the value-transfer can take place without intermediation and double use of the carbon units can be prevented. The decentralized function of DLTs with information about value-transfer validated only through consensus before it is recorded, the immutability and the distributed nature of the recorded information enable to get rid of the intermediation, while ensuring no possibility of double use.

The information in the city action plans, where it is the basis for measurement and attribution of climate contribution and the corrections that will take place after implementation, should not be alterable. This information will be directly recorded to the DLT, which will ensure it is immutable and distributed to all the node of the DLT. The nodes will be able to verify at the date when physical infrastructures were supposed to be build or renovated or retired whether this occurs. They will also be able to assess whether the expected performance in the city action plan is the achieved performance. Where the information in the investment plan is confidential, it can be translated into an emission pathway and this emission pathway will be recorded on the DLT.

9.6 Conclusions

To enable transformational changes in cities, a combination of enabling climate solutions such as innovative climate leadership, policies, technologies, financial instruments and business models is required. These climate solutions should provide to cities, their citizens but also their key enablers (financiers, private sector suppliers of good and services, technology providers):

- Direction for action
- Incentive to act
- Power to act
- Capability to act
- Opportunity to cooperate on action.

Climate solutions in cities support (i) avoided emissions through replacement of existing carbon intensive products or value-chains and (ii) the enabling of GHG emission reduction through enhanced efficiency of citizens' consumption of goods and services as well as through the decrease of the carbon footprint of products and services consumed by the citizens. They also support the enabling of GHG emission removal by citizens and the reduction of the GHG emissions or the enhancement of GHG removal of cities infrastructures.

City action plan can use top-down backcasting normative scenario, to translate their citizens' core need into enabling needs and options for infrastructures, supporting their low emission and climate resilient development. Then cities can use bottom-up forecasting exploratory scenarios as well as system analysis to identify how the different options for infrastructures serving a given core need will impact the way other core needs are satisfied. The system analysis will also identify how the options for infrastructures will impact the planet's needs. This will ensure an integrated approach to address holistically the interlinked pathways for the satisfactions of the citizens' core needs, in a connected and coordinated city. The incremental and manufacturer-focused climate action of corporates which is sector-based and problem-oriented is thus complemented with a need-based and solution-oriented approach to climate action from cities, mainstreamed into city development actions, disruptive, end-user focused and with a much broader and impactful domain of application. Cities enabled by a cluster of entities including policy makers, financiers, technology providers and individual citizens, can contribute in delivering integrated 1.5°Clow-energy demand compatible solutions, through a combination of innovative policies, financial instruments, technologies and consumption behaviors.

As part of their city action plan, financing roadmaps for the realization of new green infrastructures in cities and the upgrade of existing ones need to be developed, considering cities' specificities. Where borrowing complies with regulations and cities have guaranteed revenue streams, bonds are especially suitable for city action plans with long-term investment horizons and high-capital costs.

Over the last five years, UoP green bonds have developed and are now a source of funding for a wide variety of climate, environmental and social initiatives. To continue to develop cities while progressing toward achieving the climate goals, green bonds is a suitable financing instrument, for cities that can borrow from private financiers. They can drive markets to finance green city action plans and a sustainable world. However, cities do not always have in-house the expertise and skill to issue a bond. They also do not necessarily meet the risk/return expectations of the capital market. Furthermore, they can have difficulties to meet the minimum issue volume and absorb the transaction cost of the issuance. To address these challenges, the entity issuing the green bond could be a Local Government Green Funding Agency (LGGFA) instead of an individual city.

For green bonds to be effective in supporting low-emission development of cities, aligned with the climate goals, there must be a more impactful and more attractive framework that specifically recognizes and incentivizes the climate contribution of bond investors. Many bonds currently promote marginally green activities that could have been financed from a standard vanilla bond. Other bonds that address the Sustainable Development Goals can be appealing to cities, investors and validators/verifiers; however, their UoP often address green uses but do not contribute to the 1.5 °C climate goal as they are not enabling green activities that otherwise would not be possible. There is a dire need to accelerate the reform of the green bond framework to make it support urban development and the climate goals.

Mainstream bond investors buy bonds to meet their investment preferences. This means that a bond that funds climate improvements but delivers sub-par returns is not enough for most investors. The source of recourse in the event of default for many UoP green bonds is the bond issuer. The credit quality of the issuer is the primary security for the investors purchasing the bond. However, if investors could better visualize the credit quality of the cities (or their LGGFA) and be incentivized for investing in green city action plan, more liquidity would flow into genuinely green bonds, supporting the climate goals.

This limitation can be addressed through the proposed integrated green bond and carbon market via a new financial product: the MOS. The MOS would be akin to a warrant attached to a bond and would act as an additional source of return, tied to the amount of GHG emissions reduced by the green city action plan and the price of carbon. This is especially timely as talk of carbon pricing and carbon markets is advancing at pace. The MOS would be a commitment to deliver a unit of GHG emission reduction (a promised mitigation outcome unit) attached to the bond. It can be sold and traded immediately after the issuance of the green bond, before its maturity date, which is the date on which the actual mitigation outcome unit become due to the MOS holder. The maturity date of the MOS will be linked to the date on which the GHG emission reduction associated with the green activities of the city action plan, is expected to occur. This transforms the current carbon finance, which is an ex-post reward in the current carbon market, into an ex ante enabler in the form of additional benefit for the green bond issuer in an integrated green bond and carbon market.

To meet the conditions required for the green bond to be a financial instrument that serves the 1.5 °C climate goal, a new definition of green bond is proposed:

In the primary market:

- A green bond is a bond acquired with an appended MOS. Future repayment of the MOS is made using mitigation outcome units that are at least partly generated by activities financed by the proceeds of the bond. In exchange for the MOS, the bond buyer pays a premium or provides to the issuer another type of economic benefit compared with its baseline source of financing;
- The future mitigation outcome units intended for the repayment of the MOS and that will be generated by the assets or activities financed by the proceeds of the bond should firstly be used to compensate any climate performance gaps in the cities' investment plan. After these offsets, mitigation outcome units should be leftover for the repayment of the MOS to the bond buyer;
- As a counterpart of the economic benefit arising from the issuance of the green bond, the issuing cities fills the viability gap of the activities generating mitigation outcome units. What the proceeds of the green bond were used for becomes irrelevant;
- The shade of green of the bond is measured by the amount of MOS appended to the bond per unit of financial resources.

The proposed integrated green bond and carbon market framework has several benefits. If implemented, it will directly contribute to the climate goals by shifting mainstream finance toward effective climate solutions. It will also enhance the transparency, objectivity and environmental integrity of the green bond market by providing an objective definition of green bonds that does not provide room for misinterpretation and that is easy to assess, therefore protecting issuing cities as well as buying financiers from the accusation of 'green washing.' It will enhance the efficiency and effectiveness of climate and carbon finance by crowding in private finance. Developed countries buying MOS issued from developing countries can claim the facilitation of access to easily quantifiable private finance. Countries buying MOS, by setting as a condition for participation in their bidding processes, the disclosure by the financier of the exposure of its financial assets to climate-related risks and opportunities, can incentivize the implementation of initiatives concerning the disclosure of climate-related risks and opportunities. Financiers, that are influential stakeholders, will have additional reasons to undertake policy advocacy for enhanced climate ambition by policymakers, as this will lead to enhanced demand for MOS.

The implementation of the proposed reformed green bond framework requires the use of digital technologies such as IoT, DLT, AI and smart contract leveraging digital methodologies.

References

Alexander J, Nassiry D with Barnard S, Lindfield M, Teipelke R, Wilder M Financing the sustainable urban future Scoping a green cities development bank. https://www.odi.org/sites/odi.org.uk/files/resource-documents/12680.pdf

Eib (2015) Joint statement by the multilateral development banks at Paris, COP21 https://www.eib.org/attachments/press/joint-mdb-statement-climate_nov-28_final.pdf

Ehlers T et al (2017) Green bond finance and certification. Bank for international settlements, Basel, Switzerland. Available at: https://www.bis.org/publ/qtrpdf/r_qt1709h.pdf

EU Technical Expert Group on Sustainable Finance (2019) TEG report: EU green bond standard. Available at: https://www.politico.eu/wp-content/uploads/2019/06/EU-GREEN-BOND-STANDARD.pdf

European Commission (2016) Study on the potential of green bond finance for resource-efficient investments. European Union, Luxembourg. Available at: http://ec.europa.eu/environment/enveco/pdf/potential-green-bond.pdf

Creutzig F et al. (2016) Urban infrastructure choices structure climate solutions. Nat clim solutions 6:1054–1056

Floater G, Dowling D, Chan D, Ulterino M, Braunstein J, McMinn T (2017) Financing the urban transition: policymakers' summary. Coalition for urban transitions. London and Washington, DC. http://newclimateeconomy.net/content/cities-working-papers

KfW (2019) Research Focus on Economics Green bonds—a sustainable alternative for municipal infrastructure finance? https://www.kfw.de/PDF/Download-Center/Konzernthemen/Research/PDF-Dokumente-Fokus-Volkswirtschaft/Fokus-englische-Dateien/Fokus-2019-EN/Fokus-No.-245-March-2019-Green-Bonds.pdf

Nelson K (2020) Municipal finance: self-reliant cities generate their own revenue https://urban-links.org/insight/municipal-finance-self-reliant-cities-generate-their-own-revenue/

Nick C, Stephen B, Julian V, Darryl M (2012) Cities infrastructure: a report on sustainability. https://assets.kpmg/content/dam/kpmg/pdf/2012/05/Cities-Infrastructure-a-report-on-sustainability.pdf

The Economist (2017) What makes bonds "green"? The economist, London, UK. Available at: https://www.economist.com/the-economist-explains/2017/07/04/what-makes-bonds-green

The World Bank (2015) What are green bonds? The world bank treasury and public—private infrastructure advisory facility (PPIAF), Washington D.C., USA. Available at: http://documents.worldbank.org/curated/en/400251468187810398/pdf/99662-REVISED-WB-Green-Bond-Box393208B-PUBLIC.pdf

The World Bank (2020) Urban development. https://www.worldbank.org/en/topic/urbandevelopment

Turnbull S (June 27–30, 2002) Self-financing economic development markets and institutions network of the 14th annual meeting on socio-economics, University of Minnesota, Minnesota, USA

Thioye M (2021) Planning for climate smart cities, chapter 11 New advanced GHG accounting for cities enabled by DLT

UNDP (2016) Green bonds. UNDP, New York, USA. Available at: http://www.undp.org/content/sdfinance/en/home/solutions/green-bonds.html

Urban Development https://www.worldbank.org/en/topic/urbandevelopment/overview

World Bank (2018) The World Bank Group Announces $200 billion over Five Years for Climate Action https://www.worldbank.org/en/news/press-release/2018/12/03/world-bank-group-announces-200-billion-over-five-years-for-climate-action

Chapter 10
Market Mechanisms in the Building Sector from an Urban Perspective: Creative Building Solutions

Kwi-Gon Kim and Hee-Sun Choi

Abstract The global contribution from buildings toward energy consumption, both residential and commercial, has steadily increased reaching features between 20 and 40% in developing countries and has exceeded the other major sectors: industrial and transportation. In European countries, buildings account for 40% of total energy consumption. Growth in population, increasing demand for building services, and comfort levels, together with the rise in time spent inside buildings due to COVID-19 pandemic, assure the upward trend in energy demand will continue in the future.

Market mechanism in the building sector from an urban perspective is essential as reducing building sector emissions will have to be a critical part of any effective global effort to reduce climate risks.

This chapter explores how to build the market mechanism in the building sector from an urban perspective. The first section of this chapter overviews the issues when combining different market mechanisms within building sector and energy market-. The second section describes a framework for integrating renewable energy sources, solar water heating, lighting and cooling, centralized air conditioning, and the use of energy-efficient equipment at different component and spatial levels such as each component, a whole building and at a site or district level. The third section covers the methodologies for the market mechanisms in the building sector, while the fourth section addresses a testbed approach for a new modality and procedure of market mechanism in the building sector as a guide for a Blockchain-based community smart and resource wise market mechanism. The fifth and final section addresses the cases at residential community, site, and a whole building levels.

Keywords Building sector · Sustainable building design · Supply chain performance · Value chain · Equilibrium · Building-to-land ratio · Bioclimatic architecture · Green building standards · Smart microgrid systems · Standardized baselines · Climate smart cities · Solar microgrids · Backup system · Digital

---

K.-G. Kim (✉)
Seoul National University, 1 Gwanak-ro, Gwanak-gu, Seoul, South Korea
e-mail: kwigon@snu.ac.kr

H.-S. Choi
Korea Environment Institute, Bldg B, 370 Sichenong-daero, Sejong, South Korea
e-mail: choihs@kei.re.kr

K.-G. Kim and M. Thioye (eds.), *Planning Climate Smart and Wise Cities*,
The Urban Book Series, https://doi.org/10.1007/978-3-030-80165-6_10

sensor networks · CPA scenarios · Climate smart building · Green technology performance · Urban blue–green infrastructure

10.1 Introduction: Issues on Building Sector and Energy Market

We need to integrate new dimensions of carbon market mechanisms into the building sector.

Buildings are responsible for 39% Of global energy related CO_2 emissions. However, we have failed to accomplish buildings and renovating in an energy and resource wise way. The reasons for this failure are numerous, but include a lack of 'tools' to:

- Monitor commitments, secure transparency, build trust regarding commitments and funding;
- Promote sustainable building design for the future;
- Trace supply chain performance;
- Support the "weak link' in value chain;
- Programming capital for sustainability; and
- Establish new market mechanisms.

10.2 Framework for Market Mechanisms in the Building Sector

To deliver the new market mechanisms for building efficiency, there is a need to rethink market policies for the value given to protecting and restoring natural ecosystems, to the sustainable use of energy and resources and to improving human health. A framework for developing carbon markets for the building sector should include:

- Baseline framework: To standardize baseline scenarios;
- Digital technologies framework: To promote and invest in the necessary digital transformation and tools to secure a better tracing of GHG emissions and trading to contribute to the implementation of the SDGs and climate agreement;
- Standard and performance framework: To enforce the legislation and standards related to the energy performance of buildings;
- Governance framework: To work with stakeholders, including an open platform bringing together the buildings and construction sector, architects, and engineers and local authorities to address the barriers to renovation;
- Scoping framework: To transform the whole energy system;
- Carbon market framework: To initiate innovative financing schemes;
- Smart microgrid framework: To enable sector integration in a holistic way; and
- Design framework: To build orientation, internal zoning, ventilation methods, airtight building envelopes, daylight use, choice of building materials, recycling,

life cycle, embodied energy and toxicity; considering them can achieve marked energy savings right from the planning and design stage based on bioclimatic architecture (Fig. 10.1 and Table 10.1).

A step-by-step framework for developing market mechanisms in the building sector from a residential community perspective is shown in Fig. 10.2.

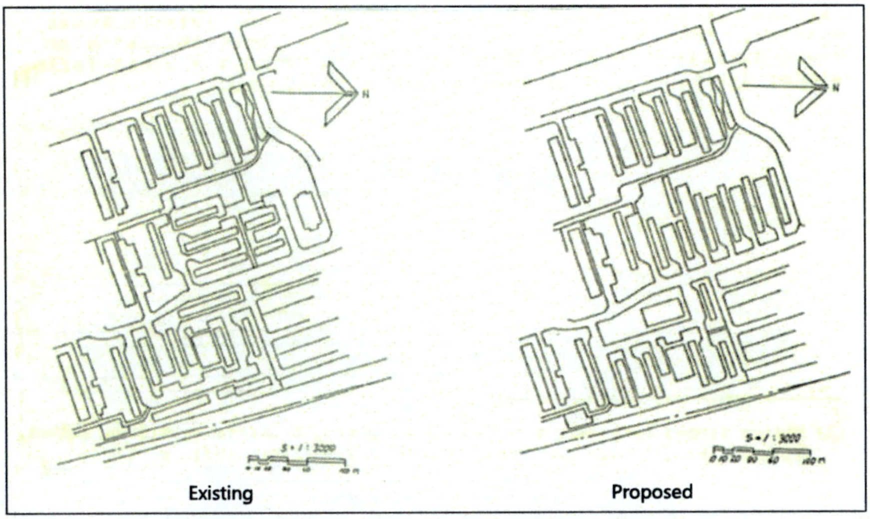

Fig. 10.1 An example of building orientation analysis at a site level

Table 10.1 Comparison of existing site layout and its alternative for orientation, distances between blocks and solar radiation on walls and roofs based on Fig. 10.1

Item	Existing	Alternative
Orientation	10 blocks: Oriented south ± 20° 7 blocks: Oriented south ± 20°	17 block: Oriented south ± 20°
Distances between blocks	1.4 h	1.4 h
Solar radiation on walls and roof	Sun data Altitude: 37.57° Season; Winter solstice Apartment block data Length: 50 m Height: 13.3 m Radiation: 225, 164x(1) 100 kcal/day	Sun data Altitude: 37.57° Season; Winter solstice Apartment block data Length: 50 m Height: 13.3 m Radiation: 293, 029x 100 kcal/day

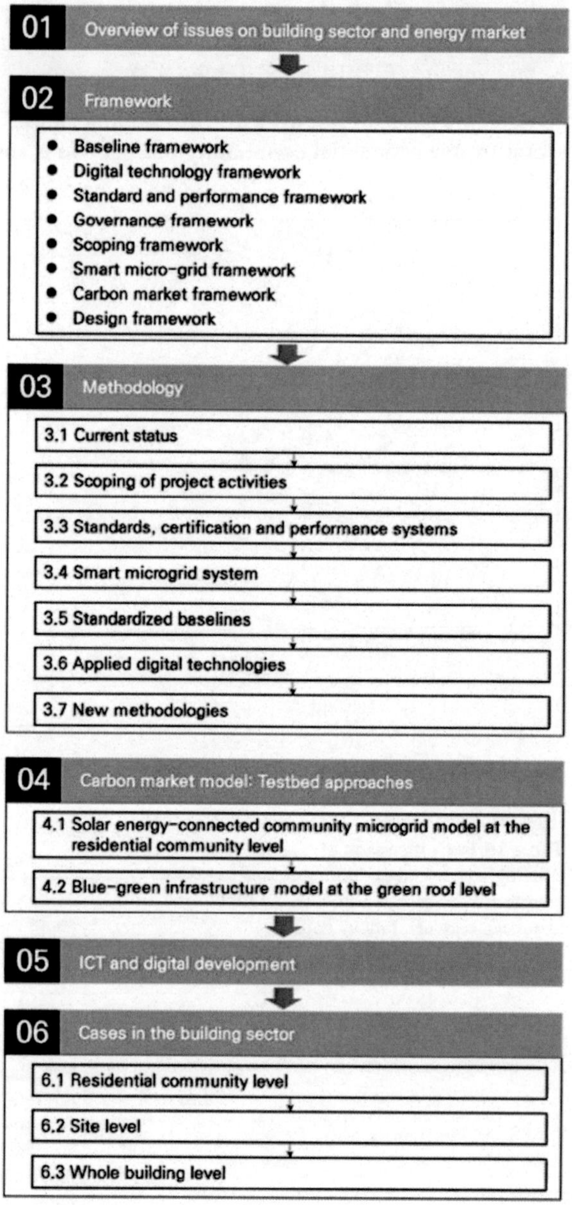

Fig. 10.2 Framework for developing market mechanisms in the building sector from a community perspective

10.3 Methodologies for the Market Mechanisms in the Building Sector

10.3.1 Current Status of Methodologies and Implementation of the Market Mechanism in the Building Sector

As described in Chap. 6, the urban market mechanism planning directs each of different sectors of a city and needs overarching considerations with their interconnections to maximize the efficiency of each of them. The authors view the market mechanism in the building sector as one of six cross-cutting components in the urban market mechanism: urban land use planning, transport, energy generation, building energy, solid waste, and waste water and water efficiency in accordance with a new tenet of the market mechanism for the environmental integrity of a whole city.

Four points are identified that need to be adhered to in the market mechanism for building projects and program of activities. The first one is that it has been reported that only a small number of registered methodologies in the building sector were implemented because of a small amount of GHG emission reductions, lack of demand, and low carbon price. The second point is that a great deal of difficulty and complexity of methodologies has caused financial burden for developing countries to hire expensive consulting firms.

The third item is that data monitoring system is designed for registering methodologies of program of activities, but proper monitoring has not really been executed as an effective vehicle of CERs. The fourth point is that lack of evidence on the data has resulted in inadequate tracing of baseline of existing technologies. From a practical standpoint, improvements are needed in the following areas:

1. Market mechanism in the building sector as integral parts of the urban methodology;
2. Market mechanism methodologies prepared for programs rather than individual projects at different level of space;
3. Full consideration of all public and private review, comments and testimonies on existing methodologies;
4. Streamlined application of smart digital technology for MRV; and
5. Better methods for co-benefits of mitigation, adaptation, biodiversity, and social cohesion, etc.

The future of the market mechanism in the building sector will be characterized by manifold activities. It can be described according to types of methodologies and technologies, green building standards, certification and performance systems, home IoT smart microgrid system, standardized baselines, and digital technologies.

Table 10.2 Comparison of existing and proposed elementary technologies for carbon markets in the building sector

Existing methodologies	Proposed methodologies
• Cook stove • Water pumping • Waste purifier • Water saving • Refrigerators/chillers • Lighting • Whole building • Other/various technologies	• Architectural elements – Direction of buildings, etc. • Facility elements – Equipment efficiency, etc. • Political elements – Policy, law, regulations, etc.

Source Kim (2019) CDM in the Building Sector, p 4

10.3.2 Inclusion of More Elementary (or Unit) Technologies in Scoping of Project Activities

An area of future importance is the use of more methodologies for technical elements, architectural elements, facility elements and political elements with data information storage and retrieval systems. There are many elements which are not fully considered for registration of methodology, as can be seen in Table 10.2. The inclusion of such elements in the building sector could help a lot to make our efforts toward ambitious resilient building and climate mitigation.

10.3.3 Incorporation of Green Building Standards, Certification, and Performance Systems into the Market Mechanism Rules and Regulations

In addition to measurable technologies (e.g., fuel switch, building insulation, etc.), extension of market mechanism to include green building standard and certification, and performance in often used in green building decisions would seem to be a particularly promising approach (Table 10.3 and Fig. 10.3).

10.3.4 AI-Driven IoT Smart Microgrid Systems

It can be proposed that the current project-based procedure of carbon market mechanisms for the building sector should be considered as an interim measure, to be eventually replaced by a more streamlined procedure closely tied to integrated settlement planning. An important area is alternative rates of energy consumption resulting from various settlement patterns described in Chap. 1. The whole energy system has to be transformed. The most innovative technique is solar energy-driven community smart microgrid systems (Fig. 10.4). The technologies for describing a solar energy-

Table 10.3 Green building standards and certification systems

Product certification	Single or multi-attribute	Type of standard or certification	Managing organization	Issue of Focus
Energy star	Single-attribute	Government certification relying on manufacture provided data or third-party testing	U.S. EPA and U.S. DOE	Energy-consuming products
WaterSense	Single-attribute	Government label based on third-party testing	U.S. EPA	Showerheads, toilets, faucets, urinals, and valves
Forest stewardship council	Single-attribute	Third-party certification	Forest stewardship council (FSC)	Forests and forestry products
SCS global services	Multi-attribute	Third-party certification	SCS global services	Wide range of products (i.e., carpets, textiles, wood products, insulation, and more)
Green seal	Multi-attribute	Third-party ISO Type 1 certification	Green seal	Wide range of sectors (paints, adhesives, lamps, electric chillers, windows, window films, occupancy sensors)
Cradle to cradle	Multi-attribute	Third-party certification, cradle to cradle certifies^{CM} product standard is managed and updated by the institute's certification standards board	Cradle to cradle products innovation institute C2CPII	Building materials, interior design products, textiles and fabrics, paper and packaging, and personal and homecare products
Greenguard	Multi-attribute	Third-party certification	UL environment	Indoor air quality, children and school focus
Green squared	Multi-attribute	Third-party ISO type 1 environmental labeling and declaration requirements (ISO 14,024)	TCNA	Tiles and tile installations

Source CDM in the building sector, p 6

Fig. 10.3 Brock environmental center achieved LEED platinum and living building certification. *Source* chesapeake bay foundation, block environmental center, bf.org/about.cbf/locations/virginia/facilities/brock-environmental-center/index.html

connected IoT microgrid system model equipped with ESS (off/on) for housing developments are as follows:

- Energy-saving houses, commercial, and office buildings which are embedded with IoT-enabled monitoring devices include:
  - Solar garden lighting
  - Solar parking lots
  - Solar water heaters
  - Solar door lockers
  - Solar EV charging stations
  - Solar cook stoves and other types of home appliances
  - Solar heat pumps (heating and cooling)
  - Solar LED room lighting, etc.
- Energy-saving public facilities include:
  - Solar parking lots
  - Solar LED street lightings
  - Solar bus stop and bus information terminals (BIT)
  - Solar smart parks

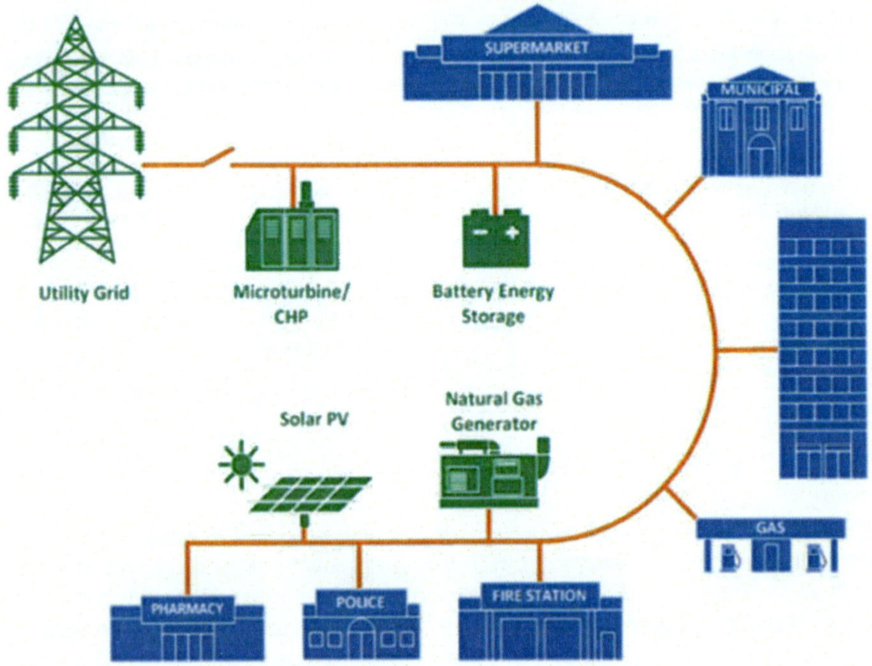

Fig. 10.4 Conceptual community smart energy microgrid. *Source* Microgrid, montgomerycountymd.gov

- Solar smart play grounds
- Solar schools
- Solar kiosks
- Solar farms
- Solar underground water table meters
- Solar benches
- Solar water desalination plants, etc.

10.3.5 Standardized Baselines

Building and house sector-specific standards or factors could be developed for baseline calculations, without measures and demonstration of additionality scenario, with measures for broad categories taking into account the specificities of sectors at international, national, regional, and urban levels. Baseline scenarios can be standardized by using historically recorded data, while prediction function of smart technologies can be developed by using statistical multiple analysis. Table 10.4 depicts the difference between CDM methodologies/tools (on the left hand side) and standardized baselines (on the right hand side).

Table 10.4 Difference between CDM methodologies/tools and standardized baselines

• International standards • To calculate emission reduction of specific applicability conditions • Specific projects boundary • Project-by-project baseline scenario determination and demonstration of additionality • Baseline using 48(a) (historical or actual), 48(b) (most attractive course of action), or 48(c) (Average of top 20%) • Project emissions • Data not monitored • Data monitored	• Sector-specific standards (Could be regional, national, or international); • Take into account the specificities of sectors; • Either calculate baseline emission factor broad class of mitigation activities (measures) taken up in the sector; or baseline emission factor for entire sector; • Baseline emission factor to be used for baseline emission calculations and demonstration of additionality; • To be used in conjunction with an approved methodology/tool

10.3.6 Digital Technologies

Big Data, AI, IoT, and Blockchain technology can be used for MRV and, in particular, accurate prediction of potential GHG emission reductions (project scenarios), as well as baseline emissions (average performance scenarios). Emission reductions are monitored for difference between baseline and climate action project buildings.

There are many IoT-enabled monitoring devices available for automated building systems. Smart lights can be connected to the Internet, allowing them to both receive data and send recorded data to a central system. This allows system management to continually optimize the bulb's performance based on historical data and predictions of future lighting needs. Smart irrigation system can do exactly the same thing with Big Data, IoT, AI, and controllers for water supply. In this regard, the authors suggest launching of Academia- Industry Cooperation Joint Call on "Climate Smart Buildings."

Blockchain technology holds much promise and other interesting opportunities for the market mechanism in the building sector as an intermediary between producers and consumers in trading energy and carbon emissions, as is shown in Fig. 10.5. This figure gives us an indication of potential for avoiding double counting and assuring transparency of trading and carbon emissions by using Blockchain technology. In this diagram, consumers can choose where to buy renewables from Blockchain technology, while producers with surplus solar energy sell the energy to their neighbors.

10.3.7 Suggestions for New Methodologies

Several suggestions can be made to allow us to look ahead into the bright future of the market mechanisms in the building sector as follows:

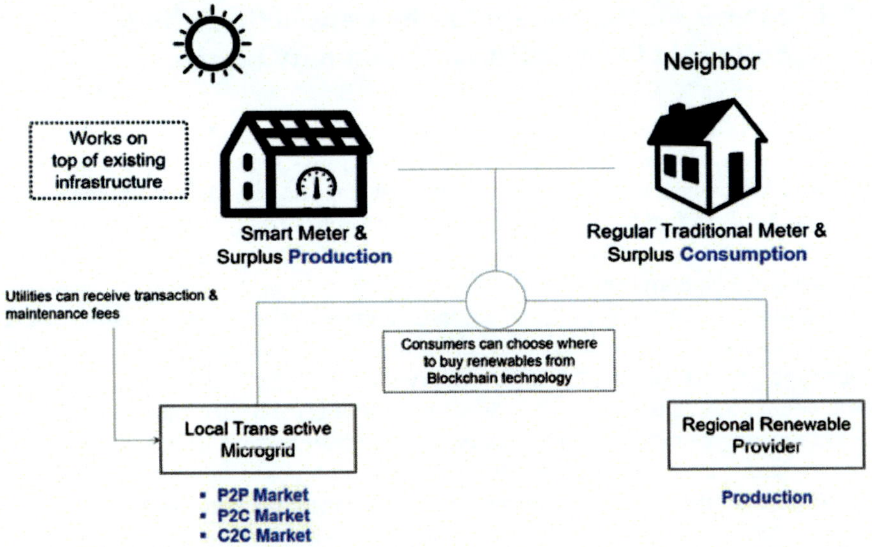

Fig. 10.5 Interacting local energy with neighbors: Blockchain is a shared database technology that processes and stores transaction data

- Methodological approaches to user needs: In terms of scoping, development of some kind of extended CDM modules for each category are recommended.
  - Individual end-uses
  - Whole building
  - Group of buildings by category (e.g., apartment buildings, condominiums, flats, single detached houses, commercial and office buildings, or districts and communities)
- Reduced transaction costs for MRV by:
  - Targeting individual end-uses
  - Using models and benchmarks
  - Using intelligent digital technologies
  - Developing baseline and additionality standards
- Providing methodologies to address specific needs, situation, and level of technology of LDCs, SIDs, African countries, underdeveloped zones in developing countries
- Integrated holistic approach versus deductive approach to the market mechanisms: "Everything is connected," This thought is particularly relevant to extended multiple PoA CDM. We have to look at different schools of thought on approaches to market mechanism, namely, integrated holistic, and deductive approaches.

10.4 Models for the Market Mechanisms in the Building Sector: At the Residential Community Level

These models are presented with the aim of offering guidance on market elements, methods, and tools for use by local authorities. Step-by-step sequences are provided to clarify key issues. It is recommended that all steps be considered, although each building needs to find the most appropriate approaches for its local situation. Even if these steps have a certain logical sequence, they should not be taken as a strictly chronological blueprint for action.

Community commitment is an important prerequisite to the process, and a core organizational body needs to be established to monitor and guide the overall marketing process. Each of the stages may require the input of different people, and each community needs to choose the type of arrangement that may best suit and reflect its administrative planning system and political context. The major actors in the community must also be committed to the structure chosen. In some communities, structures may already be in place and building on them may be feasible.

10.4.1 A Solar Energy-Connected Community Microgrid Model

Step 1: Project description and building partnerships.

The first step in modeling a solar energy-connected community microgrid involves description of follows:

- Type of building: Apartment buildings (number of dwellings, roof and veranda areas for the installation of solar panels, etc.)
- Project activities: A solar panel system for power production and backup power
- Site conditions: Location, land use, physical–chemical, biological, cultural, and socioeconomic environments. Site conditions are crucial to solar panel performance calculation and system optimization (Boxes 10.1 and 10.2).
- Partnerships: The main benefits from investing in the solar and backup power systems would be that the costs are shared among partners, renewable power is utilized by the buildings (or sold), and businesses can partially continue to operate even during blackouts.

Box 10.1 Solar energy-connected community microgrid project site, Narpio city, Finland.

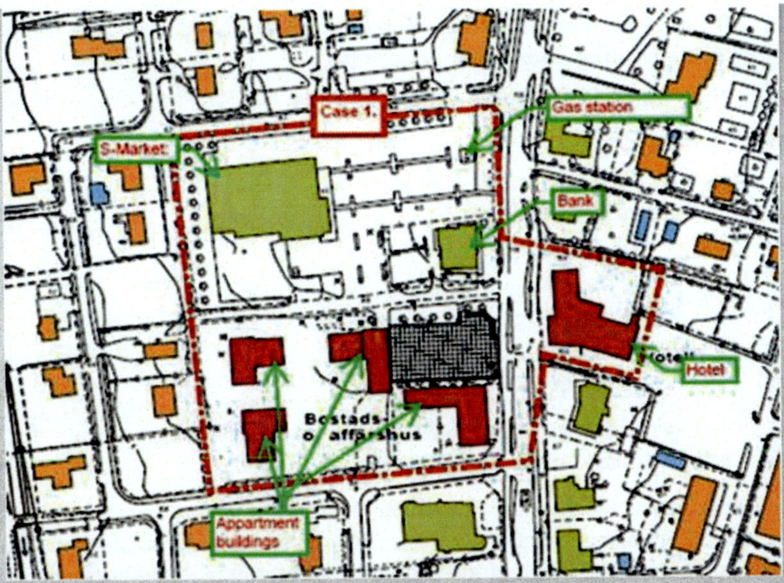

Source Pasonen and Hoang (2014)

Box 10.2 Solar microgrids in community planning, Narpio city, Finland

The area is in the center of the city and surrounded by mainly commercial and residential buildings, and the availability of renewable excess energy is limited. There are no forests, open fields, water bodies, or industrial buildings from where renewable or waste energy can be utilized. However, the roofs of the buildings could provide installation surfaces high enough for solar panels and wind turbines to avoid obstructions. The area is not adjacent to any sources of groundwater that would make geothermal heat-energy a possible option for effective space heating and cooling. In this study, we focus on the benefits of solar power for the buildings under investigation.

Step 2: Measures for climate actions_Solar energy-connected community microgrid for the residential apartment buildings.

Measures for climate actions for this model include follows:

- Roof-top solar panels
- Veranda solar panels
- Solar LED street lightings
- Solar LED park lightings
- Solar water heaters
- Solar water desalination, etc.

A solar energy-connected community microgrid model is shown in terms of modality and procedure in Fig. 10.6.

Step 3: Current energy consumption estimates or measurements for the residential apartment buildings.

During this phase, developers aim to analyze cooling, heating and electricity consumptions either from measurements or estimations, using available data. Therefore, this process should look at availability of consumption data of the buildings.

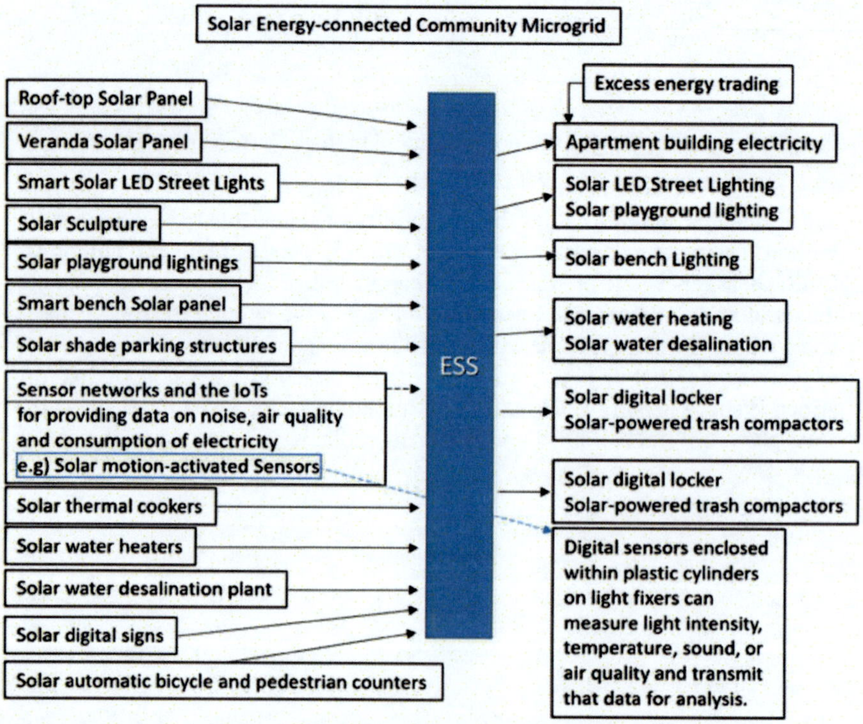

Fig. 10.6 Solar energy-connected community microgrid: modality and procedure

A long-term energy demand of the buildings should be projected for establishing baseline scenarios. Current provisions for heating, cooling, and electricity have to be described, including the district heating network for heating and the local electricity grid for all electricity. An example of energy consumption that was estimated for buildings in Narpio city, Finland, is shown in Table 10.5. Figure 10.7 illustrates a measurement sample of supermarket power consumption.

Step 4: Future energy demand of the residential apartment buildings: Baseline scenarios.

A long-term energy demand for the buildings should be projected with historical consumption data of the buildings. The projected demand can be used as baseline scenarios.

Step 5: Solar project scenarios.

Solar project scenarios are based on the number of factors, with use of solar panel performance calculation and system optimization as follows:

- Solar panel type (name, area of panel, maximum PV power of the system)
- Solar panel parameters, and digital sensor networks and the IoT used in the simulation (weather (temperature, etc.) data and other settings in the simulation)
- A solar energy-connected community microgrid operation
- Solar panel performance calculation and system optimization
- Production and storage equipment (ESS) data.

Figure 10.8 shows solar panel parameters used in the simulation, including costs, in Narpo city, Finland.

Step 6: Carbon emissions savings in a typical scenarios and carbon market.

The calculation method was used for carbon emission savings in a typical scenarios in the Narpio city case study.

When 541 gCO_2/kWh is assumed to be the average emission figure of replaced production, and 70 gCO_2eqv/kWh life-cycle emissions for the PV system, emission savings in this case can be calculated by:

$$387862\ \text{kWh} \times \left(541\ \text{gCO}_2/\text{kWh} - 70\ \text{gCO}_2\text{eqv/kWh}\right) = 182.7\ \text{tonne CO}_2\ \text{eqv}$$

Step 7: Backup system.

Having established a solar energy-connected community microgrid, we need to look at how to backup power capability. Figure 10.9 depicts a solar power system with backup power capability. Backup systems and manuals should be developed to reduce over-dependency on solar power systems.

Step 8: Big Data, data storage, and data analytics: Blockchain platforms.

Step 9: Implementation: Construction and operation.

Table 10.5 Energy consumption estimates for buildings, Narpio city, Finland

Building name	S-market		Bank		Hotel		Gas station		Residential buildings(4)	
Energy type	Heat	Power	Heat	Power	Heat	Power	Heat	Power	Heat	Power
January	27.9	86	15	4.3	53.7	37.8	0	1.6	56	22.5
February	25.5	78.9	14.8	4.3	55.4	37.8	0	1.3	57.4	22.5
Mars	17.1	86.4	9.8	4.3	44	37.8	0	1.3	48.2	22.5
April	12.1	79.4	9.5	4.3	41	37.8	0	1	45.7	22.5
May	4.9	82.1	6.7	4.3	32	37.8	0	0.7	38.4	22.5
June	3.1	80.3	4.4	4.3	26.8	37.8	0	0.6	34.2	22.5
July	2.4	88.1	3.3	4.3	19.9	37.8	0	0.7	28.6	22.5
August	3.2	87.6	3.9	4.3	22.7	37.8	0	0.8	30.8	22.5
September	4	84.7	6.5	4.3	29.1	37.8	0	1	36	22.5
October	7.2	86.5	9.8	4.3	38.5	37.8	0	1.3	43.7	22.5
November	8.5	85.1	9.2	4.3	41.1	37.8	0	1.5	45.8	22.5
December	24.2	89	14	4.3	58.3	37.8	0	1.4	59.7	22.5
Sum	140	1014	107	51	462.6	453	0	13	524	271

* *Either from measurements or estimations, using available data*
Source Pasonen and Hoang (2014) ibid., p 25

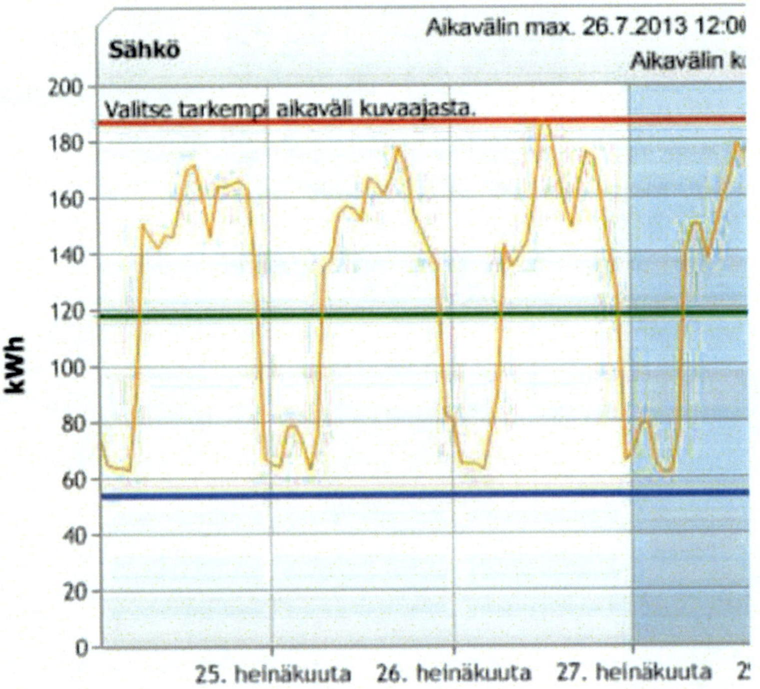

Fig. 10.7 Measurement sample of S-market power consumption. *Source* Pasonen and Hoang (2014) ibid., p. 40

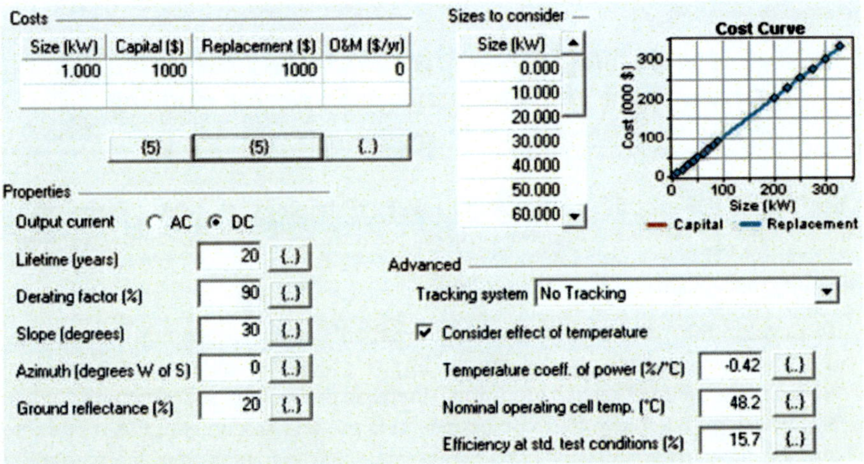

Fig. 10.8 Solar panel parameters used in the simulation. Costs (without VAT) are in euros even though the dollar sign is used in the software. *Source* Pasonen and Hoang (2014) ibid., p31

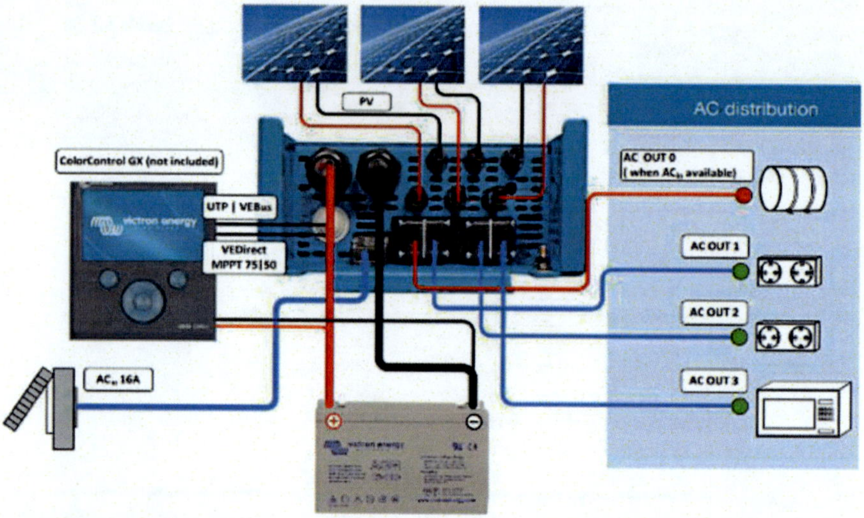

Fig. 10.9 Solar power system with backup power capability. *Source* Pasonen and Hoang (2014) ibid., p 40

Step 10: Measurement, monitoring, reporting, verification, and validation of climate actions, including citizens' contribution.

Step 11: Issuance of CERs and marketing.

10.4.2 A Model of the Market Mechanisms in Rainwater, Green Space and Solar Energy Sector for Housing Projects: Blue–Green Infrastructure

A new way to describe the relationship between technology, rainwater, green space, and solar energy sector for housing projects is the smart blue–green infrastructure.

The objectives of this model are as follows:

- To develop the ways of how to mitigate urban floods, reduce non-point pollution, promote rainwater utilization, and improve the urban microclimate and local environment, reduce carbon emissions, increase carbon sinks, and enhance urban biodiversity. To achieve these objectives, this model can choose LCA method to analyze the carbon emissions and carbon sinks in the operation of the whole life cycle of climate smart and resource-wise city projects.
- To improve global carbon market mechanisms through the implementation of a carbon emission reduction plan, and green economy and green new deal strategy.

- To develop an internationally standardized market mechanism for housing projects.

Step 1: Project description and building partnerships.

To achieve the objectives above, a pilot residential community for climate smart and wise cities can be selected as a case study area, based on which the characteristics of carbon emissions and carbon sinks from the whole life cycle can be explored and studied, and several carbon reduction measures can be proposed.

The building site and environmental setting are the starting points from which a market mechanism must be developed. However, specific details should not be given here. There should be covered in "Inventory" to describe baseline data against which baseline scenarios can be developed. The primary categories of the building site and environmental setting are as follows:

- Location and areas
- Catchment and basin distribution map
- Pervious natural land
- Impervious roofs and pavements, etc.

Figure 10.10 is a layout plan of the urban climate green roof development used by Malmo city. It shows the overall project characteristics. The green roof is a part of blue–green infrastructure at community level (Fig. 10.11). Figure 10.11 shows a schematic diagram of the climate smart and wise city which adopts renewable energy, low-carbon green infrastructure, and buildings.

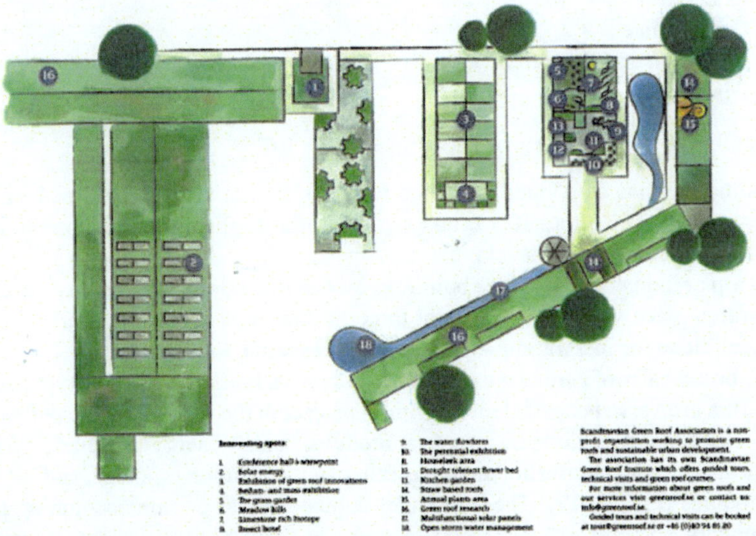

Fig. 10.10 Layout plan of the urban climate green roof development. *Source* scandinavian green roof, augustenborg botanical roof garden, undated, unpaged

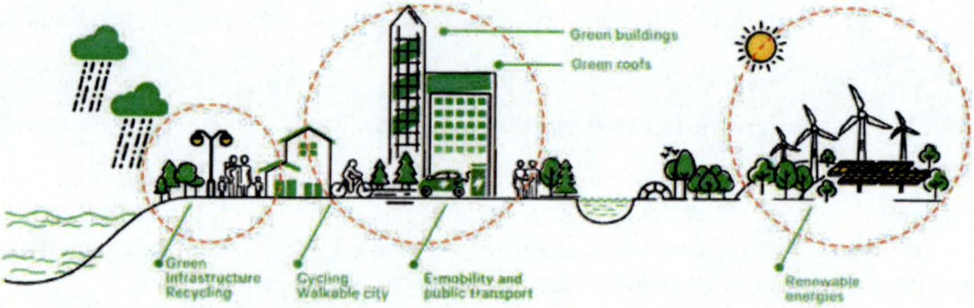

Fig. 10.11 A schematic diagram of the city which adopts renewable energy, low-carbon green infrastructure, and buildings. *Source* Zinco, Planning Guide: Green roof 4.0, Systems for roof utilization for the future, undated

Step 2: Project Activities.

The categories of technologies for the proposed model are as follows:

- Elementary/unit technologies
  - Greenin projects
  - Ecological buffering projects
  - Rainwater utilization projects
  - Green–blue systems
- Rainwater system technologies
  - Collection
  - Conveyance
  - Retention and detention
  - Discharge
- Installation of digital sensor networks and the IoT.

Malmo city has developed a method for a residential project that involves some of the technologies. Photo 10.1 is an example of the application of the methodology in residential community development.

On top of industrial and office buildings owned by Malmo city, there is a botanical garden in a place where you have not thought there was room for a park. This is a beautiful place for inspiration, education, and research.

The botanical roof garden is a part of a larger project called the Eco-City Augustenborg. This is a very successful revitalization project in the residential area of Austenborg. The project is divided into five modules: solar energy modules, greening modules, ecological buffering modules, rainwater utilization modules, and biodiversity modules (Fig. 10.12). These modules can be used as carbon accounting model components.

Figure 10.13 shows open storm water management in green roofs in combination with canals and ponds in "THE ECO-CITY AUGUSTENBORG." Visitors can use a

Photo 10.1 Walk along the path of the green roof and storm water in the Eco-City Augustenborg residential community

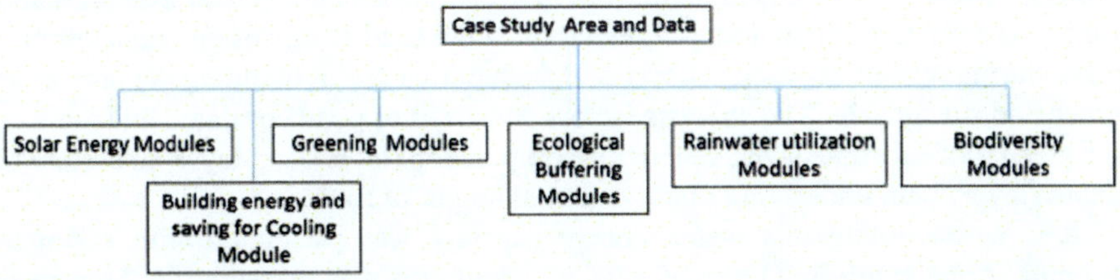

Fig. 10.12 Six modules of a green roof

Fig. 10.13 Overview of Augustenborg residential community. *Source* VASYD, the "Eco-City Augustenborg"—a walk along the path of storm water, undated, unpaged

key map to take a walk with the storm water around Augustenborg. Various techniques for open storm water solutions have been combined, resulting in technical, climate, and esthetic advantages.

The solar energy management in green roofs is combined with smart solar energy-connected microgrid in the "ECO-CITY AUGUSTENBORG," Malmo, Sweden.

Step 3: Accounting methodology for carbon emission reductions.

A new methodology and procedure of the market mechanism in water, green space, and solar energy sector for housing projects include consideration of five modules shown in Fig. 10.12. The methodology takes the characteristics and parameters of each module into consideration.

This residential community development is divided into green roof development and low impact development (LID) on the ground. An accounting methodology considers climate effects associated with green roof facilities within the project site shown in Fig. 10.11.

For the whole system of urban blue-green infrastructure at residential community level, carbon neutrality can be analyzed by life cycle assessment with empirical data or estimation for different facilities of the system.

The methodologies that exhibit the effects of green roofs include Bengtsson (2005), Villareal (2007), Quin et al. (2013), Stovin et al. (2017), and Sorensen and Emilsson (2019). In their studies, green roofs were used as one of the main components during retrofit of Augustenborg and frequently used in branding the project as green and sustainable. The analysis factors consisted of pond complex, thickness of the green roof substrate layer, vegetation composition (layer), soil layer, size, storage depth (layer) with underlying drain, and total height of roof.

It is worth noting that water storage capacity was calculated with a simple formula in the study by Qin et al. (2013), whereas it was measured in Augustenborg (Bengtsson et al. 2005). One of the key criticisms of the studies is that they have been applied to evaluation of effects on flood risk reduction. Effects on carbon emission reductions and carbon sink were not evaluated for urban climate roof. The project could be a good urban climate smart resilience pilot project at residential community level for co-benefits of climate change mitigation and adaptation.

(1) **Inventory and baseline scenario: A common roof Scenario**

Sealed surfaces of common roofs with concrete or asphalt pavements can no longer be used as an engine for plant growth and runoff delay and can therefore hardly be used for cooling evaporation and GHG emission reductions. It can thus be assumed that black roofs have no effects on carbon emission reductions and carbon sink.

(2) **Project activity scenarios: Green roof scenarios**

A comprehensive carbon accounting model for climate smart and wise city projects is developed based on the guidelines for the Intergovernmental Panel on Climate Change (IPCC) and the United Nations Framework Convention on Climate Change (UNFCCC).

A. **Components of program of activities (CPA)**

Calculation methods of the amount of net carbon emission reductions.

Due to the lack of current research on the measurement and computation-based calculation in the operation of urban climate green roof facilities, this method suggests to calculate the direct carbon sequestration from green space, rain garden, grassed swale, permeable pavement storm water storage, and the indirect carbon emission, energy saving with solar panels, evaporation with plants and water ponding in the storm water retention area beneath the green roof.

IPCC guidelines for direct emissions of CO_2 can be accounted for using measurement-based or computation-based methods.

a. **Accounting for the carbon sinks: CO_2 sinks/m^2.**

The carbon sink generated by green roof facilities basically comes from the operation phase: as follows:

- Carbon sequestration in green space by plant types such as trees, shrubs, grasses, and wetlands
- Carbon sink from rainwater utilization
- Carbon sink from runoff reductions
- Carbon sink from runoff pollutant removal
- Carbon sink from buildings' energy saving.

The assemblage of data to describe the surface areas, material consumed, unit of carbon sequestration rate, and accounting period of different facilities of the urban climate green roof in the residential community in Table 10.6 would be a way of developing a list of climate effects for consideration.

b. **The amount of evaporation cools as the indirect emission reductions: Evaporation capacity per 100 m^2.**

Evaporation can be used to optimize cooling capacity, and improve urban microclimate and building energy saving. Generally speaking, increased evaporation will always ensure greater cooling capacity in urban centers. Urban trees have a considerable cooling capacity.

Table 10.6 Format of surface areas, material consumed, unit of carbon sequestration rate, and accounting period of different facilities of the urban climate green roof in the residential community

Green roof facility	Material consumed	Unit of carbon sequestration rate (carbon emissions)	Accounting period
Rain garden (_ m^2)			
Grassed swale (_ m)		Carbon sink	
Permeable pavement (_ m^2)			
Green space (trees) (_ m^2)		Carbon sink	
Storm water storage (_ m^2)			
Solar energy panels (_ m^2)			
Urban farming (crops (_ m^2)		Shortening of food transport distance, carbon sink	

c. **The amount of runoff delay and the storm water retention for indirect carbon emission reductions.**

The storm water retention with a green roof can be used for runoff reduction, rainwater purification, and evaporation. The space required for the temporary retention of water, which enables the desired relief of all drainage systems down as far as the watercourses, can be created by installing a "space" beneath the actual green roof build-up. The additional capacity of the storm water retention is to mitigate urban floods, reduce non-point pollution, and promote rainwater utilization.

- Up to approx. 40 l/m^2 permanent water ponding in the system build-up for the irrigation of an intensive green roof
- Up to approx. 80 l/m^2 temporary water ponding in the storm water retention area beneath the green goof
- Building energy saving for cooling and saving.

Many researchers have extensively studied the energy saving of a green roof compared with a common roof. Among them, it has been found that green roofs can save a significant amount of energy, for example, 14.86 kWh/(m^2.a) compared with a cool roof.

e. **Pervious roofs for plant growth and runoff reductions.**

Actions aimed at de-sealing are becoming increasingly important, for example, cities are incorporating green roofing in development plans are rewarding the storm water retention capacity of green roofs with reduced waste water charges. This capacity to hold back water is also referred to as "retention," and it has considerable potential for carbon sinks.

f. **Urban farming for food security and food transport distance reduction.**

Current global food systems accounts for a third of total GHG emissions and are driving environmental degradation and socioeconomic inequalities all along the value chain. Cities can change that.

Urban green roofs are well suited to growing a full variety of fruits, vegetables, and herbs. They are close to the source of food demands. Therefore, farming in green roofs shortens food transport distance and thus saves energy. This in turn leads to reduce GHGs. In light of the current COVID-19 pandemic, green roofs shorten food supply chains and improve urban resilience in our food supply.

g. **Solar energy.**

Thermal solar energy panels installed on green roofs can be used to produce both heat and electricity. For example, annually each square meter of the 420 m2 of thermal solar panels at Augustenborg Botanical Roof Garden produces 350 kWh of energy. The thermal panels are connected to Malmo city's district heating system, and the photovoltaic solar panels are used to produce electricity. Combining solar panels and green roofs can have positive effects. First, the green roofs cools down the solar panels, which improves the energy production levels from the solar panels. Secondly, the effects of solar panels on a green roof are that different microhabitats are created which allows a more diverse flora and fauna (Scandinavian Green Roof, undated).

Most of the important carbon emission factors for carbon emission accounting in green roof facilities are derived from prior research. However, it is recommended to amend or adopt more appropriate values, since the selection and determination of carbon emission factors have a large impact on the accuracy and rationality of carbon emission accounting. Therefore, the model may need the uncertainty analysis using sensitivity analysis methods. The accounting period can be assumed depending on the life cycle of green roof facilities.

B. **Overall aggregation of GHG emission reductions by the proposed climate actions: Synthesis of CPA.**

An accounting method for residual emissions or emission offsets is provided after all roof emission reduction activities described above. An algebraic sum of emission reductions by each component of project activities (CPA) can be calculated and shown in the form of Table 10.7.

Step 4 ICT and Digital Development.

Blockchain or distributed ledger technologies (DLT) have received wide attention in UNFCCC COP 25 as an innovative solution to provide a reliable, incorruptible, decentralized, shared database that allows transactions between individuals who do not know and therefore do not trust each other. In recent years, blockchain technology has been tested in various international development projects. In this model, the technology is used for a shared database for monitoring of climate action, peer-to peer electricity trading, and management of housing properties.

Table 10.7 Overall aggregation of GHG emission reductions by the proposed actions (CPA)

Component of project activities (CPA)	Algebraic sum of emission reductions
CPA1: energy generation	Solar energy generation
CPA2: green space	Carbon sequestration
CPA3: building energy saving	Cooling capacity
CPA4: water saving	Reduction of waste water charges
Total emission reductions	

Unit tCO_2

10.5 The Cases for Carbon Market Mechanisms in the Building Sector

10.5.1 'A Residential Community' Level: The Case of Blue-Green Infrastructure, Augustenborg Neighborhood, Malmo City, Sweden

Sorensen and Emilsson (2019) evaluated flood risk reduction by urban green infrastructure using insurance data. The data employed in the Sorensen-Emilsson approach are insurance claims and floods reports from housing companies, and rainfall data. The insurance database covers property loss from overland flooding, groundwater intrusion through basement walls, and flooding from the drainage system, and it was used as a proxy for flood severity for different events.

This methodology compares those indicators before implementation of blue–green infrastructure and post installation data. Basically, it involves evaluation of a number of flood risk reduction effects, with use of a comparing system for flood risk evaluation. This study used datasets of five neighboring areas as baselines for Augustenborg evaluation. With the approach, urban climate effects including urban heat islands, carbon emission reductions, and carbon sink effects can be readily evaluated.

10.5.2 'A Site' Level: The Case of Alternative Urban Forms for Energy Conservation and Energy Conservation Measures, Gwachon City, Korea

Kim (1986) has developed an approach to energy flow and conservation in urban systems. A method of modeling energy flow and distribution while assessing energy demand and supply scenarios taking policy options into account against base year (1985) was used. It was found that the alternative block layouts of a selected site in terms of orientation, distances between blocks, and solar radiation on walls and roofs would increase the solar radiation by about 30%, as shown in Fig. 10.1.

10.5.3 'A Whole Building' Level: The Case of 'The Green', The Korea Land and Housing Corporation

The green technology building exhibition developed by the Korea Land and Housing Corporation, Seoul, demonstrates environmental-friendly technologies, including renewable energy and clean energy resources. It is divided into the energy source technology section, climate change response technology section, and environmental technology section. In this hall, these technologies are organically linked to show visitors an important role in realizing a climate smart building as a whole. The roof provides the opportunity to take a look at a new and renewable energy system, and the resting areas in the nature combined with green technology. The whole building exhibits a wide range of monitoring systems for green technology performance as follows:

- Smart energy, gas, and water metering which enables recording of energy, gas, and water consumption within a day;
- Remote meter reading on a scheduled and on-demand basis;
- Real-time notification of abnormal usage to the customer and/or the energy, gas, and water utility;
- Control of energy, gas, and water consumption devices within a customer's premise;
- Messaging to the customer;
- Customized targeting of segments.

10.6 Conclusions

2020 marks the fifth anniversary of the Paris Climate Agreement, and although the world has faced setbacks and extraordinary adversity, a sea of change is underway.

This chapter has examined market mechanisms in the building sector at community, site, and whole building levels.

The overall finding is that AI-driven IoT community smart microgrid systems and blue–green urban infrastructure could be employed to address the carbon market failure under the current CDM. However, a full procedure and tools for market mechanisms need to be tested in demonstration projects proposed in this chapter. The demonstration projects can be selected in both new developments and retrofit projects.

In many respects, it is still too early to draw conclusions on the economic impacts of the market mechanism, and actual contributions to emission reductions and to investment in the development and application of new, low-carbon technology in the building sector.

It is concluded that incorporation of smart green building standards, certification, and performance systems in the market mechanisms for building sector of different levels may be effective in guiding cities toward climate actions.

References

Korea Land and Housing Corporation (undated) the green: green technology building exhibition hall. Seoul, Unpaged

Kim KG (1986) Energy flow and conservation in urban systems. UNESCO, Paris, p 123

Kim KG (2019) CDM in the building sectors. Presented at COP25 side event: the role of the clean development mechanism (CDM) in decarbonizing cities and its co-benefits by UNFCCC, Madrid, Spain, 2–13 December 2019

Kim KG (2019) How digital technologies support decarbonizing cities, presented at COP 25 Side Events organized by UNFCCC, Madrid, Spain, 2–13 December 2019

Lin X et al (2018) Prediction of life cycle carbon emissions of spongy city projects: a case study in Shanghai, Chaina. Sustainability 10:3978. https://doi.org/10.3390/su10113978

Pasonen R, Hoang H (2014) Microgrids and DER in community planning: Practices, permits, and profitability, VTT Technology 189

Scandinavian Green Roof Institute (undated) Augustenborg botanical roof garden, Malmo, Unpaged

Sorensen J, Emilsson T (Feb 2019) Evaluating flood risk reduction by urban blue-green infrastructure using insurance data. J Water Resour Plann Manag https://doi.org/10.1061/(ASCE)WR.1943-5452.0001037

VASYD, The "Eco-City Augustenborg"—a walk along the path of storm water, undated, unpaged

Chapter 11
New Advanced GHG Accounting System for Cities Enabled by DLT

Massamba Thioye

Abstract CBA is the most appropriate GHG accounting system for a sustainable satisfaction of citizen's core needs. However, the current way it is operationalized, makes it more complex to implement. Access to the required information is an important barrier. There are also risks, when tracking back the carbon footprint of consumed products, of missing greenhouse emitted during their production. Indeed, the Eulerian specification of the flow of embodied GHG emissions in and out of a city cannot capture the history of the carbon embodied.

To address these issues, we propose a new GHG accounting system for city, replacing the Eulerian specification with a Lagrangian specification, thus permitting to leverage the advantages of both CBA and PBA. We call it a Lagrangian CBA. Instead of focusing on the city, it follows the products and their embodied GHG emissions through time and space, from city to city, along their value chain or value circle. It can then keep track of their trajectory and their history.

DLT is used to serve the Lagrangian description of the flow of GHG emissions embodied in products. With its immutability function, information related to each transaction of products is cryptographically sealed and recorded on distributed ledgers, with a time stamp and cannot be changed without the DLT network consensus. The smart contract function authorizes the transfer of responsibility for GHG emissions embodied in a product from its producer to its buyer alongside the transfer of the product. This provides a much higher level of granularity to the GHG accounting system, with possibility of getting detailed descriptions attached to the GHG emissions footprint of products. For each ton of GHG emissions embodied in a consumed product, it is possible to know precisely from where, in the value chains, it has been emitted as well as other relevant information. The use of digitized methodologies, IoT, AI, GPS, RFID will further automatize the GHG accounting system.

The way the baseline is set is another specificity of the GHG accounting system proposed. It departs from the current baseline framework and proposes a new type of baseline we call forward-looking baseline. This type of baseline brings out what

M. Thioye (✉)
UN Climate Change Secretariat, Haus Carstanjen, Martin-Luther-King-Strasse 8, 53175 Bonn, Germany
e-mail: mthioye@unfccc.int

K.-G. Kim and M. Thioye (eds.), *Planning Climate Smart and Wise Cities*,
The Urban Book Series, https://doi.org/10.1007/978-3-030-80165-6_11

is needed instead of (i) what is possible or desirable or (ii) what progress has been already achieved. It indicates at any point in time to cities and their climate actors, where they should be in terms of climate performance, to be aligned with the climate goals.

A framework foreseen to assign to cities accountability for GHG emissions or to credit cities for GHG emission reductions should be able to do it based on responsibilities and effort. However, a consumption-based approach for city GHG accounting, which depends on responsibilities and effort is challenging to develop. Indeed, (i) GHG emission reductions of cities can arise from a chain of climate actions, (ii) several drivers could be at the origin of an impact on city's GHG emission and (iii) for important types of climate actions, there is no direct causal link between the action of a city and the impact on its GHG emissions. A ToC, using a matrix for system analysis, establishing the linkages between the actions of a city as change agent, and the impact it is claiming, is proposed to address these challenges. It also enables the ex-ante determination of the climate contribution of a city action plan, useful for many reasons:

- It establishes the link between the actions that are to be introduced in a city action plan and the impact that is expected. Impact cannot be claimed without action;
- It enables city government to prioritize the city action plan that addresses the development needs of the city while being aligned with the climate goals;
- It enables a city to join a Local Government Green Funding Agency (LGGFA) which has as criteria for participation, forward-looking green activities in a green city action plan;
- Determining the emissions of future actions in a city action plan enables influential stakeholders such as central government, financiers and citizens that are constituencies, to proactively encourage greener greenfield and renovation plans;
- In case a city issues a green bond for the implementation of its city action plan, the ex-ante determination of the impact of the city action plan may be required.

In most of the papers, addressing the topic of categorization of consumption-based GHG emissions, a sector-based approach to categorization is used. An approach to categorization of consumption-based GHG emissions that depends on need is proposed. It builds the consumption categories around the core basic needs citizens must satisfy.

Keywords Lagrangian CBA · DLT · Transfer of responsibility for GHG emissions · Forward-looking baselines · ToC · Matrix for system analysis · Ex-ante determination of impact

Disclaimer The content of this chapter has no relationship whatsoever with the UNFCCC process or the work of the secretariat. Its findings, interpretations and conclusions do not reflect the views of the secretariat. The secretariat does not vouch for the accuracy, currency or completeness of the data or information contained in the article and excludes all liability for any loss or damage arising (including through negligence) in connection with and as a result of any reliance on the data and information contained in the article and/or its use. Questions concerning the analysis and views expressed in the article shall be addressed to the author.

11.1 Introduction

At the current level of GHG emission, the world remaining GHG emissions budget to stay below a 1.5 °C temperature increase will be consumed by 2030. The stake is the survival of humankind. Therefore, there is a need to scale up and accelerate the pace of climate action, and complement incremental improvements with breakthrough transformative changes. The net-zero commitment of Parties and non-Parties stakeholders including sub-nationals is an important move in the right direction. However, the significance and the urgency of the response require doing more. A GHG emissions budget as well as a pathway for its consumption between now and the date of achievement of the net-zero emission commitment or pledge is also needed. This is particularly relevant to cities. Indeed, they have a high potential for climate action as described at Chap. 9 of this book (Thioye 2021).

Incentivizing climate action by cities requires the development of a framework for the measurement of the impact of their climate actions. This is a complex task that raises several new and challenging issues, most of them related to the development of a suitable GHG accounting system:

- That can support the alignment of cities with the 1.5 °C goal.
- Able to attribute responsibility of emission and credit emission reduction effort to the right stakeholder.
- Able to determine the climate contribution of a city action plan both ex-ante and ex-post.
- That can address the issue of data availability.
- That can set suitable baselines.
- That enables to incentivize climate action by producers as well as consumers in the city and determine the impacts of the different types of climate action.
- That is need-based.

There are several initiatives aiming at developing a framework for the measurement of the impact of climate action by non-Party stakeholders in general and cities more specifically. Guided by the desire to make the framework immediately operational, some of these initiatives are constrained by what is currently understood as technically possible, including in relation with data availability. This is for example the main reason why PBA to GHG accounting is used for cities, although it has obvious limitations. Other initiatives put emphasis only on the usability of the framework and the extent to which the targeted users will be willing to adopt it. Their attempt is to build a framework agreeable to the future users. There is a lot of merit in ensuring the absence of technical and acceptability barriers to facilitate the future operationalization of a GHG accounting framework. However, this should not be at the detriment of its relevance or its environmental integrity. Indeed, the starting point should be to define what is needed to ensure relevance, with regard to the objective the framework is pursuing, the issue it is intended to address. Then, the next step is to explore what is currently possible and what is not, what is acceptable to the potential users and what is not. The gap between what is needed and what is currently possible and/or is acceptable to potential users will facilitate the identification and the formulation of the issues that need to be addressed to design a relevant, usable and easy to operationalize framework. This is what guided the approach taken here, in this chapter of the book. It identifies and formulates the issues that need to be addressed and proposes possible solutions. It is expected to trigger further thinking and serve as basis for further work.

The views expressed in this chapter are the expert view of the author. They do not reflect the view of the UNFCCC secretariat.

11.2 Terms and Definitions

The definitions of scope proposed in the GHG protocol and the GPC (WRI, C40 cities and ICLEI; 2014) are suitable for production-based approach (PBA) to GHG accounting for city. They are less appropriate for a consumption-based approach (CBA) to GHG accounting. However, when the GHG emission reduction of a specific city's stakeholder (city government, citizen, business within the boundary of the city) is considered, we will adopt the definition of scope from the GPC.

Climate contribution of a city.

The climate contribution of a city is defined as the impact of its climate actions. The contribution can be a surplus or a deficit and will be measured in terms of GHG emission below the baseline (climate contribution surplus) or above the baseline (climate contribution deficit).

Normative climate scenario.

They are backcasting top-down scenario whose starting point is a desirable future goal. Normative scenarios describe various options available for adoption starting from now in order to reach that desirable future goal. The starting point of a normative scenario in this chapter is an international long-term consensual global temperature goal such as the 1.5 °C temperature increase above pre-industrial levels.

11.3 Accounting of GHG Emissions in Cities

11.3.1 Features of the GHG Emissions Accounting System in Cities

A GHG emission accounting framework is needed to measure and recognize the impact of climate actions in cities with the view to incentivizing their scaling up. The framework should be designed to have the maximum effectiveness, meaning here the capability to succeed in making cities aligned with the climate goals. For that purpose, the framework should be able to drive climate action from the key targeted actors, where it is the most needed and using the most suitable lever, by defining:

- What to measure?
- Where to measure it?
- How to measure it?

While it is widely accepted that there is no GHG accounting framework for city in absolute better than the others, it is possible to find one which is the most suitable for the achievement of a specific objective such as the one set above: making cities aligned with the climate goals. Thus, the first step in the design of a GHG accounting framework for a city is the identification of the stakeholders which need to be targeted for climate action, those that have the responsibility, power and capability to act. The main stakeholder identified is the city government. Indeed, it is in a unique position to conduct a variety of climate actions either directly reducing GHG emission or enabling other entities to do so. Cities government can be reducer of the GHG emissions related to the operation of their own infrastructures, enabler of GHG emission reduction of the citizens and businesses in the city and disruptor of GHG emissions intensive products and markets. Examples of their climate actions include:

- Directly reducing GHG emissions of cities' building, heating system, waste management facilities, car fleet).
- Enabling GHG emission reduction by citizens, businesses and industries through the infrastructure they build (GHG emissions reduction achieved through transport network, EV charging station, facilitation of more efficient consumption).
- Incentivizing climate action through the policies they implement.
- Exercising their influencing and enabling power on their citizens (information and awareness campaigns on consumption of sustainable goods and services, capacity building, providing access to low-cost finance for energy efficiency improvement and enhanced deployment of renewable energies) or on their suppliers.
- Facilitating the disruption of products and markets with high net GHG emissions footprint (green procurement, compact and complete city).

The other stakeholders to target for action include:

- The entities enabling the city government including the central government, financiers, technology providers.
- The entities enabled by the city government including citizens, producers of goods and services within the administrative boundaries of the city.

11.3.2 Most Used Approaches for GHG Accounting System in Cities

If city governments are the key stakeholders targeted for city climate action, what they can act on and what they cannot act on is important in designing the GHG accounting system for a city. There are GHG emissions within the boundary of cities on which they cannot act (e.g., implementation of a cement plant authorized by the central government which cement is sold in the entire country and even exported). There are also GHG emissions that are out of the boundary of cities they can influence. The largest potential of GHG emissions on which cities can act is not necessarily emitted within the geographical boundaries of cities. This includes all the GHG emissions reductions related to enhanced circularity, more efficient consumption, switch to less GHG emissions intensive products, change of behavior of individual citizens.

There are currently three main approaches for city's GHG accounting (Balouktsi 2020; Broekhoff et al. 2019):

- Production-Based Approach or PBA.
- Consumption-Based Approach or CBA.
- Community-wide Infrastructure Footprint Approach (CIFA).

The first two approaches are binary approaches. They either look at a city as a producer or as a consumer. These are simplifications that can lead to important inaccuracy, for example, when PBA is used for a 'consumer city' or CBA for a 'producer city.' The third approach considers a city as a prosumer responsible for the GHG emissions related to its production and its consumption. Therefore, CIFA does not allow aggregation at a region level, for example, to track global progress because of the double counting of GHG emissions and GHG emission reduction.

All the three approaches describe the flow of embodied GHG emissions from and/or in a city. This is a field approach to GHG accounting.

PBA describes the flow of GHG emissions from activities taking place in a city, embodied within the products they deliver. It is simpler to implement because the required data are more widely available, with the use of satellite observations to support the monitoring of GHG (Stephen Hardwick and Heather Graven 2016). However, it does not put accountability and credit where it is the most relevant. Indeed, it does not capture a broad range of mitigation action impacts that can arise from city government mitigation action including changing the consumption pattern of the citizens or green procurement by the city government if the related goods and services are imported. On the other hand, PBA addresses GHG emissions the city government may have no influence on.

CBA facilitates the translation of global GHG emissions budgets into city GHG emissions budget. However, it is not enough granular to inform about the history of the GHG emissions embodied in the product. This is the issue highlighted by C40 (C40 Cities 2018) related to the limited granularity of the EEIO modeling which does not inform cities on where in the supply chain GHG emissions arise. This is particularly challenging to ensure that all the GHG emissions related to the production activities along the value chain are captured by the GHG emissions considered embodied in the product consumed. For example, the GHG emission of defective products at some stage in the value chain that does not reach the end consumers may not be accounted for. The accounting system may also miss some products that are consumed in the city.

CIFA addresses all climate actions in the city related to the key infrastructure, which is useful for measuring the impact of enhanced circularity.

None of these approaches consider the effort of the city. Let's take an example where a city A set a green Venture Capital Fund seeded with its public finance at a first loss position and give access to low-cost financing to its cleantech startup. Consequently, cleantech startups develop in the city and one of them innovates, reducing the cost and enhancing the performances of a low GHG emissions footprint product which become attractive for the citizens of city B. Even if city B has a CBA accounting, should the related reduction of GHG emissions footprint be attributed to city B? Conversely, if thanks to awareness raising from its city government, citizens of city B accept to pay a premium price for a low GHG emissions footprint product from city A, thus enabling its import by city B from city A. Even if city B has a PBA accounting, wouldn't it be fair to attribute the related reduction of GHG emissions footprint to city B?

Finally, for the PBA, CBA, and CIFA, the approach defines simultaneously two things: what to look at (GHG emissions footprint of consumed products or GHG emissions related to production) and to whom it is attributed (respectively, the consumer or the producer).

11.3.3 Moving from Eulerian to Lagrangian Description of the Flow of GHG Emission Embodied in Products for Enhanced Granularity of the GHG Accounting System

The main purpose of a GHG accounting framework is to facilitate the effectiveness of a climate mitigation scheme. Therefore, the selection of what to account for is directly tied to the objective of the scheme. As evidenced in Chap. 9 of this book, there are two main approaches to climate mitigation actions. One is sector-based and problem-oriented. It is consistent with a PBA GHG accounting. The other one is need-based and solution-oriented. It requires for GHG accounting a CBA. So far focus has been put on sector-based and problem-oriented approaches to climate mitigation actions, the main entities targeted for climate action being the big polluters. This led to an incremental and manufacturer-focused approach to climate action by corporates. This is urgent to complement this approach with a need-based and solution-oriented approach to climate action which is more transformative, and cities, as consumption centers are in the best position to do that. Indeed, by influencing end-user's consumption behavior, cities can disrupt markets and entire value chains or decrease their GHG emissions footprints. While the incremental and sector-based framework for climate mitigation action reduces the GHG emission of the polluters, mainly through the development and deployment of new climate technologies and production processes, the need-based and solution-oriented approach of cities would explore alternative solutions for a sustainable satisfaction of the core need of their citizens. CBA is the GHG accounting system that promotes the need-based and solution-oriented approach to climate mitigation action. It will therefore be preferred to the PBA. Indeed, the development and implementation of a green city action plan including green urban planning for a compact, complete, connected and coordinated city (adapted from Floater 2017) is likely to be among the most impactful type of climate action in a city. As its impact is reflected more in the GHG emissions footprint of the products consumed within the boundary of the city and less in the GHG emissions of the producers of good and services within the city, the approach proposed will take and will look at the GHG emission footprint of products as the primary responsibility/source of credit of the city. The most suitable approach for the determination of accountability/credit for GHG emissions/emission reductions of a city is a CBA. This supports a 'low GHG emissions footprint products' pull policy. However, it needs to be improved.

All the three approaches currently available are using, according to classical field theory, a Eulerian description of the flows of embodied GHG emissions. They are looking at them with a focus on a given location in the space: a city. This can be visualized if the city is considered as a circle in front of us, and we are watching the flow of embodied GHG emissions getting in and out of the circle. This Eulerian specification of the flow of embodied GHG emissions in and out of a city cannot capture the history of the GHG emissions embodied. It can only inform about the volume of GHG emission embodied in products through processes operating within the city. For the volume of GHG emission flowing in the city, embodied in products, made outside the city and the volume of GHG emission embodied in product leaving the city, information related to the history of the products is needed. The flow of GHG emissions in and out of the city are required to enable the calculation of the volume of GHG emissions embodied in products consumed in the city.

We propose to use a combined PBA and CBA, made possible by a Lagrangian description of the flow of embodied GHG emissions. We call it a Lagrangian CBA GHG accounting system. Instead of focusing on the city, it follows the products and their embodied GHG emissions through time and space, from city to city, along their value chain or value circle. It is possible then to keep track of their trajectory and their history. Each ton of embodied GHG emission will carry its own properties, the different products in which it has been embodied changing with time until its product carrier is consumed by an end consumer.

The approach also delinks the ‘what to look at’ from the ‘to whom to attribute the responsibility/credit of the impact noticed on what was looked at.’ The ‘what to look at’ (GHG emissions footprint of consumption vs GHG emissions of production) is determined in a way that makes visible the impact of the most prominent climate actions undertaken in the city. However, it does not mean that the city will always have the responsibility/credit for its consumption. The credit related to the consumption of low carbon footprint products can be transferred to the city they are imported from, in case that city is the enabler that made its consumption possible, for example, by making it be the most competitive product for the need it satisfies, without any action from the importing city. Recognizing and giving credit to the exporting city for its effort to develop competitive low GHG emissions footprint products supports a ‘low GHG emission products’ push policy. It can incentivize cities to implement incubators and accelerators to promote the development of startup working on innovative climate solutions. The example provided below illustrates how CBA can be adjusted to give credit to an exporting city which develops competitive low GHG emission footprint products, therefore enabling their consumption by an importing city.

If i represents the products which consumption by other cities is enabled by a city A (e.g., by making each of them the most competitive for the satisfaction of a given need), and j represents the products which consumption by other cities is not enabled by a city A:

$$
\begin{aligned}
\text{City A GHG emissions} &= \text{GHG emissions related to consumption of city A} \\
&= \text{embodied GHG emissions in products imported by A} \\
&\quad + \text{GHG emissions within the city A} \\
&\quad - \text{embodied GHG emissions in products exported by A} \\
&\quad + \text{embodied GHG emissions in products } i \text{ exported by A} \\
&\quad - \text{embodied GHG emissions in products} \\
&\quad \text{displaced by the products } i \text{ in the importing cities}
\end{aligned}
$$

For any importing city B receiving products *i* which consumption is enabled by the city A, the GHG emissions are calculated as:

$$
\begin{aligned}
\text{City B GHG emissions} &= \text{GHG emissions related to consumption of city B} \\
&= \text{embodied GHG emissions in products imported by B} \\
&\quad + \text{GHG emissions within the city B} \\
&\quad - \text{embodied GHG emissions in products exported by B} \\
&\quad - \text{embodied GHG emissions in products } i \text{ imported by B} \\
&\quad + \text{embodied GHG emissions in products displaced} \\
&\quad \text{by the products I in city B consumption.}
\end{aligned}
$$

The displaced products are the most likely products that would be consumed by city B if the consumption of products *i* was not enabled by city A. It could be the products consumed at the time products *i* were made available, and later, any products that would have been consumed in the absence of products *i*. It can be either the most attractive alternative to products *i* or the benchmark products consumed in the absence products *i*.

This proposed Lagrangian CBA incentivizes action by all entities with capability and power to act and recognizes the impact of their climate action regardless of whether they are acting as producer or consumers. It is inviting cities to not only promote the consumption but also the development within their cities of low GHG emission products. City government can incentivize cleantech startup to act, stimulating sector innovation for the development of low-cost and high-performance low GHG emission products.

11.3.4 Addressing the Simplicity or Accuracy Trade-Off

As mentioned above, the specificity of the PBA is its simplicity and the availability of data under that approach. The PBA also ensures the avoidance of double counting. These are very interesting features of the PBA, and it would be good to leverage when using CBA for GHG accounting. The shift from the Eulerian description of the flow of GHG emissions embodied in products to a Lagrangian description will enable to benefit from these great features of PBA while using CBA.

The accounting of GHG emissions will begin as a PBA. GHG emissions related to production of an entity in a city are first allocated to that entity and that city by default. And during its entire journey along the value chain or the value circle, the GHG emissions embodied in products will only be removed from the account of one city if it is recorded in the account of another city. This ensures nothing fall through the cracks. Indeed, GHG emissions embodied in products exported by a city are validated and deducted from its GHG emissions account once they are accepted by an importing city as GHG emissions embodied in its imported products. After acceptation, the GHG emissions embodied in the products related to that transaction is simultaneously removed from the account of the exporting city and added to the account of the importing city. In that way, the total cities emissions determined through PBA will always be equal to the total cities' emissions determined through CBA. Data being available for PBA, the initial step that allocates to producer and their city the GHG emissions related to their production is easy to implement. The way the GHG emissions are distributed among transacting cities will change from PBA to this Lagrangian CBA, but the total GHG emissions of the transacting cities will be the same. In addition, the GHG emission embodied in products that are not consumed for whatever reasons (limited quality, unsold, loss between production and consumption) is attributed automatically to their producer. Therefore, there is no single ton of CO_2 lost when passing from a PBA accounting at the creation of the embodied GHG emission to a CBA when it is definitely attributed to an end consumer. And, this approach does not provide any room for double counting. The use of a Lagrangian description for the GHG emission accounting enables to combine the benefits of PBA and CBA and address their limitations.

11.4 Ex-post Determination of the Climate Contribution of a City

The Lagrangian CBA for cities' GHG accounting provided at 11.3.3 is used for the measurement of the climate contribution of a city. Indeed, it gives, during a given year y_i, the GHG emissions of a city that can then be directly compared to its GHG emissions budget for that same year y_i to assess its climate performance. But, to address possible variation of GHG emissions related to lower population growth compared to the baseline population growth, actual GHG emission per capita is compared to baseline GHG emission per capita for the determination of the climate contribution of a city.

11.4.1 Leveraging Digital Technologies to Implement a Lagrangian Description of the Flow of GHG Emission Embodied in Products

The determination of the GHG emissions related to the consumption of products can be very simple if digital technologies such as Distributed Ledger Technologies (DLT), Internet of Things (IoT) and Artificial Intelligence (AI), Global Position Sensors (GPS) and Radio Frequency Identification (RFID) are leveraged to serve a Lagrangian description of the flow of GHG emissions embodied in products. These digital technologies can facilitate the connection of dispersed entities within different cities, transacting through complex networks of value circles/chains. The immutability and consensus building functionalities of DLT, its distributed nature as well as its smart contract is particularly of relevance (Evry financial services 2015). With the immutability function, information related to each transaction is cryptographically sealed and recorded on distributed ledgers, with a time stamp. This appends only features of ledgers guarantee that the recorded information related to a transaction between entities cannot be changed without the DLT network consensus. The smart contract function includes coded agreements between the transacting entities and can self-execute agreed operations. It authorizes the transfer of responsibility for GHG emissions embodied in a product from its producer to its buyer alongside the transfer of the product. The same smart contract that recognizes the buyer of a product as the new owner will also recognize simultaneously that it inherits the responsibility of the embodied GHG emissions. With the consensus building functionality, the entities involved in the value circles or value chains can verify the transactions before they are recorded and vote to ensure the reliability of the records. GPS connected to the DLT informs about the position of the products. The combination of these functionalities secures for the nodes of the DLTs reliable information, enabling the tracking of the products transacted among them as well as the flows of embodied GHG emissions. The availability to all the stakeholders involved in the value circles/chains of real-time data about the transfers of responsibilities for GHG emissions prevents fraud and falsification. For each product, the history of its embodied GHG emission is provided to all the node of the DLT, with full traceability about the place in the value circle/chain from where each ton of GHG emission was emitted and all the relevant information related to the conditions of emissions. RFID attached to the product and linked to DLTs can further increase the granularity of the accounting system by tracking the product through the value circle/chain all the way to the point of sale and give insight about the category of end users to whom the embodied GHG emission should be attributed. This can enable targeted climate action by the city government. The proposed Lagrangian description of the flow of GHG emissions embodied in product supported by digital technologies enhances drastically the level of granularity of the GHG accounting system of cities while reducing its complexity and preventing any double counting as well as any ton of GHG emission from falling through the cracks during the transfer of responsibility for GHG emissions from producers to buyers of products.

To operationalize this Lagrangian CBA, the entities in the value circles/chains should participate as a node to the private DLTs of their value circles/chains.

Where possible, Internet of Thing enabled devices connected to the DLT can further automatize the data value chain by collecting automatically the information required for the calculation of the GHG emissions by a digital methodology. The large volume of data collected, and data calculated by the digital methodologies, is processed with an AI technology trained to provide insight on possible climate action to reduce the GHG emission related to the satisfaction of a given need of a defined stakeholder in the city.

11.4.2 Stepwise Approach for the Ex-post Determination of Cities' Climate Contribution

Baseline setting

The steps below can be followed to set a baseline for the ex-post calculation of the climate contribution of a city:

- Translate the global GHG emissions budget (at the year the city starts implementing its pledge) derived from the global 1.5 °C temperature increase goal into a city GHG emissions budget and a date for the achievement of net-zero emissions
- Translate the city GHG emissions budget into GHG emission pathway by attributing GHG emission budget to the different years between the start year

y_1 and the year y_{end} before the one when the net-zero GHG emission is pledged to be achieved

- Make a projection of the population in the city for the years y_1 to y_{end}.
- For the years y_1 to y_{end}, determine the ex-ante baseline GHG emission per capita EI_{yi} for y_i by dividing for each year the GHG emissions budget allocated to that year by the population anticipated for that year $P_{ant,yi}$.

Determination of the GHG emissions during year y_i

- Entities record daily on the DLT of their value circle/chain, using PBA, their production and all the information allowing the calculation by the relevant digitized methodology of their GHG emissions. The place from where the production took place is indicated by a GPS connected to the DLT, and the time is provided by the DLT time stamp. Where several products are generated following different processes, the GHG emissions arising from each process are attributed to its products. Where several products are generated from the same process, the GHG emissions are attributed to these products according to their value by the digitized methodology. The GHG emissions per unit of quantity of product are determined. At the production stage, the responsibility for the GHG emissions embodied in the products is first attributed to the producer, and accordingly recorded in the DLT ledger as GHG emissions of the city where the producing entity is located, just like a mined bitcoin is assigned to its owner.
- For each transaction between entities, the city of the selling entity as well as the city of the buying entity is recorded, with their populations. The buying entity (including visitors to a city) acknowledges through smart contract the transfer of accountability related to the GHG emissions embodied in the transacted products from the account of the selling entity to its account with a time stamp. With the information recorded on the DLT, it is possible to backtrack and not only have the total GHG emissions embodied in a product held by an end consumer but also how this GHG emissions footprint was built up from cradle to cradle or from cradle to grave, across the value circles or the value chains. The transfer of accountability works the same way and in the same direction as a transfer of value (such as bitcoin). However, while the code of agreement of the smart contract for product transfer is a recognition by the seller of the ownership by the buyer of the product from now on, for the transfer of GHG emission responsibility, the buyer recognizes that it inherited the responsibility from the seller.
- In real time, digital methodologies related to the DLT transform the data into GHG emissions and per capita emissions following a CBA, for the cities which entities are nodes of the DLTs. They also determine how the GHG emissions of a city decompose across products and across position in the products' value circle/chain. All this information is directly recorded and distributed to all the nodes. Just like blockchain record the entire history of a bitcoin since it is mined with the different transfer that occurred, the DLT will record the history of products and their embodied GHG emissions from cradle to cradle or from cradle to grave. This gives to a city the information required to identify where the GHG

emissions embodied in its consumption are coming from, and what actions could be undertaken to reduce them by reducing either the GHG emission footprint of the products, or the volume of products it is consuming. The per capita emission for year y_i is calculated by dividing the total GHG emissions of the city during year y_i with the population anticipated for that year $P_{\text{ant},yi}$.
- Within the city, citizens can be encouraged to participate to the DLT as end consumer. The DLT would provide information about what category of end consumer is consuming which type of products, where and when. This enables importing cities, with the information made available, to limit its consumption of products with high GHG emissions footprint, through the implementation of relevant city action plans, reducing the scope 3 and 4 emissions of the city government.

Measurement of the annual climate contribution of a city.

The climate contribution CC_{yi} of a city during year y_i is determined by:

$$CC_{yi} = \left(EI_{yi} - \text{per capita emission year } y_i\right) \times P_{\text{ant},yi}$$

However, this value needs to be corrected in case the population is less than what was anticipated, to avoid a situation where a city claims GHG emission reductions simply because its population has decreased.

$$CC_{yi} = \left(EI_{yi} - \text{per capita emission year } y_i \times P_{\text{ant},yi}/P_{\text{act},\,yi}\right) \times P_{\text{ant},yi}$$

If $CC_{yi} > 0$, there is an excess of climate contribution, the achievement being above the reference achievement.

If $CC_{yi} < 0$, there is a deficit of climate contribution, the achievement being below the reference achievement.

11.4.3 Specificities of the Consumption-Based Approach to GHG Accounting and Limitation in Determining the Climate Contribution with no Linkage to Actions

A framework foreseen to assign to cities accountability for GHG emissions or to credit cities for GHG emission reductions should be able to do it based on responsibilities and effort. However, there are challenges related to the development of a consumption-based approach for city GHG accounting, which depends on responsibilities and effort:

- For important types of climate actions, there is no direct causal link between the action of a city and the impact on its GHG emissions. These types of climate actions include, for example, the city exporting competitive low GHG emission

products. In these cases, the city is not reducing the GHG emission related to its consumption, but it is enabling the reduction of GHG emissions related to the consumption of another city;
- GHG emission reductions of cities can arise from a chain of climate actions. For example, actions from the central government can empower a city which then enables citizens' actions that directly reduce GHG emissions;
- Several drivers could be at the origin of an impact on city's GHG emissions. What entails the consumption of some high/low GHG emission footprint products could be out of the city boundaries, with the city having no responsibility/credit for consuming them.

To avoid multiple claims and double counting, it is important to disentangle the different chains of action-impact at the origin of the GHG emission of a city. An ex-post determination of the climate contribution of a city alone cannot address such complexity. It also does not permit the evaluation of the climate contribution of a specific city action plan. An approach combining a theory of change (ToC) and a system analysis is proposed to tackle that challenge through an ex-ante identification of the actions the city needs to undertake to be able to claim the achievement of a defined climate goal, as well as the determination of how these actions are driving the achievement of the climate objectives through their outputs and their outcomes. The ex-post determination of the climate contribution of a city, combined with a ToC established ex-ante and linking a climate contribution target with concrete actions through outputs and outcomes, gives to a city a strong argument to claim the climate contribution it targeted if it is confirmed by the GHG emissions of the city measured after the implementation of the actions identified through the ToC. The actions demonstrate the effort. The ToC is developed in the context of the management of the cities' GHG emissions.

11.5 Management of Cities' GHG Emissions Through a Need-Based Solution-Oriented Approach

Cities have more potential for GHG emission reductions, acting on consumption. Indeed, GHG emissions related to the operation of cities' infrastructures are just a little part of the total GHG emissions of a city, and city governments have limited influence on GHG emissions related to business activities in their city. The need to mobilize the important GHG emission reduction potential related to consumption positions the CBA as the most suitable GHG accounting approach for cities.

Under the CBA, the GHG emissions of cities are the GHG emissions embodied in the products consumed within the boundary of the city. Therefore, managing the GHG emissions of a city means managing the GHG emissions embodied in the products it consumes. To be effective, it requires going one level of granularity up and managing the GHG emissions within the city related to categories of consumptions and categories of consumers the city government can influence.

11.5.1 Theory of Change to Link Action and Impact

A ToC establishing the linkages between the actions of a city as change agent, and the impact it is claiming, can help address the challenges identified under 11.4.3. In particular, it can unravel a situation whereby the same credit is claimed by both producers and consumers.

The ToC describes and illustrates, ex-ante and in detail, how and why a city will achieve a GHG emission budget target in a given situation. It connects within the city action plan, a city's current level of GHG emissions, its GHG emission budget and the climate mitigation actions it plans to implement, describing the missing bridge between these latter. It does that by first deriving from the global GHG emission budget, the city's GHG emission budget, then the city emission budgets for the satisfaction of different core needs. These variables are the dependent variables that form the elements of the dashboard. From these variables, working back using top-down backcasting normative scenario, all the conditions that must be in place for the achievements of the objectives related to the dependent variables are identified as outcomes variables or relay variables. The relations among these variables are also determined. Working further back, the outcome variables enable the identification of the outputs and actions required to produce the outcomes. This gives an in-depth understanding of the change process and a tool to track progress toward the city GHG emission budget goal.

Through a need-based solution-oriented approach to climate action by cities, they are expected to mainstream their climate mitigation actions into the city development actions. This can make the ToC more complex. In this case, it is recommended to use a system analysis matrix as described in 11.5.6, to support the ToC.

A ToC is important to understand the chain of climate mitigation actions in city, and the related outputs and outcomes required to achieve a given impact on the GHG emissions of cities for its replication.

Finally, where green bond is issued to finance a city action plan, the ex-ante evaluation of the action plan impact on the city's GHG emission is needed. This requires determining the impact of the different actions in the city action plan. To be effective, the ToC needs critical thinking and the assumptions used for the backcasting normative scenario should be explicit.

11.5.2 Target Setting for Cities

11.5.2.1 From What is Possible or Desirable to What is Necessary

The scale and the urgency of needed climate actions demand bold actions from cities. It starts with the setting of an ambitious target. It should not be a target set ex-post based on climate actions considered as possible: cases where action planning precedes target setting as indicated by Balouktsi (Balouktsi 2020). It should rather be a moonshot target set ex-ante, based on what is needed for the city to effectively contribute to the achievement of the 1.5 °C temperature increase goal. The gap between the target set for an alignment with the climate goal and the current GHG emission pathways of the city defines the action the city should take, including the development of innovative climate solutions on technology but also on policy, financing instruments, governance, leadership, business model and consumption behaviors. Setting a comprehensive target for a city should include by when it intends to achieve a net-zero emissions, a GHG emissions budget as well as a pathway for its consumption.

11.5.2.2 Scope of the Target

City governments are expected to conduct scope 1, 2, 3 and 4 climate change mitigation actions. While their scope 1 and 2 GHG emissions are more under their influence, they also have the power and capability to undertake impactful scope 3 and 4 climate change mitigation actions. Indeed, the potential of their climate contribution covers substantial scope 3 and 4 climate contributions, that is important to mobilize. This requires the net-zero emissions and GHG emissions budget pledges of cities to go beyond scope 1 and 2 GHG emissions and cover both scope 3 and 4 GHG emissions. With a global goal that encompasses all their GHG emissions, cities can also set specific goals in the form of GHG emissions budget for categories of consumption or in the form of GHG emission reduction for each scope of GHG emission, both in a way that maximize the mitigation impact achieved.

11.5.2.3 Baseline Setting

There are three categories of baselines in the current baseline framework for measurement of climate contributions of an entity:

- Backward-looking baselines, which are references to measure progress, compared to the projection in the future of historical climate performance; they are named Business as Usual baselines; determining the climate performances of a city using these baselines highlights the level of climate effort already deployed by the city
- Capability based baselines, which are references for the comparison of achieved performance with the top X% of the best possible performances. This category includes performance of best peers or performance of the best available climate solutions. Comparing actual performances with these baseline reference levels brings out how a city is performing compared to what is possible
- Readiness-based baselines, which are references for the comparison of achieved performance with the performance of the most attractive course of action.

Although all these types of baselines can provide interesting insights, they do not address one important function: The tracking of progress toward the climate goals. Improvement determined by comparing the situation after the implementation of a city's climate action with what would happen in its absence is important to bring out the level of effort already deployed. However, determining the additional amount of effort required to join the targeted GHG emission pathway or the extent to which a city was able to go beyond its targeted emission pathway is more important. Where the target is to achieve net-zero GHG emissions by a defined date with a defined GHG emissions budget and a pathway, then this pathway should become the baseline, the reference level against which progress toward the goal is measured. The need of this new category of baselines arises from the availability of long-term global climate goals, each city can translate into its own specific goals for the short medium and long term. We have named these types of baselines forward-looking baselines. They point to what is needed instead of (i) what is possible or desirable or (ii) what progress has been already achieved. They indicate at any point in time to cities and their climate actors where they should be, in terms of climate performance, to be aligned with the climate goals, and bring out any climate performance gap between where cities are and where they should be. We have all experienced a situation where we are traveling for an appointment we do not want to miss, with very tight timelines. In the middle of the journey, our focus is never how much time we have already spent in the journey or where we would be if we were driving a faster car. Our focus is on how much time is left to get there and how to ensure that we arrive on time.

For the operationalization of the GHG accounting framework, it is advisable to translate the global GHG emission goal of the city into GHG emission goals for categories of consumptions.

11.5.2.4 Reference Unit

The target is first in the form of GHG emissions budget. However, to be able to track progress and measure performance, it must be translated into an intensity metric a city can use to assess its climate performance. One approach is to translate the GHG emissions budget into GHG emission pathway for the city. This is then combined with projections on population and economic growth to derive intensity metrics. The most widely used reference units are tCO_2 eq per (capita*year), $tCO_{2\,eq}$ per gross city product and $tCO_{2\,eq}$ per gross domestic product. tCO_2 eq per (capita*year) is particularly suitable for CBA. This is the metric used.

11.5.3 Categorization of the City Consumers

In a city, there are three important categories of consumers:

- The city government.
- The citizens.
- The businesses in the city.

All these three city stakeholders can contribute in reducing the GHG emissions footprint of products consumed in the city, including imported products through three main categories of outcomes.

Other key stakeholders are the central government and cities' possible financiers.

11.5.4 Outcomes for the Management of City Consumption

There are several outcomes city governments can target to influence the GHG emissions footprint of products consumed within a city. These are developed in Chap. 8 of this book (Thioye 2021). These includes:

- Outcome O_{ce1}: avoiding emissions through disruption of existing GHG emissions intensive products or value chains by a product from a value circle/chain different from the one which generated the disrupted product, for example, through urban planning and the development of smart and wise cities, compact, complete, connected and coordinated (adapted from Floater 2017) or through urban forestry (reduction of heat load and the need for air conditioning).
- Outcome O_{ce2}: enabling GHG emission reduction through enhanced efficiency of citizens' consumption of goods and services. The impact is emission reductions arising, for example, from the city investing in infrastructures and equipment that enable enhanced circularity of the economy, transforming value chains into value circles, extending the time products stay in a high value state, circulating in these value circles and facilitating the reduction of waste.
- Outcome O_{ce3}: enabling GHG emission reduction through the decrease of the GHG emissions footprint of products and services consumed by the citizens. The impact is emission reductions arising, for example, from information and awareness campaigns on consumption of sustainable products, capacity building and access to low-cost finance for energy efficiency improvement and enhanced deployment of renewable energies.

These outcomes to manage the cities' emissions are used integrated with the outcomes O_{cdi}, relevant to cities' development.

11.5.5 Categorization of the City Consumptions

CBA promotes a need-based and solution-oriented approach to climate mitigation action, which is more suitable than a sector-based and problem-oriented approach, particularly for cities. Indeed, it has several advantages:

- It focuses on what cities need to work on: the well-being of their citizens through the satisfaction of their core needs.

- It enables the mainstreaming of city's climate mitigation action into the city development action and urban planning.
- It facilitates transformative changes in consumption behaviors that can disrupt high GHG emissions footprint products and markets and complement the current incremental improvement at the sector level.

As CBA has been used for GHG accounting very recently, there is no broadly agreed approach for the categorization of consumption-based GHG emissions (Broekhoff et al. 2019). However, in most of the papers addressing the topic of categorization of consumption-based GHG emissions, sector-based approach to categorization is used, despite the many benefits a need-based approach to categorization of consumption-based GHG emissions can offer. It is recommended to build the consumption categories around the core basic needs citizens have to satisfy. The criteria for two products to be in the same category is their use for the satisfaction of the same core basic needs. Fig. 11.1 below gives the core basic needs, some enabling needs and how they are linked to inputs and processes.

Five core basic needs have been identified here to which a sixth one representing needs that do not fall in any of these categories was added:

- Access.
- Nutrition/Health.
- Lifestyle.
- Clothing.
- Shelter.
- Other.

However, cities can classify the core basic needs differently.

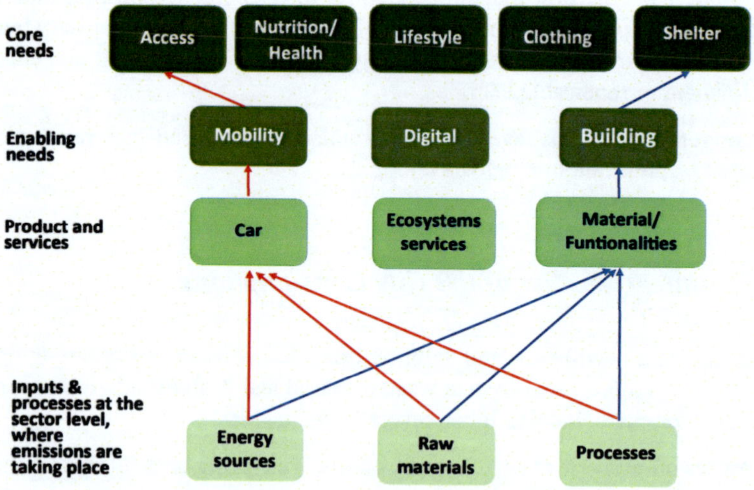

Fig. 11.1 The core basic needs

As it can be seen from Fig. 11.1 above, a same product can contribute in satisfying two different core basic needs. Power is, for example, needed for the satisfaction of all the five basic needs. It has then to be disaggregated into different types of power, e.g.: power for access: P_A, power for nutrition: P_N.

The 6 main parameters the city will monitor to track its climate performances will be the GHG emissions E_A, E_{NH}, E_L, E_C, E_S and E_O related to products consumed to satisfy, respectively, the core basic needs of access, nutrition/health, lifestyle, clothing, shelter and other. These parameters will be monitored alongside the other parameters D_{evi}, related to the development objectives of the city.

11.5.6 *Matrix for the Management of city's Consumption*

A management of cities' GHG emissions mainstreamed in the city development plan can benefit in borrowing from the system approach, the matrix for system analysis in Fig. 11.2 below to support the ToC, particularly when the number of variables involved in the city climate and development action plan is high. A top-down backcasting normative scenario derives the actions A_i from (i) the emissions budgets targeted for the different core basic needs $E_A, E_{NH}, E_L, E_C, E_S$ and E_O and (ii) the development objectives of the city D_{evk}. Doing so, it enables to build the matrix through the identification of the outcomes, the outputs and the actions. Indeed, the matrix links the dependent variables which track the impact of the development and climate actions, to the outcomes, then to the outputs and then to the actions identified to drive the different dependent variables. The actors that need to be involved in the different actions are also identified. Used to identify normative scenario, the

	A_i		O_{pi}		O_{cdj}		O_{ce1}	O_{ce2}	O_{ce3}		D_{evk}		E_O	E_A	E_{NH}	E_L	E_C	E_S
A_i			***					***										
O_{pi}					***		***		***									
O_{cdj}														***				
O_{ce1}															***			
O_{ce2}																		
O_{ce3}																***		
D_{evk}																		
E_O																		
E_A																		
E_{NH}																		
E_L																		
E_C																		
E_S																		

Fig. 11.2 Matrix of system analysis linking the impact on GHG emission budget and actions

matrix responds to the question what should be done from now on to keep the GHG emissions of a city below a given GHG emission budget?

Once it is built, the matrix is used again but now bottom-up to identify possible pathways of forecasting exploratory scenario, responding to the question: What will happen if a given set of actions are implemented under given conditions? Because an action can have unintended impact beyond the output and outcome which enabled its identifications, there are interactions that make the pathways of forecasting exploratory scenario different from the pathways of the backcasting normative scenario.

When they are at the first column, the variables are considered as influencing variables while at the first row they are dependent variables. This means that filling the cell (e.g., row A_i, column O_{ce2}) should be read as the action A_i has influence on the outcome O_{ce2} (in black in the matrix). The matrix visualizes the different exploratory scenario pathways. One example of pathway represented in blue in the matrix is A_i influences Opi. Then, Opi influences O_{cdj}, O_{ce1}, O_{ce3}. Then, O_{cdj}, O_{ce1}, O_{ce3}, influence, respectively, E_A, E_{NH} and E_L. In a city action plan, the number of key dependent variables, outcomes, outputs and actions (influencing variables) can easily be more than 50. Hence, the importance of the matrix to identify the pathways including the side effects (side unintended benefits as well as undesirable side effects).

A row brings out all the variables impacted by the variable in that row and at the first column, acting as an influencing variable. A column shows all the variables influencing the variable at that column and at the first row, acting as influenced variables. The diagonal is empty. The rows at the lowest part of the matrix are also likely to be empty as these variables are influenced variables and are in general not influencing any other variables. The first columns of the matrix are also likely to be empty as these variables are influencing variables and are in general not influenced by any other variables.

11.5.7 Identification of Mitigation Actions for the City Action Plan

There are several normative climate scenarios developed by IPCC, IEA, IIASA or PIK. They translate the climate mitigation goal of the Paris Agreement into global emission budget and attribute this global emission budget to the different productive sectors in a way that minimizes the cost of achievement of the global mitigation goal. Because of their focus on production, these scenarios are not appropriate for the CBA, as this GHG accounting approach does not address GHG emissions related to production at the sector level but GHG emissions related to consumption. For a comprehensive pledge, cities are expected to define a net-zero emissions achievement date, a GHG emissions budget and an emission pathway. Then, they can determine (i) how the GHG emissions budget could be allocated to the satisfaction of the different

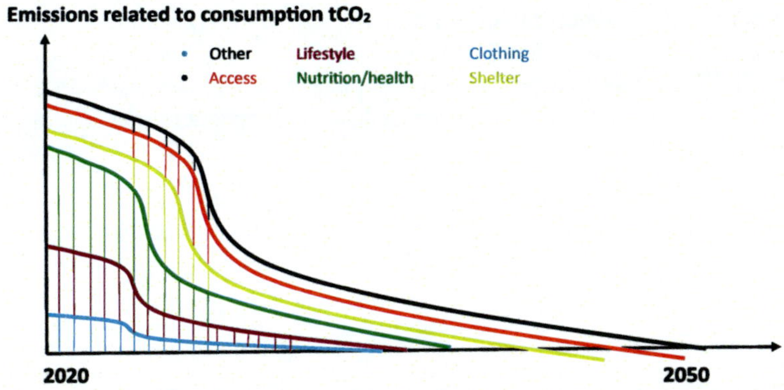

Fig. 11.3 Translation of the city GHG emission budget into emission pathway for consumption to satisfy the core needs

core needs, in a way that minimizes the cost of maintaining the GHG emissions within the GHG emission budget, and (ii) the related core need specific emission pathways as represented below in Fig. 11.3.

GHG emission targets are set for the different categories of cities' consumption established based on the citizens' core needs. Cities can then through backcasting normative scenario, translate the emission budget related to each of the citizens' core needs into emission budgets for enabling needs, emission budget for products and options for actions supporting the satisfaction of the core needs while limiting the related GHG emissions below the GHG emission target. In this exercise, it is worth putting effort in exploring transformative actions that can (i) disrupt entire high GHG emissions footprint value chains and products or (ii) transform value chains into value circles.

The climate actions and the development actions form the integrated climate and development city action plan. They enable the city development while keeping the GHG emissions of the city below its forward-looking baseline emissions, which is the emission budget of the city. The climate action plan serves as an action roadmap for the achievement of the city's climate contribution. These actions can include greenfield actions such as the construction of new infrastructures, renovation action such as improvements measures to be applied on existing physical infrastructures as well as engagement actions with citizens. They indicate the actions that need to be deployed in the city during the timeline of the city action plan, to limit the city GHG emissions below the city GHG emissions budget each year y_i. Their implementation is important for the city to demonstrate effort and its prominent role in the achievement of the climate contribution that will be measured ex-post.

The top-down backcasting normative scenario will further decompose for each year y_i the GHG emission budget for consumption to satisfy a specific core need, into more operational parameters that facilitate the identification of the climate mitigation actions required to bring the GHG emissions related to the consumption satisfying a specific core need below its baseline value.

The GHG emission budget or baseline emissions $E_{k,yi}$ for each core need k (e.g., shelter) during year y_i, is decomposed for each product j contributing to the satisfaction of that core need k into:

- Average GHG emission footprints of the products j $EFP_{k,j,yi}$
- The quantities of products j required for the satisfaction of the core need k $Q_{k,j,yi}$

$$E_{k,yi} = \sum_{j} EFP_{k,j,yi} \times Q_{k,j,yi}$$

To reduce the GHG emission footprint related to the consumption of a core need k, cities can act on any of the two variables $EFP_{k,j,yi}$ or $Q_{k,j,yi}$ or on both.

Decarbonization of products

City governments can reduce the GHG emission footprint $EFP_{k,j,yi}$ of one or several products j contributing in the satisfaction of a core need k during year y_i. They can decarbonize products generated by their own infrastructures. They can also engage with producers located at any stage of the supply chains within or out of the city boundaries and require actions from them to reduce the GHG emissions footprint of some of the products their city consumes. The GHG emissions footprint of products can also be reduced by recycling part of the product components or through refurbishment. This can be facilitated by city governments providing equipment and infrastructures enabling the recycling. Figure 11.4 below visualize the impact of this decarbonization strategy.

The products targeted for decarbonization, their current GHG emissions footprint, their expected future GHG emissions footprint and the date by when the lower GHG emissions footprint products will be available need to be defined as part of the theory

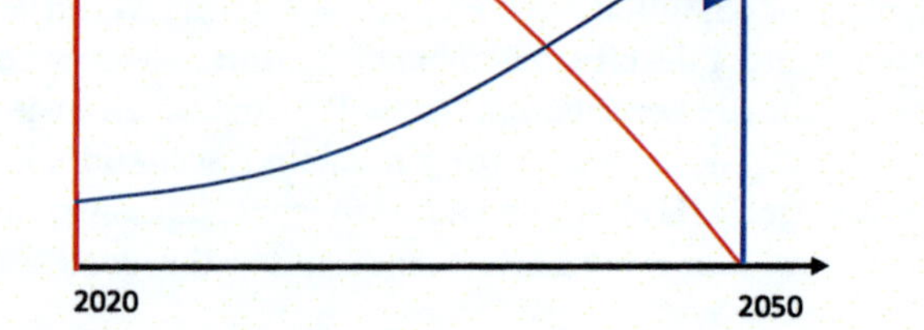

Fig. 11.4 Diagram visualizing the outcome O_{ce3} of a decarbonization strategy

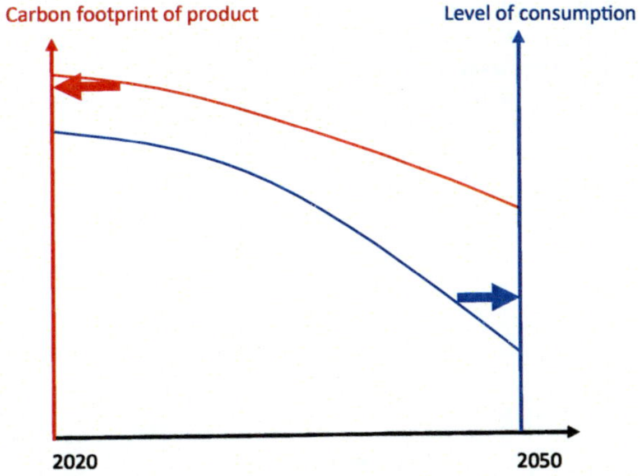

Fig. 11.5 Diagram visualizing the outcome O_{ce2} of a strategy aiming at keeping product in use

of change. The actions the city government needs to take for the decarbonization of each targeted product are also defined, with their expected outputs. Table 3 of Chap. 6 of this book gives a broad range of decarbonization climate actions relevant to the energy for building sector addressed by the Clean Development Mechanism (CDM) methodologies (Gajanana Hegde 2021).

Keeping products in use

City governments can contribute in reducing $Q_{k,j,yi}$, the quantity of one or several products j consumed during year *yi*, contributing to the satisfaction of a core need *k*. They can promote and facilitate the sharing, location, maintenance, repair and upgrade of products in use. They can also promote and facilitate reselling and donating, transforming value chains into value circles. Figure 11.5 below visualize the impact of a strategy aiming at keeping product in use and reducing their quantity consumed.

The products targeted for lifetime extension, their current lifetime, their expected future lifetime and how the product consumption will evolve in the future because of the life extension will need to be defined as part of the theory of change. This is the outcome of city government actions. The actions the city government needs to take to facilitate the lifetime extension of each targeted product are also defined, with their expected outputs. This type of climate actions has not been covered by the CDM methodologies.

Disrupting products with high GHG emission footprint

City governments can contribute in disrupting one or several products *j* with high GHG emission footprint consumed during year *yi*, by making them become redundant, reducing drastically the consumption of the quantity $Q_{k,j,yi}$ of these products, contributing to the satisfaction of a core need *k*. They can promote the development and wide use of low GHG emissions footprint products which can satisfy the same core needs while being produced by a value circle/chain different from the one

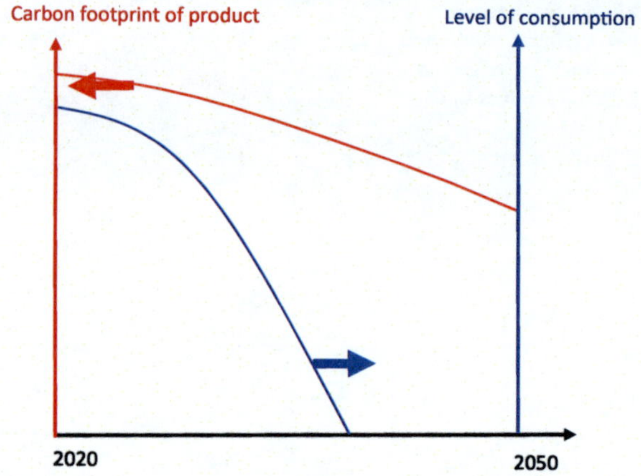

Fig. 11.6 Diagram visualizing the outcome O_{ce1} of strategy aiming at replacing a product with another with lower GHG emission footprint and from a different value circle/chain

which generated the disrupted product. Figure 11.6 below visualizes the impact of the strategy aiming at disrupting products with high GHG emission footprint.

The products targeted for disruption, the current volume of consumption of each of these products, how these volumes are expected to evolve with time, will need to be defined as part of the theory of change. This is the outcome. The actions the city government needs to take to facilitate the disruption of each targeted product are also defined, with their expected outputs. This category of climate actions is not covered by existing methodologies. This is because of the lack of direct causality between the climate action (taking place in a given value circle/chain) and its impact (taking place in a different value circle/chain). The ToC and the matrix for system analysis are proposed to fill that gap. A map identifying actions and linking actions, outputs, outcomes and impacts is provided below (Fig. 11.7).

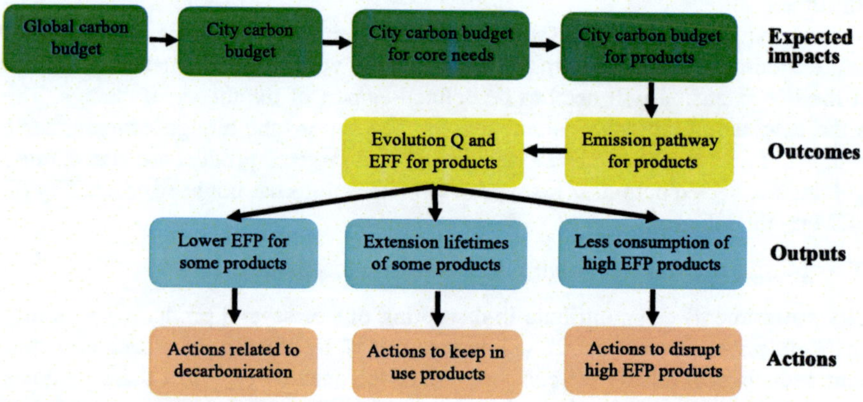

Fig. 11.7 Map identifying actions and linking action, outputs, outcomes and impacts

11.5.8 Ex-ante Determination of the Impact of a City Action Plan

Existing GHG accounting frameworks can only measure and reward ex-post the impact of climate action. If it is related to an infrastructure, it has first to be built and operated. Indeed, the recognition and the reward are result-based and done ex-post upon the implementation of the physical assets and based on their climate performance during their operation. Emission reductions below a baseline can be recognized and rewarded. As it can only be an ex-post reward for result achieved and not an ex-ante enabler to make possible the expected GHG emission reductions, it creates a signal that can enhance the willingness to act of the city but not its capability to respond to the signal. This result-based GHG accounting can only be reactive and backward-looking in rewarding climate action. Unlike such result-based GHG accounting, a GHG accounting based on the city action plan would be proactive, forward-looking and an enabler of climate action. Also, the ex-post attribution to a city of climate contribution below expectation, arising from the implementation and operation of the infrastructures in its action plan and resulting in GHG emissions intensive activities by its citizens, would look like a punitive measure. Indeed, in most cases, there will be no more possibility of economically acceptable remedy. The implementation of infrastructures conducive to high GHG emissions would lock the cities in a high GHG emission pathway for decades. Instead, if the GHG emissions are accounted and attributed to a city based on its action plan, it will be encouraged to change the plan and strive to improve its climate performances. This is a preventive measure to the development and implementation of GHG emissions intensive city action plan. The GHG accounting framework proposed measures ex-ante the GHG emissions of cities related to their city action plans.

The ex-ante determination of the impact of a city action plan is useful for many reasons:

- It establishes the link between the actions that are to be introduced in a city action plan and the impact that is expected. Impact cannot be claimed without action.
- It enables city government to prioritize the city action plan that addresses the development needs of the city while being aligned with the climate goals.
- It enables a city to join a Local Government Green Funding Agency (LGGFA) which has as criteria for participation, forward-looking green activities in a green city action plan.
- Determining the emissions of future actions in a city action plan enables influential stakeholders such as central government, financiers and citizens that are constituencies, to proactively encourage greener greenfield and renovation plans.
- In case a city issues a green bond for the implementation of its city action plan, the ex-ante determination of the impact of the city action plan may be required.

The ex-ante determination of the impact of a city action plan takes the opposite direction of the top-down normative scenario, starting from the actions to be undertaken and determining the expected outputs, the expected outcomes and the expected impact from the implementation of the actions.

Example of actions

- Enabling the use of sustainable wood for construction with the recycling of the wood at its end lifetime.
- Enabling the development of ESCO in the city.
- Making available low-cost loan for building upgrade.
- Growing the city in a compact, complete, coordinated and connected manner.

Example of outputs

- Reduction by X during year y_i compared to year y_{i-1} of the consumption volume of product j with GHG emissions footprint $EFP_{k,j,yi}$ and increase by Y of the consumption volume of product 1 with GHG emissions footprint $EFP_{k,1,yi} < EFP_{k,j,yi}$ $Y < X$.
- Extension of the lifetime of product j from 1 year to 1.5 year.
- Reduction by X% of the GHG emissions footprint $EFP_{k,j,yi}$ of product j consumed during year y_i.

Example of outcomes

- Decrease of the GHG emission footprint $EFP_{k,j,yi}$ and/or the quantities $Q_{k,j,yi}$ for the products j consumed.
- Emission pathways of the products affected by the actions are shifted downward.

Impact

- Reduction of the GHG emission of the city related to consumption.

Scenarios that enable the satisfaction of all the core needs with the lowest level of GHG emissions and the lowest costs can be identified. The system analysis ensures an integrated approach to address holistically the interconnected pathways for the satisfactions of the citizens' core needs.

The actions that will be implemented will not necessarily be exactly the actions identified by the city through the ToC process. But even if the same actions identified through the ToC were the ones put in the climate action plan, the impact can be different from the expected impact which is the city GHG emission budget that started the ToC, due to the unintended effects. There may be also actions the city decides not to undertake because the matrix of system analysis brought out their harmful impact on development objectives. This is due to the interactions among the pathways. Where the aggregated impact of the different actions does not allow to fit within the GHG emissions budget, additional measures will need to be taken to further reduce the GHG emissions so that they fit within the GHG emission budget of the city. These measures could be scaling up the deployment of existing climate solutions or developing new more transformative climate solutions.

Recognizing cities' climate contribution based on information in a city action plan (e.g., in the context of issuance of green bond which proceeds will finance the city action plan) raises some new challenges. The implemented greenfield and renovation actions may be different from the ones that were described in the city action plan or may be less impactful than what was expected in the city action plan. It means that the climate contributions measured based on registered actions introduced in or removed from the city action plan will need to be corrected depending on the gap between the performance of the actions registered in the city action plan (that determined the climate performance of the city) and the actual performance of the actions implemented. However, while a physical infrastructure cannot be altered and is transparent, a simple information in a plan can be altered. To ensure that the information that was in the registered city action plan and that has been the basis for measurement and attribution of climate contribution will not be altered, the immutability function of DLT can be used.

11.5.9 Policy Instrument Enabled by the New GHG Accounting System

As it is possible with the proposed GHG accounting system, to follow the history of each ton of GHG emission embodied in transacted products, cities accountability can be defined much simpler. A city will be responsible for each ton of embodied GHG emissions which history ends within its boundary, with the product it is embodied in being consumed by one of the city stakeholders. As GHG emissions embodied in exported products continue their history, they are not covered by this definition. The city will also have credit for each ton of GHG emission that would have been embodied in a product it has avoided the production.

The capability of tracking the history of the embodied GHG emission provided by the new GHG accounting system opens important new perspectives in the fight against climate change. It enables city government and other sub-national authorities to levy a carbon border tax, where they are empowered to do so by their central government. Such tax would make a lot of sense. As city government is inheriting the responsibility of the GHG emissions embodied in products consumed within their city, regulatory instruments to act on such GHG emission should be made available to them. Raising awareness of their citizens will not suffice as the only instrument to influence their consumption.

Such a tax would be easy to operationalize with the proposed new GHG accounting system. By accepting through the DLT, the responsibility of the GHG emissions embodied in an imported product, an importer in city A triggers the crediting of its GHG emission account with these specific GHG emissions. If the imported product is used for the manufacturing of another product, its embodied GHG emissions are transferred to that new product. In case an imported product is re-exported as is or used to manufacture another product exported, as soon as the buyer from another city accepts the responsibility of the GHG emissions embodied in the products, they are removed from the account of the importer in city A. The GHG emissions left in the account of the importing entity in city A are those that end their history in that city. They will be reflected in the city's GHG emissions and therefore are dutiable. The DLT can automatically collect the proceeds of the tax based on the quantity of GHG emissions left in the importer's account.

To ensure that the tax will not compromise the satisfaction of a core need, its proceeds can be used to subsidize another product satisfying the same need but with lower GHG emission footprint. It can also be used as seed money at the first loss position, for a Venture Capital Fund financing the development of innovative products satisfying the same core need as the taxed product, but with lower GHG emission footprint.

11.6 Conclusions

CBA is more appropriate as GHG accounting system for city. However, it faces important barriers including access to the required information, risks of missing GHG emitted but not embodied in the accounted consumed products with the limited granularity of the EEIO modeling.

To address these issues, a new GHG accounting system for city is proposed, replacing the Eulerian specification with a Lagrangian specification, thus leveraging the advantages of both CBA and PBA. We call it a Lagrangian CBA. Instead of focusing on the city, it follows the products and their embodied GHG emissions through time and space, from city to city, along their value chain or value circle. It can then keep track of their trajectory and their history. It also fully benefits at its initial stage from the simplicity of the PBA and the availability of data.

Another specificity of the Lagrangian CBA is that it enables to incentivize and recognize action by all entities with capability and power to act regardless of whether they are acting as producer or consumers. It can encourage in the same city a citizen to change its consumption behavior but also a cleantech startup to innovate for low GHG emission products.

DLT is leveraged to serve the Lagrangian description of the flow of GHG emissions embodied in products. This provides a much higher level of granularity to the GHG accounting system, with possibility of getting detailed descriptions attached to the GHG emissions footprint of products. For each ton of GHG emissions embodied in a consumed product, it is possible to know precisely from where, in the value chains, it has been emitted as well as other relevant information. The use of other digital technologies such as IoT, AI, GPS and RFID linked to the DLT can further automatized the GHG accounting system.

The way that baseline is set is another specificity of the GHG accounting system proposed. It departs from the current baseline framework and proposes a new type of baseline we call forward-looking baseline. It brings out what is needed instead of (i) what is possible or desirable or (ii) what progress has been already achieved. It indicates at any point in time to cities and their climate actors where they should be, in terms of climate performance, to be aligned with the climate goals.

To be able to assign to cities accountability for GHG emissions or to credit cities for GHG emission reductions based on responsibilities and effort, we propose a ToC, using a matrix for system analysis, establishing the linkages between the actions of a city as change agent, and the impact it is claiming. This also enables the ex-ante determination of the climate contribution of a city action plan which links actions and impact, facilitate prioritization, enables influential stakeholders such as central government, financiers and citizens that are constituencies, to proactively encourage greener greenfield and renovation plans and facilitate the issuance of green bonds to finance city action plan.

The consumption categories are built around the core basic needs citizens must satisfy.

The proposed new GHG accounting opens important new perspectives in the fight against climate change. It enables city government and other sub-national authorities to levy a carbon border tax where they are empowered to do so by their central government. The tax proceeds can be used effectively to ensure that the tax does not compromise the satisfaction of any core basic need.

The proposed GHG emission accounting is a tool available to cities for an effective management of their GHG emissions. However, it can even do more. Its implementation nationwide, in all the cities or sub-nationals of a country, with their empowerment to use carbon pricing to manage their GHG emissions could be an alternative to a carbon pricing applied at the sector level, based on production. At least it addresses the issue of lack of competitivity at the export which has been one of the main barriers which deterred some countries from implementing carbon pricing. Future researches could address that aspect.

References

Balouktsi M (2020) Carbon metrics for cities: production and consumption implications for policies. Build Cities 1(1):233–259. https://doi.org/10.5334/bc.33

Broekhoff D, Erickson P, Piggot G (2019) Estimating consumption-based greenhouse gas emissions at the city scale (SEI Report February 2019). Retrieved December 2, 2019, from https://www.sei.org/wp-content/uploads/2019/03/estimating-consumption-based-greenhouse-gas-emissions.pdf

C40 Cities (2018) Consumption-based GHG emissions of C40 Cities (Cities climate leadership group (C40) report in partnership with the University of leeds (UK), the University of New South Wales (Australia), and Arup. Retrieved December 2, 2019, from https://www.c40.org/researches/consumption-based-emissions. Consumption-based GHG emissions of C40 cities. https://www.c40.org/researches/consumption-based-emissions

Evry Financial services (2015) Blockchain powering the internet of value

Floater G, Dowling D, Chan D, Ulterino M, Braunstein J, McMinn T (2017) Financing the urban transition: policymakers' summary. Coalition for urban transitions. London and Washington, DC. http://newclimateeconomy.net/content/cities-working-papers

Froystad P, Holm J (2015) Blockchain: powering the internet of value. https://www.finyear.com/attachment/637653/
Hegde G (2021) Planning for climate smart cities, chapter 6 possible climate actions in cities
Stephen H, Graven H (2016) Imperial college London, grantham institute briefing paper No 16: satellite observations to support monitoring of greenhouse gas emissions. https://www.imperial.ac.uk/media/imperial-college/grantham-institute/public/publications/briefing-papers/Satellite-observations-to-support-monitoring-of-greenhouse-gas-emissions-Grantham-BP-16.pdf
Thioye M (2021) Planning for climate smart cities, chapter 9 mitigation action enablers for climate smart cities
WRI (2014) C40 CITIES, ICLEI. Global protocol for community-scale greenhouse gas emission inventories. https://ghgprotocol.org/sites/default/files/standards/GHGP_GPC_0.pdf

Appendix
A Guide to Carbon Neutral City Plan 2050: An Urban Integration Platform Modeling Approach

Kwi-Gon Kim

Introduction

A new era for climate smart and wise city planning with carbon neutrality strategies has arrived. The goal of NDCs can be achieved through cities. The carbon neutrality is a matter of survival for humankind and other living things.

The objective of "A Guide to Carbon Neutral City Plan 2050" is to assist cities to fulfill their indirect commitment to the Paris Climate Agreement by aligning with international frameworks and protocols with stakeholder engagement. Cities are sub-national and do not make direct commitment to the Paris Agreement. The Guide includes the neutrality planning process in order.

At the global level, a host of countries have recently announced major commitments to significantly cut their carbon emissions, promising to reach "net zero" in the coming years. The term is becoming a global rallying cry, frequently cited as a necessary step to successfully beat back climate change, and the devastation it is causing. According to United Nations News (December 12, 2020), so far 186 parties to the Paris Agreement have developed nationally determined contributions (NDCs), and 75 parties submitted new or updated NDCs (UNFCCC 2021). Net-zero commitments and NDCs of the countries are not directly linked to carbon neutrality strategies of cities, although objectives are the same.

Practically every country has joined the Paris Agreement on Climate Change, which calls for keeping the global temperature to 1.5 °C above pre-industrial era levels. If we continue to pump out the emissions that cause climate change, however, temperatures will continue to rise well beyond 1.5 °C, to levels that threaten the lives and livelihoods of people everywhere.

K.-G. Kim (✉)
Seoul National University, Seoul, Republic of Korea
e-mail: kwigon@snu.ac.kr

K.-G. Kim and M. Thioye (eds.), *Planning Climate Smart and Wise Cities*, The Urban Book Series, https://doi.org/10.1007/978-3-030-80165-6

This is why a growing number of countries are making commitments to achieve carbon neutrality or "net-zero" emissions within the next few decades. It is a big task, requiring ambitious actions starting right now.

Net zero by 2050 is the goal. But countries also need to demonstrate how they will get there. Efforts to reach net zero must be complemented with adaptation and resilience measures, and the mobilization of climate financing for developing countries. The good news is that the technology exists to reach net zero, and it is affordable.

At the city level, on the other hand, greenhouse gas accounting and target-setting protocols such as the Global Protocol for Community-Scale Greenhouse Gas Emissions (GPC) refer to "carbon neutrality" and "net zero," but they do not clearly define these terms or provide the detailed guidance that cities need to develop citywide carbon neutrality implementation strategies. While ambitious cities agree that carbon neutrality should be achieved at the citywide scale, there is a continuing debate about how residual emissions should be defined (e.g., the scopes, boundaries and sectors), canceled out (e.g., with carbon credits) or removed (e.g., with negative emissions technologies) to achieve carbon neutrality (C40 Cities Climate Leadership Group 2019).

Carbon neutral city planning can make a tremendous difference in environmental sustainability and integrity and the quality of life and well-being of people living in cities by setting an ambitious target of net-zero emissions by 2050. Net-zero emissions can be achieved by four broad strategies: innovative urban/spatial planning, smart cutting-edge technologies, greening power supply and carbon offsetting in alignment with emerging global mechanisms. However, carbon offsetting should be separate from other strategies because there is ongoing debate on safeguards that should apply to use this option as a complementary option referring to a perspective of environmental integrity.

Although a growing number of cities have developed strategies for carbon neutrality, no universal carbon neutrality analysis methodology has been developed to date. This guide has examined several methodologies and their comparative features and suggests innovative new methodologies. It is intended to assist cities to develop strategies, measures and project activities for carbon neutrality in response to global requirements to keep the global temperature to 1.5 °C above pre-industrial era levels. It is clear that such carbon neutral strategies will help explore the design of the market mechanisms under Article 6. Under the market mechanism, a city's entire carbon neutrality plan and its implementation likely qualify.

There are several reasons why we have to look at the importance and value of "Carbon Neutral City Plan 2050" as a part of NDCs-national climate action plans. They are as follows:

1. Even though cities are sub-national and non-state actors, the objectives of the cities are the same as net-zero commitments and NDCs of the countries.
2. In about 30 years, close to 70% of the world population will be living in cities. Therefore, the cities are even more crucial to the accomplishment of carbon neutrality

3. Cities could be important game changers for implementation of NDCs on the ground based on their spatial/digital world planning.
4. Cities and municipalities have more strong sense of belongings to their carbon-free communities and their own local practices than the countries. Therefore, many cities have more ambitious emissions reduction targets than the countries.
5. The relationships between cities and nations could be much more "closer", "robust" and "enduring", not "in conflict" as partners to combat climate emergency. The identification of measures for scope 1 (city GHG emissions boundary), scope 2 (national GHG emissions boundary) and scope 3 (global GHG emissions boundary) is not explicitly described in NDCs.
6. Digital transformation to climate action is easier in cities to provide connectivity in the untact era.
7. Cities need to reboot economy from COVID-19 pandemic crisis under the Green Deal for green investment or stimulus plans.

The guide provides guidance on

- Framework for logical sequence to carbon neutrality strategy development
- Definition and principles that should be exhibited when choosing strategies measures and projects to cut carbon emissions
- Dynamic methodologies for carbon neutrality analysis based on an integration platform modeling approach
- City profile and analysis for climate, energy and nature
- Climate actions and timing of strategies including climate targets, measures and projects taken to achieve carbon neutrality
- Pathways to accomplish the climate actions and reporting on progress
- Policy implementation roadmap that includes roles for the city and partners and wider project benefits available to cities
- Project deliverables that can help people responsible for overall carbon neutrality planning in cities
- Co-benefits of carbon neutral city plan 2050 which should be an essential document for planning climate and wise cities.

Framework for Developing the Guide

The proposed guide is developed into seven phases. Phase 5 consists of four steps. Each phase is illustrated by examples of current city activity that can act as models to shape future practice. Three appendices complement the text:

Phase 1: Introduction.
Phase 2: Framework for developing the guide.
Phase 3: Definition and principles of carbon neutrality in cities.
Phase 4: Dynamic methodologies for carbon neutrality analysis.
Phase 5: The process as a whole.

5.1 City profile.
5.2 Integrated climate actions.
5.3 Pathways to accomplish the climate actions.
5.4 Policy implementation roadmap.

Phase 6: Essential project deliverables for carbon neutral city 2050 plan.
Phase 7: Co-benefits of carbon neutral city plan 2050.

Definition and Principles of Carbon Neutrality in Cities

Definition

Carbon neutrality means having a balance between emitting carbon and absorbing the equivalent amount of carbon from the atmosphere in carbon sinks. Removing carbon dioxide from the atmosphere and then storing it is known as carbon sequestration. In order to achieve net-zero emissions, all worldwide greenhouse gas emissions will have to be counterbalanced by carbon sequestration with the appropriate timescale. It is important to make good historic emissions and absorptions by 2050.

Carbon sink is any system that absorbs more carbon than it emits. The main natural carbon sinks are soil, forests and oceans. According to estimates, natural sinks remove between 9.5 and 11 Gt of CO_2 per year. Annual global CO_2 emissions reached 38.0 Gt in 2019.

To date, no artificial carbon sinks are able to remove carbon from the atmosphere on the necessary scale to fight global warming.

Part of the carbon stored in natural sinks such as forests is released into the atmosphere through forest fires, changes in land use or logging. This is why it is essential to reduce carbon emissions in order to reach climate neutrality. Another way to reduce emissions and to pursue carbon neutrality is to offset emissions made in one sector by reducing them somewhere else. This can be done through investment in renewable energy, energy efficiency or other clean, low-carbon technologies and carbon sequestration. The EU's emissions trading system (ETS) is an example of a carbon offsetting system (European Commission 2019).

Principles

The following principles should be exhibited when planning carbon neutral cities.

- Gradual target principles: Additionality issues
  - Prepare city's roadmap to cut emissions and reduce remaining emissions by local carbon sequestration and carbon offset purchases outside the city to ensure the carbon neutrality by 2050.

- Set up annual tree-planting goal to increase percentage of tree canopy cover by 2030 and 2050, respectively.
- Calculate additionality in terms of carbon sequestration.
  Additionality aims at measuring the net sequestration difference deriving from the implementation of the proposed project by estimating the change between the situation without the project and the situation with the project at the city level. Timescale should be considered.
- Adopt BAU projection approach at the city level to consider population and economic growth, technology innovation, behavioral change of people, etc., over the time.
- Determine the additionality of a project: "project-specific" and "standardized." Each of these approaches has strengths and weaknesses (Broekhoff et al. 2019). Carbon neutrality projects in cities as a whole may need different types of additionality methodologies/protocols. Each project which is included in the carbon neutral city plan 2050 should be evaluated on an analysis of an individual project's characteristics and circumstance or against a set of pre-determined eligibility (e.g., performance benchmarks that—in principle—distinguish additional from non-additional projects).

- Social-spatial-digital integration platform principle: Sector cohesion issues
  - Try social inclusion and socio-spatial integration, by preventing social segregation and mitigating social imbalances between neighborhoods.
- Land tenure security principle: Ownership issues
  - Consider land tenure security which is a requirement for forest carbon (C) finance investment, as a way to ensure that C emission reductions can be legally delivered to buyers and to promote permanence of emission reductions from forest activities. Areas under unsecure land tenure are regarded as less attractive.
- Carbon offset principles: Nature-based environmental integrity issues including achieving land degradation neutrality
  - Take the idea of offsetting carbon emissions which is a bold step inventing a way in which credits for such offsets can be traded like commodities. The credits from carbon sequestration by forests are in various forms such as CDM and ETS called these credits certified tradable offsets. They can include investments in both carbon sequestration and clean technology outside the cities.
- Digital transformation principles: Transparency issues
  - Accelerate digital transformation which is the pivotal shift that has made carbon neutral city possible. To develop a carbon neutral city, we need to adapt much of our infrastructure and buildings to the digital transformation. Digital retrofits will be a fundamental part of this process but are currently a step

that is frequently overlooked even at a city level. Climate smart digital design, operations and management of our energy and carbon offset systems in every aspects of carbon neutral cities need to be incorporated from the outsets. This will require a digital first approach and investment to create various smart grids, smart land use, smart buildings, smart IoT homes, smart transport, smart workplace, smart factories, and smart urban infrastructure which are components of carbon neutral cities.

- Financing mechanisms access to capital: Innovative financing mechanism issues
  - Play a role creating or expanding financing options and improving access to such financing, both for city-led projects (i.e., public sector) and for those projects that initiate on the private sector side.

In 2021 and beyond, the digital journey needs to continue to deliver much needed smartness, inclusiveness, resilience and resource wisdom, creating green jobs and economic growth while simultaneously helping to dramatically reduce our impact on the city and ultimately halting climate change.

Box 1

Guiding principles for the carbon neutrality strategy, Washington DC, USA

- Advance Equity: Ensure that the benefits of climate action extend to all district residents.
- Transparency and Accountability: The process to develop the strategy should be inclusive, and the information provided should be accessible. The strategy should establish clear milestones and metrics that will be incorporated into our regular public reporting.
- Goal-oriented: Achieving carbon neutrality will require tough choices and bold action. The strategy should provide help the district evaluate and make those choices.

Source: Department of Energy and Environment, Washington DC, 2018.

Dynamic Methodologies for Carbon Neutrality Analysis: An Urban Integration Platform Modeling Approach

Different methodologies for carbon neutrality analysis identify various pathways or scenarios under different assumptions. Various integration platforms would be required for each methodology. A climate integration platform means integration of information from the disparate systems of climate actions into a unified whole to provide a holistic view of the overall performance and state of the city and its various functions by using knowledge-based platforms for sensing, for communications, for integration and for intelligent decision making.

Critique of Current Approaches to City Carbon Neutrality Planning

In many cases, there are no clear indication of relationships between urban planning and carbon neutrality planning. Carbon neutrality strategies in cities are currently separate from sustainable urban development and strategic planning for two reasons. The first reason is that urban land use/spatial planning has not been pivotal sources of carbon emissions and solutions for carbon neutrality strategies. The second one is, therefore, that carbon neutrality strategies have not been integrated into existing urban planning process. Even though some cities such as the City of Cape Town consider spatial planning (e.g., transit-oriented, densified and diversified urban development and growth, well-located and affordable housing) as part of their commitments, carbon neutrality strategies need to be implemented under the broad framework of sustainable urban development and strategic planning.

Today, there is a new emphasis on an intersectoral planning approach incorporating spatial and climate aspects as well as new health pandemic, social, cultural, economic and digital elements. The new methodologies for carbon neutrality analysis should be characterized by intersectoral coordination and financial feasibility.

Types of Methodology

Five methodologies can be suggested for carbon neutrality analysis, which show different degrees of net-zero commitments across the broad spectrum of sustainability from strong sustainability to weak sustainability. The types of methodology that have evolved in each city have depended on the city's data availability, impact of Green New Deal exposed, level of technology, legal system and institutional framework and the relative role of the different actors in the development process. Method 1 has developed for net-zero emission reductions. Other methods can assist in specifying the range of emission reductions that may occur.

Method 1

The first method provides a means for reaching net-zero city/community emissions with the business-as-usual emission projection 2050. In 2009, the City of Melbourne updated the Municipality Business Strategy Map contained within Zero Net 2003 with the revised 2020 projections. To reach zero net community emissions, reductions considered achievable via the strategies contained in Zero Net Update 2008 have been determined as well as the contribution required from decarbonizing the energy supply and offsetting. Figure 1 illustrates implementation of the strategies under the Zero Net Update 2008, City of Melbourne.

Nature-based solutions (NBS) to address city carbon neutrality 2050, including management of forest, soil, freshwater, river, wetland, and estuarine protected areas, should be part of dynamic carbon neutrality models.

Figure 2 shows City of Melbourne—the municipality business strategy map which contains the same information shown in Fig. 1 in a graphic form. The relative contributions toward zero net emissions are shown in Fig. 2. It includes 8,014,600 t Business-as-usual emissions 2020, 1,153,387 t residential and commercial, 96,580 t passenger transport, 1,455,200 t decarbonising the energy supply, 94,600 t industrial and manufacturing (10% reduction) and 5,214,833 t offsetting which contribute to zero net emissions.

The equation model for GHG balance sheet indicates strong sustainability goal. The carbon offsetting has the following features:

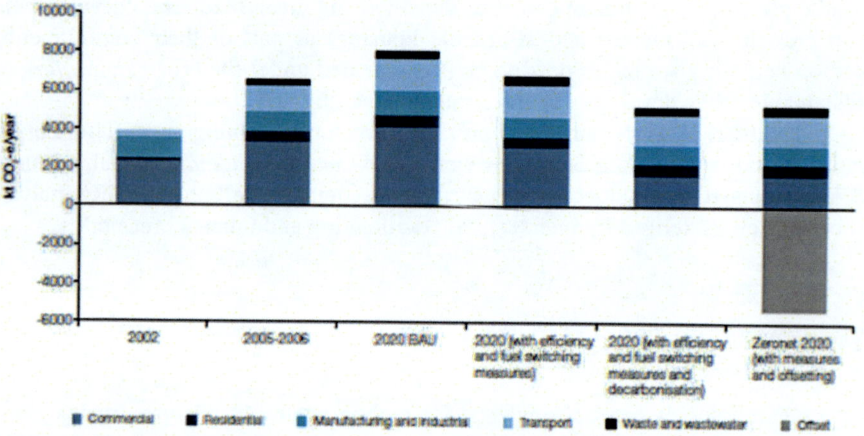

Fig. 1 Greenhouse gas emissions (kilotonnes carbon dioxide equivalents) for the City of Melbourne from 2002, 2005–06 to 2020 (under a business-as-usual and *Zero Net Update 2008* scenario). *Source* City of Melbourne (2009) Zero Net Emissions by 2020: Update 2008, p 28

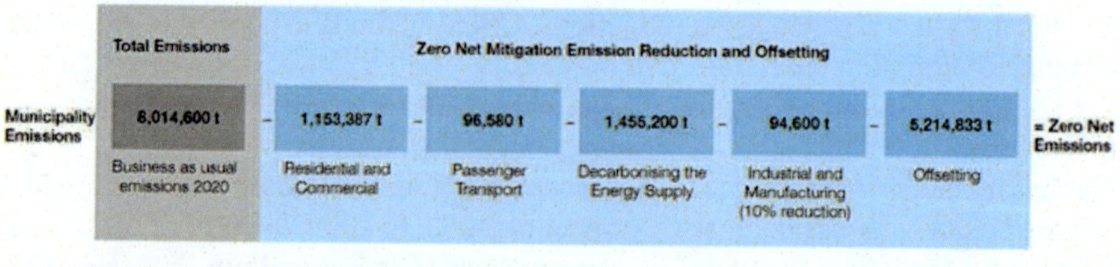

Fig. 2 City of Melbourne—the municipality business strategy map—an equation for strong sustainability. *Source* City of Melbourne (2009) Zero net emissions by 2020: update 2008, p 29

- Carbon offsets were an intrinsic feature of the Australian Government's National Emission Trading Scheme.
- A carbon credit purchasing scheme for tenants was part of the pilot municipal emission trading market.
- The City of Melbourne planned to invest in blue mallee eucalyptus plantations as feedstock for renewable power generation, with eucalyptus oil as a by-product.

As another example, New York states that, by 2050, the city must deliver on its commitment to directly reduce GHG emissions 80 percent below 2005 levels. To achieve a 1.5 degree Celsius outcome, NYC, along with cities around the world, must also reach carbon neutrality, meaning zero net emissions citywide. The city will lead in the development of a global protocol for carbon neutrality that may include large-scale renewables, as well as carbon sequestration and carbon offsets to account for residual GHG emissions after all technically feasible reductions are achieved (Fig. 3).

The term carbon offset broadly refers to a reduction in GHG emissions—or an increase in carbon sequestration (e.g., through land restoration or the planting of trees)—that is used to compensate for emissions that occur elsewhere.

However, the carbon offsets after 2020 would be different from the Kyoto Protocol. Under Kyoto, for companies committed to taking climate action, it has not been hard to find low-cost carbon offsets in developing countries. The 2015 Paris Agreement could change this. For the first time, nearly every country (186 parties as of the end of last year) in the world has identified explicit actions (i.e., "contributions" through NDCs) they agree to make to reduce GHG emissions and adapt to climate

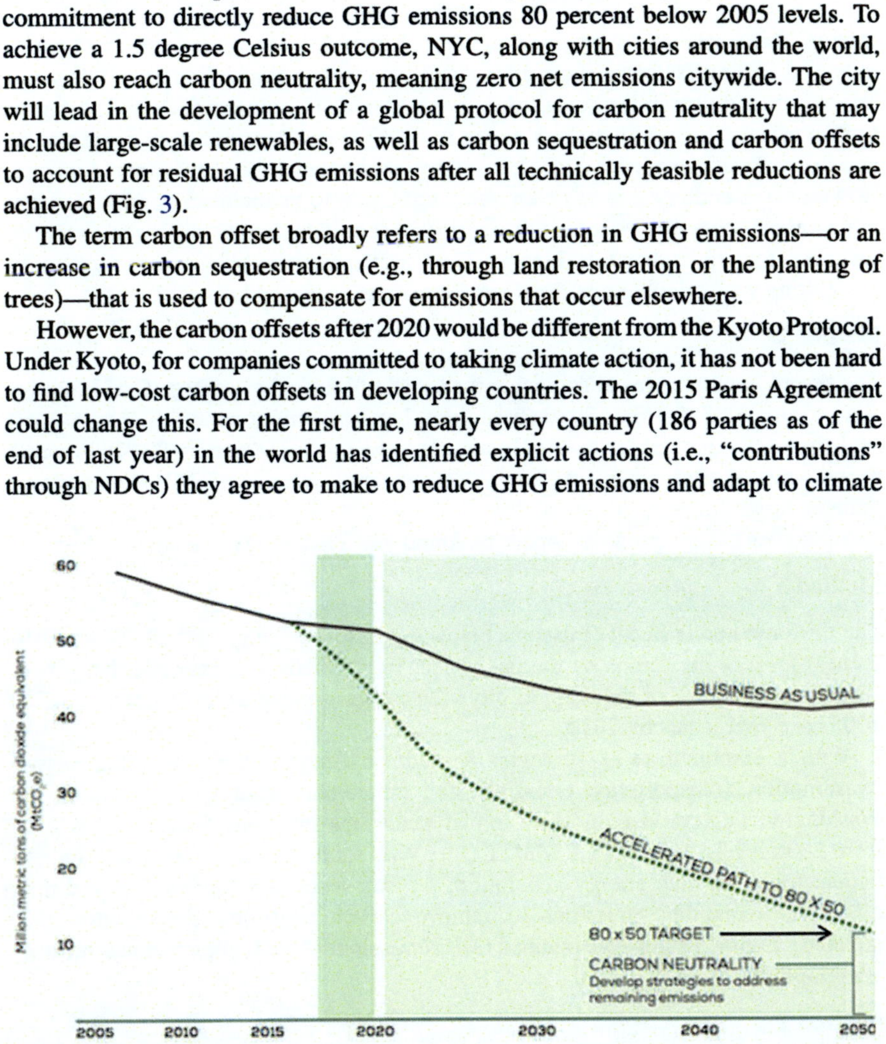

Fig. 3 Long-term plan of New York City. *Source* 1.5C: aligning New York city with the Paris climate agreement

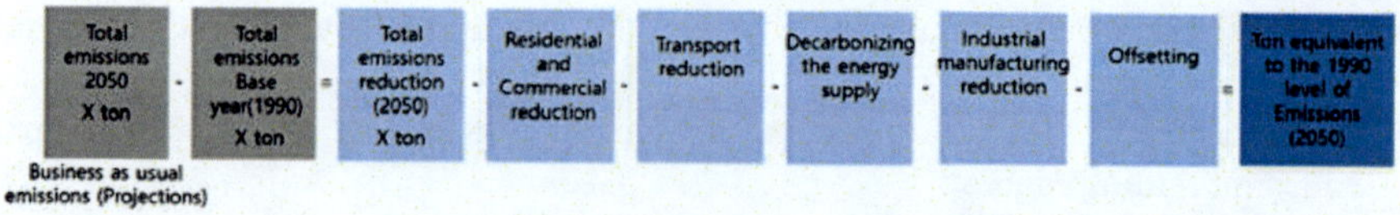

Fig. 4 Carbon neutral city 2050—the city base year (1990) strategy map: realistic target for weak sustainability

change. This approach is a major change from Kyoto, where only industrialized countries could find offset projects in developing countries, providing them with needed investment and promoting sustainable development. This does not mean the end of carbon offsets. In fact, Article 6 of the Paris Agreement explicitly recognizes the possibility for international cooperation through the transfer of emission reductions. However, if a country allows an emission reduction to be claimed by another party (either another country or some other entity), it should no longer be able to count the reduction toward its own GHG target. The Paris Agreement has language expressly prohibiting such "double counting" among countries (Broekhoff et al. 2019).

Method 2

The second method aims at achieving the target of kilotonnes carbon dioxide equivalent to the 2005, 2006 or 2010 (base year) level of emissions by 2050 with 2050 projections under business-as-usual scenario (projected emissions in 2050). Figure 4 shows a model of carbon neutral city 2050—the city base year (1990) strategy map: realistic target for weak sustainability.

Examples of the approach cannot be found yet. This option is a new method.

Method 3

The third method is to cut emissions below the base year (e.g., 2005, 2006 or 2010 tonnes) level of emissions by the target year (e.g., 2050). For example, New York City has set the goal of reducing its citywide carbon emissions by 30 percent below 2005 base year levels by 2030.

With a combination of strategies to address energy supply, building energy consumption, transportation emissions and municipal waste and wastewater, the Washington DC could reduce its annual emissions by more than 89% by 2050 from 2006 levels, to reach 1.2 MMTCO2e. With more aggressive state and federal targets for renewable energy generation, we can reach 0.5 MMTCO2e and then offset the remaining emissions to achieve carbon neutrality (Washington DC, undated). Figure 5 shows the planned GHG emissions—gradual path to neutrality of Washington DC.

Method 4

The fourth method is to reach carbon emissions in the target year per capita below the national average of tones per capita in the same year. Per capita emissions per year can be used for a need-based solution-oriented approach to climate contribution

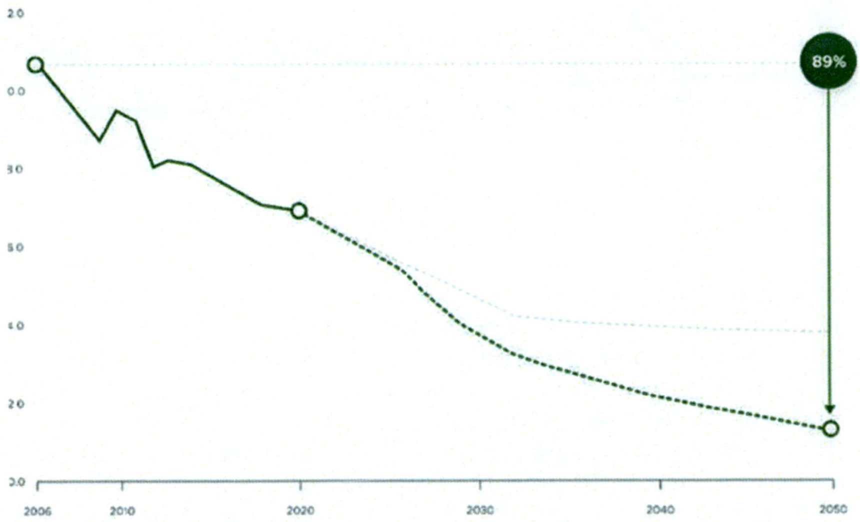

Fig. 5 Citywide GHG emissions—gradual path to neutrality

of a city. It builds the consumption categories around the core basic needs citizens must satisfy (Thioye 2021).

Method 5

The fifth method can use the base and target year and emission reduction percentage similar to NDCs (2050) and LEDS (2030) at the national level when mitigating climate change of cities.

At the national level, there is no hard and fast rule for the base year in the NDCs. But relatively few use 1990 as the base year (there are only three now: UNFCCC site provides more information.) (Hegde 2021). Recent NDCs use other base years such as 2010, 2013 and 2017. In the case of Republic of Korea, the latest submission of December 2020 takes 2017 as the base year. It is stated that the Republic of Korea has set the ambitious target of reducing one-fourth of the total national GHG emissions within approximately 10 years from the base year 2017. Target year in the case of Korea is 2030 for the NDC. The interesting measures in the Korea NDC include buildings, transport and energy. It may be useful for cities to consider the measures in the NDCs to reach and exceed if possible (Hegde 2021).

Box 2

Summary of measures included in the Korea NDC

- With regard to new buildings, including private-owned buildings, more buildings will be required to be certified as zero-energy buildings that minimize energy consumption, and energy efficiency standard will be strengthened for home appliances and office equipment.

- Smart meters will be provided for 5 million household units of apartment.
- For the building sector, a total of KRW 20 trillion will be invested in green remodeling of 225,000 public rental homes and 2000 public buildings (day care centers, healthcare centers, etc.) used by vulnerable populations and transformation of 2890 units of school into green smart school.
- Create six hydrogen-based cities.
- Ban on constructing new coal-fired power plants.
- Additional reduction of existing coal power plants
- Conversion to eco-friendly fuels such as LNG (from coal)
- Scale up renewable energy (to increase the share of renewable energy up to 20% by 2030 and 30–35% by 2040).
- The solar and wind power capacity is expected to increase three-fold from the 2019 level by 2025 (from 12.7 GW to 42.7 GW).
- In the transportation sector, the Republic of Korea has set the target of deploying 3 million units of electric vehicles and 850,000 hydrogen vehicles by 2030 with a view to scaling up the deployment of zero-emission vehicles.
- Model shift in freight transportation from road to rail and shipping

  – Expanding eco-friendly ships fueled by LNG.

Source: Submission Under the Paris Agreement: The Republic of Korea's Update of the First Nationally Determined Contribution, December 30, 2020.

The Process as a Whole

The components and planning process of the carbon neutrality plan 2050 are proposed as a whole. Even if the following steps have a certain logical steps, they should not be taken as a strictly chronological blueprint for action. A cyclic process for all steps is recommended, although each city needs to find the most appropriate approaches for its local situation.

City Profile

To serve as an "evidence baseline" for activities supporting climate neutrality, it is necessary to draft a city profile for climate, energy and nature based on an inventory (e.g., 1996 IPCC Inventory Guidelines) and analysis of the city's existing energy balance, climate footprint, GHG balance sheet, vulnerabilities and opportunities.

The profile can be supported by geographic information systems (GIS) to identify priority areas for planning climate mitigation and adaptation measures and energy efficiency actions or to identify areas with better potential for renewable energy (Figs. 10, 11, and 12).

Box 3
Inventory of greenhouse gas emissions, New York City, USA

It included CO_2 emissions (2007) resulted from the city in 1995, 2000 and 2005. The base year 2005 was chosen to align with the larger framework of PlaNYC. While consistency with the city's larger planning process was deemed a higher priority, it was accepted trend analysis and comparison with historical years was of value. To accommodate both purposes, inventories were also conducted for two background analysis years, 1995 and 2000. New York published citywide CO_2 emissions in 1995, 2000, 2005 and projected 2030 trend under business-as-usual conditions based on these figures.

Source: New York City Mayor's Office of Long-Term Planning and Sustainability, Inventory of New York City Greenhouse Gas Inventory, 2007.

Figure 6 illustrates the connections between data mining and other technical terms, including machine learning, AI, data management (database), big data and cloud computing. Figure 7 illustrates an application of the GIS/GPS-based smart technology for predicting the future ranking of real estate value using knowledge from multiple datasets. Many smart meters have been installed to remotely monitor the electricity, water and gas consumption in a household. The data can help understand a household's living patterns and estimate a family's economic ability, which is useful in precision marketing. When used aggregately, the data can infer the economic prosperity of a neighborhood and can be used to predict the future value of real estate in the neighborhood (Zheng 2018). The same procedure can be utilized for historical trend analysis and projections under business-as-usual conditions of carbon emissions. It should be noted that anonymization of data may have to be highlighted to address data protection issue.

Fig. 7 shows a city's spatial features which affect carbon emissions resulting from land use, transport, energy, buildings, water and waste services. The spatial features determine the physical and economic opportunities and constraints, and parameters within which GHG emission inventories will need to be conducted. GHG emissions are directly related to the choice, location and design of infrastructure systems and service delivery technologies. For instance, transit-oriented development is financially viable only at certain threshold urban densities (e.g., compact cities) and is optimized under certain horizontal and vertical spatial layouts and land use patterns.

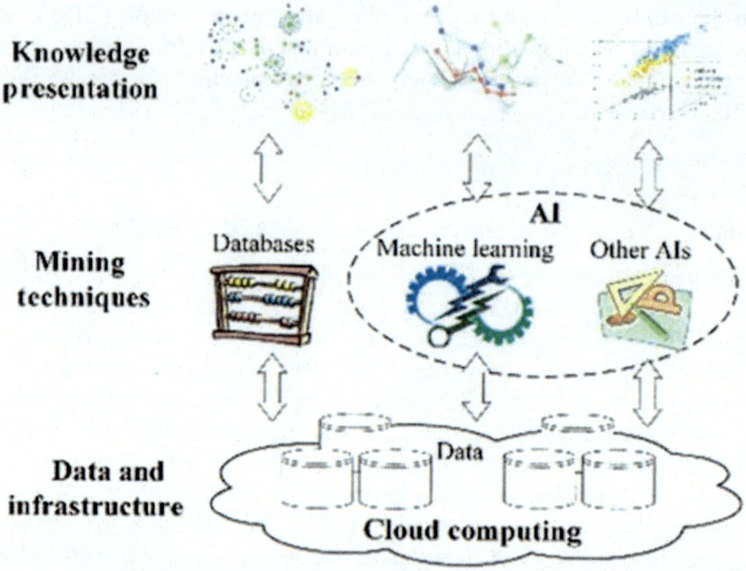

Fig. 6 Connection between data mining and other terms

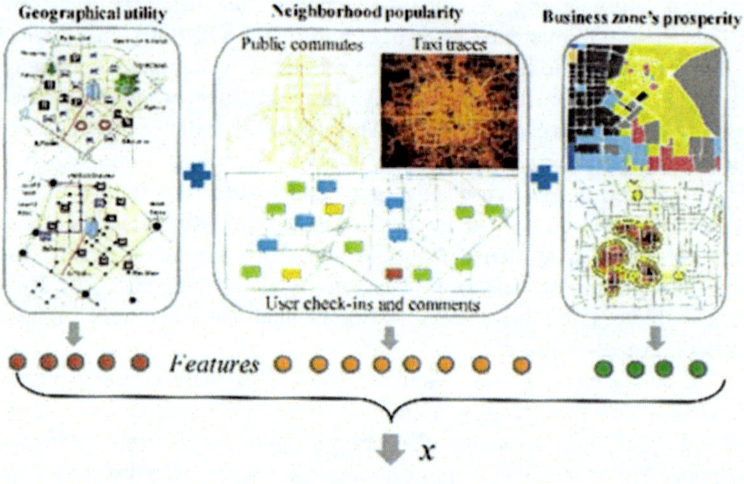

Fig. 7 Ranking real estate using knowledge from multiple datasets. *Source* Zheng (2018)

Therefore, these spatial parameters should be a part of an inventory and analysis of cities.

Box 4
Setting the baseline for residential buildings in Republic of Korea

Republic of Korea was the first to develop a standardized baseline for the residential buildings,[1] i.e., emission intensity values in tCO_{2e}/m^2 were determined by applying the "TOOL31: Determination of Standardized Baselines for Energy Efficiency Measures in Residential, Commercial and Institutional Buildings" (TOOL31).[2]

The residential and commercial building sectors in the country accounted for about 23% of the total energy consumed and 28% of the GHG emissions in 2018 as per the information from IEA.[3] The NDC of Korea envisaged a 67% reduction in building sector emissions in comparison with business-as-usual emissions by 2030.

Policy measures for mitigation in the country included the Korea Emission Trading System (K-ETS), the GHG & Energy Target Management System (TMS) and the use of Clean Development Mechanism and Korea offset credits. Specifically, for the building sector, regulatory measures to improve the building energy efficiency include the Building Energy Efficiency Certification (BEEC) that provides a minimum requirement for specific energy consumption per $m^2$of new buildings. The National Building Energy Database (NBED) is a comprehensive database developed by the Ministry of Land, Infrastructure and Transport (MOLIT) with support of the Korean Appraisal Board (KAB). The NBED contains information on different building characteristics (such as location, type of building, year of construction, gross floor area, height, number of stories) and energy consumed (electricity, gas and district heating). Building characteristics data from 7.2 million building units and monthly energy consumption data from 57 million meters aggregating to more than 4.4 billion data points were collected (Fig. 8).

[1] The documents related to this standardized baseline can be accessed at <https://cdm.unfccc.int/methodologies/standard_base/2015/sb151.html>.

[2] https://cdm.unfccc.int/methodologies/PAmethodologies/tools/am-tool-31-v1.pdf/history_view.

[3] https://www.iea.org/data-and-statistics?country=KOREA&fuel=Energy%20consumption&indicator=CO2Industry.

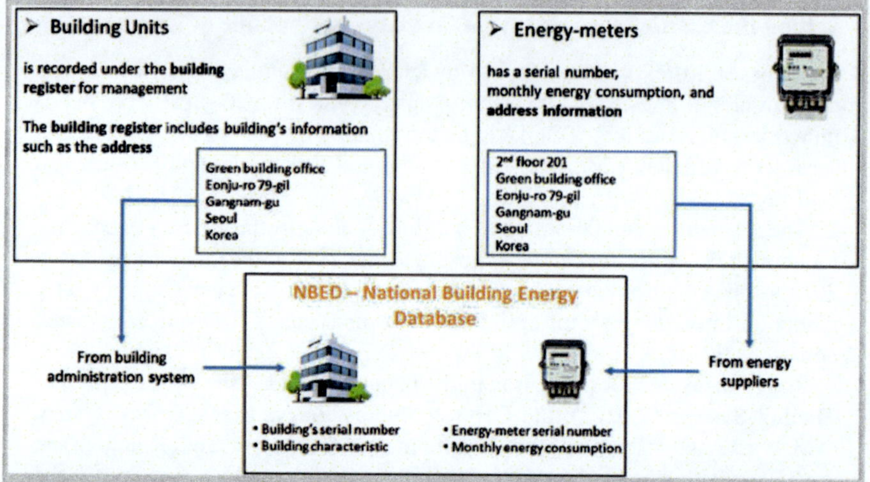

Fig. 8 Integration principle of the NBED. *Source* "Development of Korean Standardized baselines in the building sector", presentation made by Korea Appraisal Board (KAB) and Ecoeye at the workshop "CDM standards and tools for GHG emissions in the building sector", held at the 81st meeting of the CDM methodologies panel on 17/02/2020

For the standardized baseline development, target buildings were prioritized based on the emission reduction potential and data availability. Residential apartments that receive electricity from grid and that are not connected to district heating units were selected (non-residential buildings and single family housing units were excluded).

Initially, 120 categories of data/parameters were identified for new and existing buildings considering the factors that affect energy consumption and emission reduction, data coverage under the NBED and the requirements of the TOOL 31. These variables included climatic zones (i.e., Southern, Central and Jeju), administrative provinces (e.g., Daegu City, Daejeon City, Gwangju City, Seoul and Ulsan City) and gross floor area based on the "Building Act of the Republic of Korea" (i.e., A60,[4] B85[5] and C85[6]). A data filtering process was applied to remove outliers, such as buildings with gross floor area below 5 m^2, building units with zero energy consumed in any month.

[4] Gross floor area $\leq$ 60 m^2.

[5] Gross floor area > 60 m^2 and $\leq$ 85 m^2.

[6] Gross floor area > 85 m^2.

Given the number of parameters for the standardized baselines was high, the data was statistically assessed to observe which of the criteria did not have a significant impact in the results and hence could be omitted. After a careful review, aggregating the data by the climatic region and the area of the household showed that it could provide a representative and useful result; therefore, the criteria "Province" were removed, and the number of standardized baseline values decreased from 120 to 18 (9 for new buildings and 9 for existing buildings) (Fig. 9).

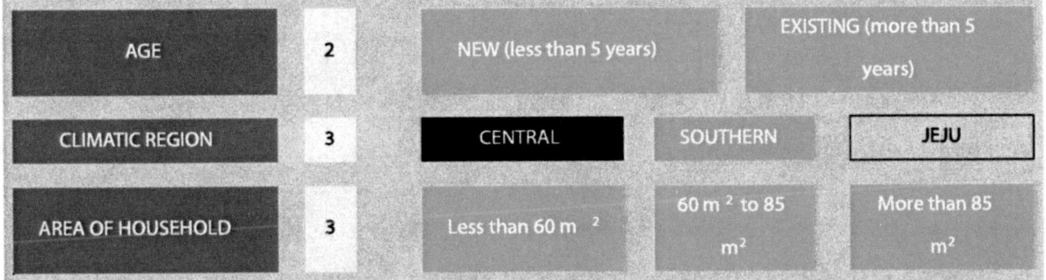

Fig. 9 Criteria applied to determine the number of standardized baseline values

The specific CO_2 emissions of building units were calculated based on the procedures and equations provided by the TOOL31, i.e., the electricity and gas consumption multiplied by respective emission factor divided by the gross floor area of the building unit. The results were separated for each building criteria (e.g., new A60 building units located in the Southern region) and sorted from the lowest specific CO_2 emissions to the highest specific CO_2 emissions. The average of the top 20% buildings by performance was set as the standardized baseline value for those building unit types.

The final results were as follows:

The specific energy consumption levels for residential building units specified by the BEEC are reproduced below, and new residential buildings must achieve at least BEEC levels 1. As can be inferred from standardized baseline values in Tables 1 and 2 above, they are almost always above BEEC level 1 + which means mitigation projects should aim for 1 + + or higher level. Such a robust level of baseline values is expected to trigger ambitious climate action encompassing construction projects but also supply chain of building materials to achieve transformative changes in the building sector consistent with the national and international climate goals (Table 3).

Table 1 Baselines for EXISTING building units

Parameter	Climatic region	Residential building unit GFA category	Specific energy consumption (kWh/m^2.year)	Specific CO_2 emission factor (tCO_2/m^2.year)
$SE_{CO2,\ Top20\%,\ i}$	Central	A60	119.2261	0.0378
		B85	103.4301	0.0329
		C85	85.0041	0.0271
	Southern	A60	98.5367	0.0336
		B85	85.8994	0.0294
		C85	71.6434	0.0243
	Jeju	A60	40.5255	0.0346
		B85	33.2806	0.0292
		C85	29.3337	0.0264

Table 2 Baselines for NEW building units

Parameter	Climatic region	Residential building unit GFA category	Specific energy consumption (kWh/m^2.year)	Specific CO_2 emission factor (tCO_2/m^2.year)
$SE_{CO2,\ Top20\%,\ i}$	Central	A60	119.2261	0.0378
		B85	96.2420	0.0314
		C85	85.0041	0.0271
	Southern	A60	98.5367	0.0336
		B85	85.8994	0.0294
		C85	71.6434	0.0243
	Jeju	A60	40.5255	0.0346
		B85	31.8987	0.0285
		C85	29.3337	0.0264

Table 3 "Building energy efficiency certification" (BEEC) levels associated with the specific energy consumption ranges

BEEC Level	kWh/m^2.year
1+++	<60
1++	60 ~ 90
1+	90 ~ 120
1	120 ~ 150
2	150 ~ 190
3	190 ~ 230
4	230 ~ 270
5	270 ~ 320
6	320 ~ 370
7	370 ~ 420

Integrated Climate Actions in Practice: A Target and Indicator Approach

Climate Targets: Optimal Scenarios for Achieving Carbon Neutrality by 2050

Normative scenarios describe various options available for adoption in order to reach achievable, potential and socially desirable future target starting from now.

The city needs to analyze the strengths, weaknesses, opportunities and threats (known as SWOT) based on the city profile described in city profile for evaluating impacts of changing climate and emission reduction potential. Emission reduction modeling and benchmarking under the key categories (e.g., residential, commercial, industrial and transport sector, decarbonizing the energy supply, carbon sequestration, etc.) can be conducted to determine the total emission reduction achievable, the energy demand reductions achievable and the change in energy demand from electricity sourced from the grid for each category.

It can be expected that there will be achievable or potential emission reductions in residential, commercial, passenger transport, manufacturing and infrastructure (water and waste treatment) in terms of demand management. The additional emission reductions are expected from decarbonizing or reforming the energy supply (energy switching to solar PV, wind, CHP, etc.) and the level of offsetting. From the modeled potential emission reductions, targets can be set up for both energy reductions and GHG reductions based on percentage reductions from base year levels prior to offsetting. Or the emission and energy reductions of all sectors resulting from emission reduction modeling can be subtracted from total emissions (e.g., 2050 BAU scenario) to produce carbon offsetting target as described in "Method 1". Target conflicts should be identified and minimized as far as possible.

Box 5
City climate targets

Example 1: Berlin

Base year: 1990

2020: 40%

2030: 60%

2050: 85%.

Example 2: Washington, DC

Base year: 2006

2032: 50%.

Emissions by source (22~75%): To cut emission 50% by 2032 by reducing energy use and the use of renewable energy, with actions for government, local business and residents:

- Transport: 22%
- Waste: 3%
- Buildings: 75%
- Tree canopy cover: 40% (trees planted annually: 14,137 trees)
  2050: 89% to become climate neutral, compared to the base year 2006 (BAU scenario).

Example 3: Amsterdam

Baseline: 1990

2050: 95%.

Strategies

City strategies should be designed to achieve cities' climate targets to become carbon neutral by 2050. The strategies for achieving the targets are one of the best starting points to develop integrated climate actions. It is important to note that there are some umbrella considerations for developing the strategies as follows:

- **Policy integration strategies: Sector cohesion**

Policy needs to be integrated at the highest strategic levels, moving away from sectoral approaches and working for intersectoral climate action or cross-over climate action directed toward carbon neutrality.

- **Methodological strategies: Weak sustainability vs. strong sustainability**

Weak sustainability is characterized by a non-declining combined stock of capital and assumes that man-made capital can be replaced with natural capital for carbon sequestration. Strong sustainability, on the other hand, implies that natural capital cannot be replaced by any other capital. Ecological or carbon footprint can be used as an indicator of strong sustainability. We need to gradually move from a path of weak to strong sustainability, which can be reflected in ecological footprint or carbon emissions per capita, to achieve carbon neutrality in cities, as described in "Definition of Carbon Neutrality."

- **Connectivity strategies: Total system approach vs. sectoral approach**

The urban matrix is assembled by superimposing the different sectors and elements. The systems for land use, water, transport, buildings, forests and the strategy for energy can work together. The physical/territorial spatial system provides a tool that serves to guide land and other resource use patterns, mobility patterns, forms of urbanization and community and green and park systems.

For example, smart grid technologies have emerged from earlier attempts to use electronic controlling, metering and monitoring in electricity infrastructure. In recent years, many smart meters and sensors have been installed in electrical grids, generating data about electricity consumption, transmission and distribution. The electricity consumption of an apartment or a building can be used to optimize residential energy usage, shifting peak loads to periods of low demand. The sensor data from electrical networks can help optimize energy transmission and distribution. DLT/ blockchain technology can be used for energy and carbon emissions reduction tracking, accounting and trading.

- **Technology strategies: City digitalization strategies during the global climate emergency and COVID-19 pandemic**

The recent biggest climate survey shows majority believe climate is a global emergency (Euronews, January 28, 2021).

The climate change and COVID-19 pandemic underline the need for fast, intelligent and sophisticated decision making in city government. Now, as cities look to the future, they are harnessing the power of innovative smart technology and interactive 3D virtual twins to help the cities plan, develop and test strategies to support their recovery and build resilience for meeting future carbon neutrality of cities.

- **Additionality assessment strategies: To convert assessment of GHG emissions resulting from alternative policies, strategies, measures and activities to bring out carbon neutrality into an urban methodology for carbon financing in a holistic manner (Urban CDM)**

Aggregating various technological and offsetting options at the city level would achieve a scale of emission reductions suitable to access newly emerging carbon market mechanisms. Additionality means the change between carbon emissions without strategies under business-as-usual and carbon emissions with the strategies at the city level. Testing and calculating the additionality can be done by cross-sectoral programmatic approach and single-project assessments. In the context of international market mechanism, according to the Paris Agreement, environmental integrity means that the transfer of units does not result in higher global emissions than the NDCs had been achieved only through domestic action. Environmental integrity should be central to both the effectiveness and credibility of urban market mechanisms under Article 6. The combined strategies, measures and project activities would have to be bundled into one activity. This would probably include a variety of different technical and organizational solutions and also support plans, programmes and regulatory activities.

Box 6
Carbon neutrality strategies in cities

Example 1: Berlin, Germany

Berlin's new energy and climate protection strategy, which will develop the capital into a climate-neutral and climate-adapted city by 2050, is an excellent example of integrated, sustainable urban development.

- Phase out coal and markedly reduce oil consumption.
- Massively expand solar power.
- Increase average energy-related renovation rate to 2% per year by 2030.
- Higher standards for new buildings.
- Gear climate protection agreements and public procurement to climate neutrality.
- Further change in the transport system (modal split) in favor of the ecomobility alliance.
- Strengthen climate-friendly consumption through information, pilot projects role models.
- Network, strengthen and permanently establish climate education.
- Overarching aspects:
  (1) Create institutional structures.
  (2) Build up funding architecture.
  (3) Establish climate protection as a cross-cutting issue.
  (4) Develop monitoring.
  (5) Permanently establish participation.

Example 2: LA County, USA

To drive action toward net-zero greenhouse gas emissions: More than 150 strategies.

- Health including initiatives around food systems.
- Transit-oriented development.
- Housing affordability.
- Zoning policies to "promote thriving places for current and future generations".
- Input from stakeholders throughout the county.

Example 3: Glasgow, UK

Glasgow City of Council has announced that the city has been set a target of becoming carbon neutral by 2030 with 61 recommendations.

- A new local transport strategy.
- An electric vehicle strategy.

- Development of a plastic reduction strategy.
- A circular economy route map.
- A food-growing strategy.

Measures

A list of potential measures for key categories necessary to achieve the above strategies is as follows:

Land use planning

- Green land use patterns: Soil organic carbon (SOC) sequestration associated with the use of soil, including crop management practices
- Zoning (vulnerable zones, risk-prone zones, low-emission zones, etc.)
- Change in forest areas due to changes in land use and forest fires (reduction of carbon sink): % of tree canopy cover
- Smart platform model for land use (Tables 4).

Spatial planning: Enabling smart, wise, compact, complete, connected and coordinated cities

- Urban layout (limit car dependency, etc.): Table 5.
- Density (mixed-use compact communities, etc.).
- Untact public space.
- Infill development.
- Transit-oriented development.
- Rehabilitation and regeneration of problem areas (former industrial sites, etc.).
- Encouragement and support for smart eco-villages, smart eco-towns or smart sustainable settlements.
- Enhancing of social-spatial integration.
- Mitigating social imbalances between neighborhoods (Figs. 10, 11 and 12).

Table 4 Interaction matrix between GHG reduction activities and land use types: An illustration

Area		Residential	Commercial	Private buildings	Public buildings	Urban Infra			Industry	Agriculture	Forest	Others
Sector						Water work	Sewage	Landfill				
Energy	Renewable (biomass)	Solar energy	Solar energy	Solar energy	Solar energy				Solar energy			
	Distribution	District heating and power system, digital data	Digital data	Digital data	Digital data	Digital data	Digital data	Digital data	Digital data	Digital data	Digital data	
	Energy saving (at home and at work city size)	Optimizing energy & resource use	Use of an area energy network for the efficient use of energy									
Solid waste management	Collection											
	Management (Heat recovery)											
Sewage	Treatment											
	Reuse											
	Heat recovery											
Waterworks	Improvement of water supply					Smart water grid						
	Water saving											
Rainwater	Collection	• Rainwater penetration and storage rate (↑) • Rainwater evaporation rate (↑) • Rate of direct runoff of rainwater (↓)										
	Use											

(continued)

Table 4 (continued)

Area / Sector		Residential	Commercial	Private buildings	Public buildings	Urban Infra: Water work	Urban Infra: Sewage	Urban Infra: Landfill	Industry	Agriculture	Forest	Others
	Rainwater garden					Blue-green infrastructure						
Lighting	Replacement (including LED)											
	Newly set	Solar–LED lighting	Solar-LED lighting									
New public transport		Electric and hydrogen cars	Smart parking						Transit-oriented development			
Land use		Compact city:land form, zoning reform				Risk-prone zones					NBS	
Spatial planning		High density development (population density)			Untact public space							
Designing complex	Ventilation											
	Orientation											
	Landscaping											
	Passive design											
Wetland restoration	Marsh										Integrated system of natural infrastructure	
	Swamp											

(continued)

Table 4 (continued)

Area		Residential	Commercial	Private buildings	Public buildings	Urban Infra			Industry	Agriculture	Forest	Others
Sector						Water work	Sewage	Landfill				
Urban form	Horizontal city		Standards, code of practices									
	Vertical city											
Open space (ratio of open green space)											Carbon forest	
Buildings		Low energy retrofitting	Low energy retrofitting	Low energy retrofitting	Low energy retrofitting							
Low-carbon lifestyles		Social-spatial integration										
Carbon-oriented urban regeneration		IoT smart homes	Infill development						Low carbon smart zone			
Low-carbon business practices									Smart factory, carbon neutrality plan	Smart farming		
Use of natural climate control systems										Crop management	Carbon sink	
Smart grid-system(use of an area energy network for the efficient use of energy)		Platform	Platform	Platform	Platform	Platform	Platform	Platform	Platform	Platform	Platform	
Composting								Recycling		Urban farming		

Table 5 Spatial dimension of urban sectoral interactions: potential for climate policy complementarity and tradeoffs: an illustration

Impact → (read horizontally)	Land use zoning	Transportation	Natural resources	Building	Renewable energy	Waste and water
Land use zoning Land use zoning determines the density, height of buildings and proportion of undeveloped land on each property		Segregation of land uses impacts travel distances and frequency; transit-oriented development zones encourage use of mass transportation	Zoning designates natural resource areas that may be set aside to reduce vulnerability to flooding or urban heat island effects	Zoning impacts placement and density of buildings, which in turn impacts building energy efficiency and vulnerability to flooding and urban heat effects	Zoning density can constrain on-site renewable energy production but can also increase efficiency of service delivery	Zoning density can determine the efficacy of delivery of waste, recycling and composting services, and the energy required for and efficacy of delivery of water services
Transportation Transportation policies determine the development and extension of road and mass transportation networks	Transportation infrastructure policies shape demand for land, and acceptance of density increases		Transportation system impacts natural resource and preserved zones		Transportation policies can require renewable energy sources for mass transportation systems	
Natural resources Natural resource policies determine which areas are preserved from development and what uses are acceptable on them	Natural resource policies determine the limits of developed land use zones	Natural resource policies affect the placement of road and mass transportation infrastructure	Natural resource policies can contribute not only to the carbon and biodiversity benefits, but also to the well-being and prosperity of people		Natural resources endowment makes certain renewable energies possible	

(continued)

Table 5 (continued)

Impact → (read horizontally)	Land use zoning	Transportation	Natural resources	Building	Renewable energy	Waste and water
Building Building policies, including building codes, affect building materials, construction types and other physical conditions	Building codes can increase acceptability of high-density zones by requiring design features to improve quality of high-density structures				Building codes can require the on-site generation of renewable energy	Building codes can require design and building materials that produce less construction waste and reduce water consumption in buildings
Renewable energy Renewable energy policies can increase on-site renewable energy production and share of energy produced by renewable sources				Renewable energy-connected smart grids can help citizens manage their own community and home energy schemes and ultimately reduce their carbon footprint		Renewable energy production can involve high water consumption
Waste and water Waste policies determine the means and extent of waste disposal. Water policies determine service extent, pricing and water sources		Disposal of refuse in an optimal way can cut down wasted journey and unnecessary CO_2 emissions				Smart water grid can help to solve a number of problems related to water shortage, climate change and integrated management of water

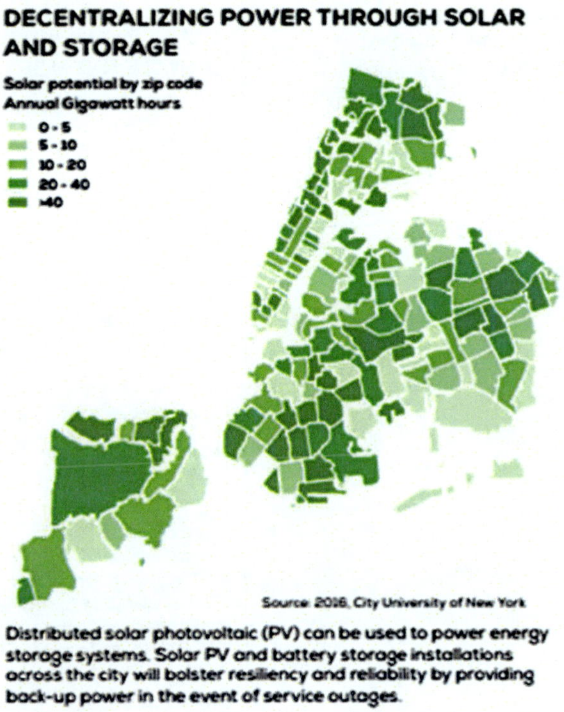

Fig. 10 Decentralizing power through solar and storage

Urban energy

- Integrated energy infrastructure platform for renewable sources of energy, district heat-cooling-electricity systems and waste-to-energy systems.
- Public transition provision.
- Renewable energy-connected smart community microgrid.
- Decarbonizing energy supply.
  - Solar, wind, CHP, etc.

Transportation

- Organizing and improving transport flow and interconnections (bus, train, river and sea transport and bike and pedestrian lanes).
- Public transport (bus rapid transit: BRT) and non-motorized transport options.

Buildings

- Smart building platform.
- Building renovations.
- IoT smart homes.

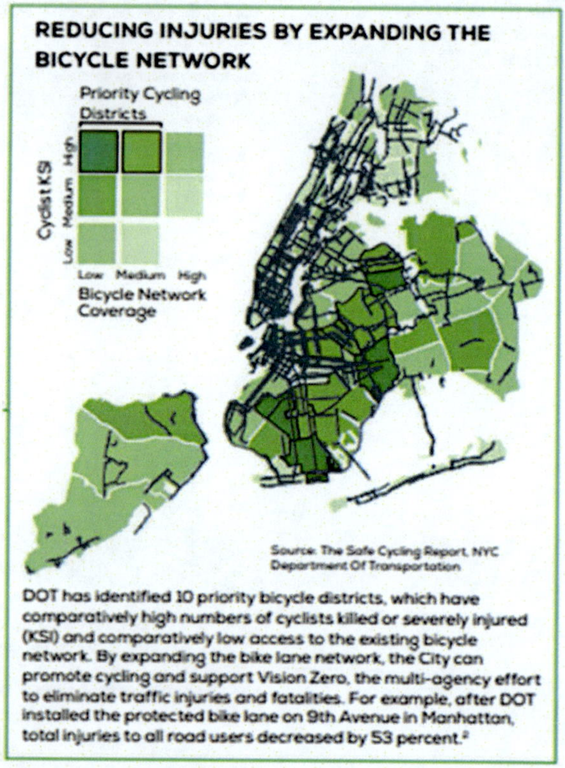

Fig. 11 Reducing injuries by expanding the bicycle network

Green spaces and other natural infrastructure

- Integrated system for a continuous structure of green spaces and other natural infrastructure (forest, water, wetland, wind, soil, farms, etc.).
  - From clean air to provide natural cooling and carbon sequestration, forests support our health in ways that may not be visible, but their value is far from zero (UNDESA 2021).

Water

- Integrated water resources management (IWRM).
- Smart water grid platform (Fig. 13).

Industry

- Smart factory platform.
- Smart and clean industrial activities.
- Climate-friendly products.

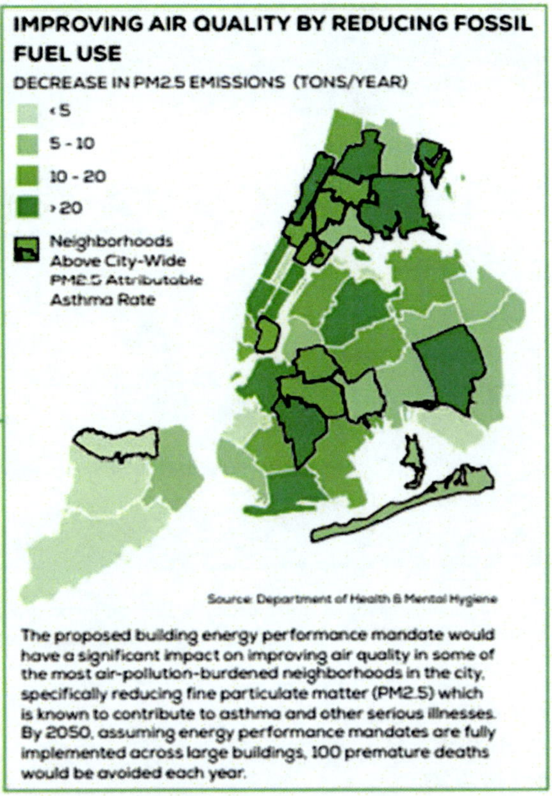

Fig. 12 Improving air quality by reducing fossil fuel use

Waste

- Smart waste treatment platform: Integrated waste management system.
- Taking the entire cycle of waste.
- Waste-to-energy.
- Reduction of waste.

Urban farming

- Growing local food.
- Food services and ICT equipment.
- Food security.

Disruptive digital technology and governance

- Digital transformation.
- eGovernance.

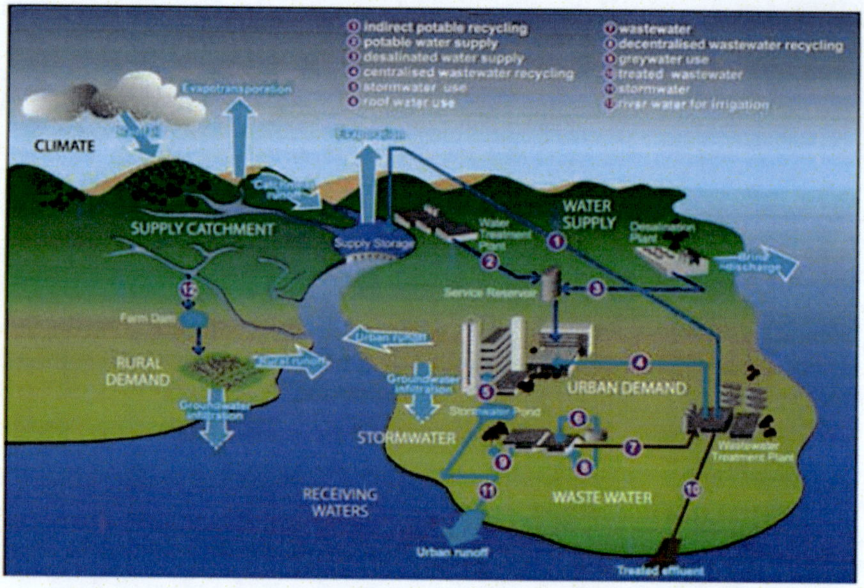

Fig. 13 Components of total urban water cycle in urban context. *Source* Water Research Foundation, CSIRO (2010) Integrated Urban Water Management Planning Manual, p 31

- Carbon financing.
- FinTech.
- LocalPay.

Standards and code of practices

- Better standards for energy efficiency.
- Explicit definition and code of practices for climate smart sustainable settlements, including smart eco-villages and eco-towns.

Box 7
Carbon Neutrality Measures in Helsinki, Finland

Making Helsinki carbon neutral image: The city is committed to cutting 4 million tons of CO_2 emissions over the next 10 years through the following various measures:

One fifth of Helsinki's emissions comes from traffic. The city's ambitious goal is to reduce traffic emissions by 69% from the 2005 level by 2035. Finland's national goal is 50%.

One key requirement for reaching this goal is that people reduce their kilometers traveled by car and choose low-emission methods of transport. Becoming carbon neutral also requires further development in vehicle technology.

Among the procedures are:

- Promoting the use of electric cars by means such as extending the charging network.
- Reducing emissions from heavy traffic with equipment choices and environmental criteria.
- Reducing emissions from port operations.
- Taking part in creating follow-up surveys regarding the pricing system for vehicle traffic.
- Raising parking fees and extending parking fee zones.
- Promoting walking, cycling and the use of public transport.
- Implementing new mobility services.
- Reducing distances with dense urban and traffic planning.

By 2035, approximately 14 million floor square meters' worth of new buildings will have been completed in Helsinki. This is equal to almost 600 Parliament Houses. The heating of buildings causes more than half of Helsinki's emissions. The greatest emission reduction potential lies in energy renovations: For example, when a building is renovated, it can be made significantly more energy-efficient than before. Emissions from buildings can be reduced by 80%. Because only a small percentage of all buildings located in Helsinki are owned by the city, it is important to encourage residents and organizations to take part in reducing emissions. The buildings owned by the city hold 11% of the emission reduction potential of the entire building stock of Helsinki. The majority of the measures to reduce emissions are financially attractive to building owners in the long term. They often improve liveability as well.

Among the procedures are:

- Providing advisory services to support Helsinki residents' energy renovations and increased use of renewable energy.
- Steering district planning more toward carbon neutrality than before (74% reduction:2015–2035).
- Steering people toward energy-efficient solutions and renewable energy through building control services.
- Improving energy efficiency and increasing the use of renewable energy in the city's service and residential buildings.
- Taking the entire carbon footprint of construction into account and promoting wooden construction.
- Replacing outdoor lights with more energy-efficient alternatives.
- Making provisions for emission-free thermal and wind energy.

The value of the city's annual procurements is €2 billion, half of which comes from construction. The development programme of Helen Oy, the energy company owned by the city, is responsible for emission reductions in energy production.

Helen Oy's procedures will reduce Helsinki residents' district heating emissions by 74% by 2035. Reducing the emissions from electricity production affects the emissions of the entire nation. The goal is to stop using coal entirely in the 2030s at the latest. The Hanasaari coal plant will be closed in 2024.

Among the procedures are:

- Replacing fossil fuels by building heating plants that run on renewable energy.
- Utilizing waste heat.
- Implementing heat pumps.
- Utilizing the demand response for heat and electricity.
- Switching to wind and solar energy in electricity production.
- Using electricity storage facilities.
- Applying energy solutions of the future clean energy production.

The procurements form more than 40% of the city's expenses. Helsinki's goal is to reduce emissions from its own procurements with environmental criteria regarding aspects such as construction materials, food services and ICT equipment. Furthermore, the city creates platforms and opportunities for developing smart & clean business activities.

Among the procedures are:

- Making the development of climate-friendly products and services in Helsinki as easy as possible.
- Developing climate-friendly solutions together with businesses and residents.
- Daring to try new things and create pilots and sharing what we have learned from them.
- Developing the city's procurement criteria so that they encourage operators to provide climate-friendly products and services.
- Taking the entire lifecycle of the products and services we procure into account.
- Promoting a sharing and circular economy in terms of the use of the city's facilities, libraries, food management, etc. Monitoring our procurements and reporting them in a more methodical manner (Figs. 14, 15 and 16).

Fig. 14 A carbon neutral city image: Walking along carbon free street

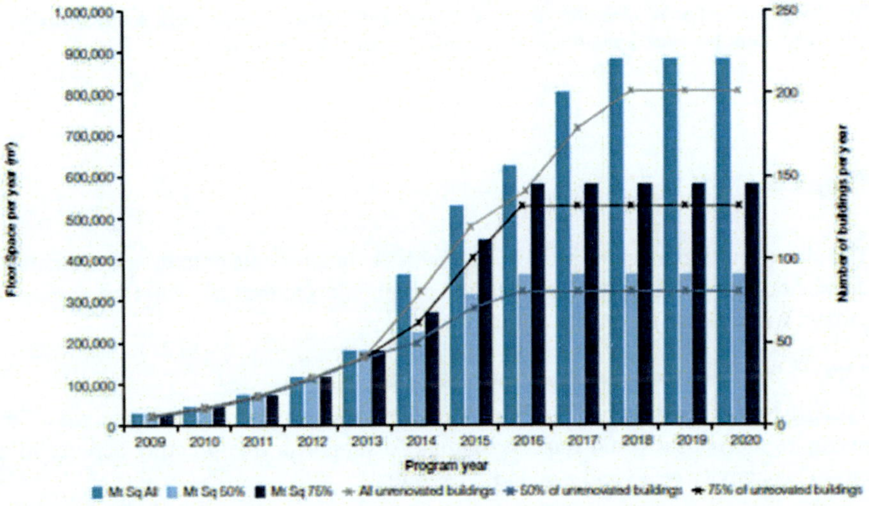

Fig. 15 Options for a program to retrofit City of Melbourne unrenovated buildings. *Source* City of Melbourne (2009)

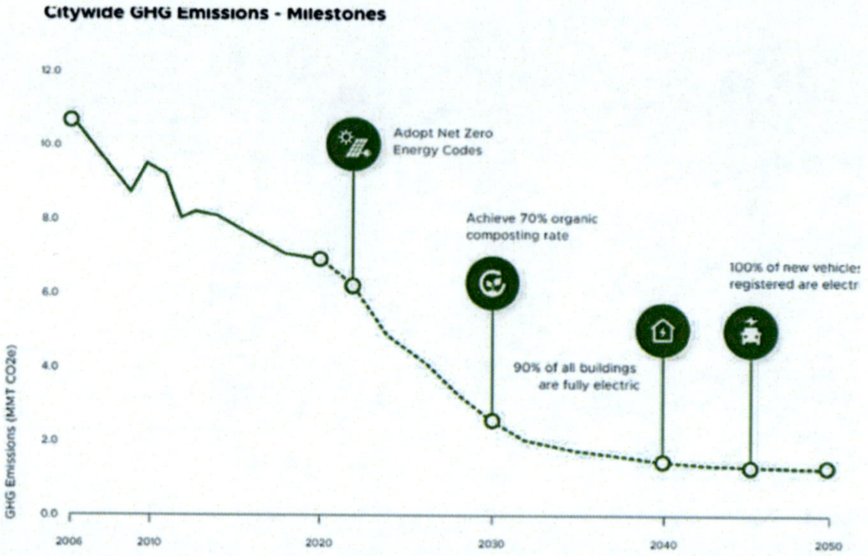

Fig. 16 Identification of measures for scope 1 (city GHG emissions boundary), scope 2 (national GHG emissions boundary) and scope 3 (global GHG emissions boundary)

Projects

The plan should connect the proposed activities to quantifiable outputs and ultimately desired outcomes. This enables cities and stakeholders to evaluate what the proposed project will achieve if funded.

Project activities

Specific types of project can be undertaken to achieve the outputs and outcomes. For example, a proposal could request funding to support a tree-planting activity in a city. This activity will yield a quantifiable output.

Box 8
Climate project activities in Glasgow, UK

- One street project which plans to include district heating.
- An innovative roof mounted solar PV canopy.
- Ducted wind turbines.
- Energy arbitrage.
- Power storage.
- EV charging stations.

- Smart grid controls.
- All council vehicles to be zero emission by 2029.
- A low-emission zone for Glasgow City center.
- A wide range of active travel projects.

Outputs

An output is a short-term result achieved as a result of an activity. For example, at the end of project period, after several sessions of tree-planting, X number of trees will be in the city. This quantifiable output will yield various project outcomes.

- 2032 goal: 40% tree canopy cover.
- Trees planted annually: 14,137 trees.

Outcomes

A project outcome is a medium to long-term result that occurs and/or continues after the project ends. For example, after several sessions of tree-planting activities, this project yielded an output of X number of trees in the city. Medium to long-term outcomes of this project include

- Carbon sequestration.
- Beautification of the city.
- Increased environmental awareness of residents.
- Biodiversity enhancement.
- Improved air quality.

Gradual Pathways to Carbon Neutrality: Organizational Actions

The pathways to reductions show an approach that the city takes to improving energy performance and carbon sequestration in existing and new developments by sector. For instance, the potential roll-out of the building energy program is illustrated in Fig. 15. A suitable ramp-up period is required to provide sufficient time for building owners to be signed up and the energy performance contracting industry to be prepared for significant increases in workload. Possible pathways can include

- Mapping the carbon neutrality process based on city documents and national statistics and validating it through process-owner interviews.

- Assessment of scenarios and assumptions of the future which weaken the reliability of the carbon neutrality plan (Figs. 16 and 17).

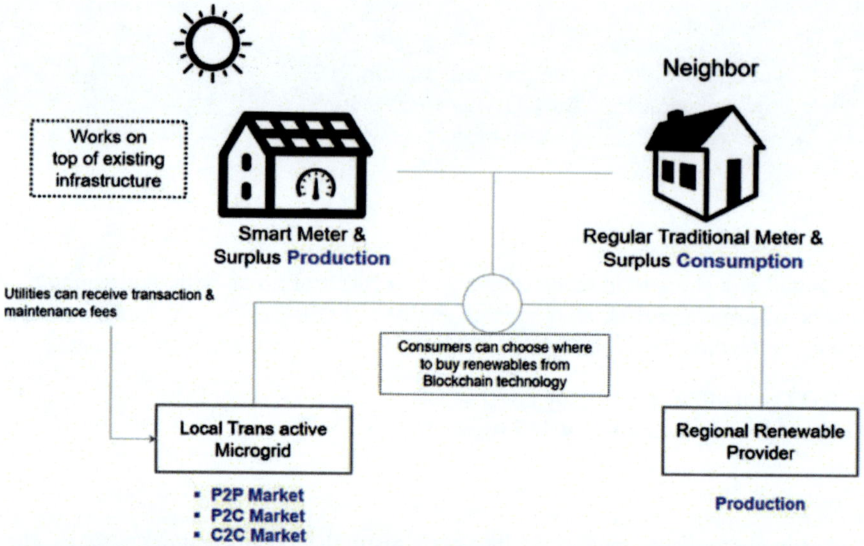

Fig. 17 Interacting local energy with neighbors: bockchain is a shared database technology that processes and stores transaction data

Box 9

Basic plan for climate change response, the Republic of Korea

Prior to the legislation, the Republic of Korea incorporated the updated target into the 2nd Basic Plan for Climate Change Response, the overarching national plan for climate actions established in October 2019. The 2nd Basic Plan for Climate Change Response envisions a sustainable and low-carbon green society. Under this vision, the Republic of Korea aims to limit the national GHG emissions to 536 MtCO2eq by 2030 and enhance capacity of all sectors to implement the Paris Agreement. To achieve these objectives, the Republic of Korea set out climate mitigation and adaptation measures by sector under its core strategies, transition to a low-carbon society, establishment of a robust adaptation system and enhancement of a climate change response framework.

Source: Submission under the Paris Agreement, the Republic of Korea's Update of its First Nationally Determined Contribution, December 30, 2020.

- Major measures: The city's direct and mandatory measures.
- Out-of-city boundary energy production.

(1) Energy industry: Carbon neutral electricity production.
(2) National government.

(3) Real estate owners.
(4) International organizations.

- Urban planning will be pivotal for a sustainable energy future. Urban energy planning should be integrated more into existing urban planning process.
- Centralized electricity production should be allocated for cities.
- Consumption-based energy efficiency measures would be taken into account in parallel when securing the transition to carbon-free energy production.
- Enabling GHG emissions reduction through enhanced efficiency of citizens' consumption of goods and services.

  (1) Enhanced circularity of economy.
  (2) Promotion of sharing economy (e.g., public parking for car sharing).

- Enabling GHG emissions reduction through the decrease of the carbon footprint of products and services consumed by the citizens.

  (1) The use of electric vehicle charging stations for e-mobility.
  (2) Information and awareness campaigns on consumption of sustainable goods and services.

- Enabling GHG removal by the citizens.

  (1) Promotion of urban trees.
  (2) Increase of urban farming.

- Emissions reduction related to the operation of cities' infrastructures.

  (1) Operation of public buildings.
  (2) Social housings.
  (3) District heating and cooling systems, etc.

- While the shares of fossil fuels are decreasing, consumption from grid-supplied energy systems is likely to increase, which changes the carbon neutrality requirements accordingly.
- Proven annual carbon compensation mechanisms related to energy production funds, green cards and carbon accounting should be included to make sure that the annual net-carbon balance is zero or negative.
- For transparent statistics and carbon accounting, allocated energy production should be separated in the statistics so that actual carbon emissions can be calculated for the sectors and cities. Without this separation, double counting will exist. DLT/blockchain technology could help this process a lot.
- Blockchain technology-based carbon tracking, accounting and trading.

  - A framework for measurement of the impact of climate actions enabled by blockchain/DLT (Thioye 2021).
  - Appendix C to promote the use of digital technologies.

- It is proposed that in municipal carbon accounting, municipal energy production for all scopes should be allocated to the city, and scope definition in carbon neutrality should justify this more clearly.

Implementation Roadmap

Given all the principles, methodologies and process, how can city climate planners do to implement the carbon neutrality plans of their cities? This section provides examples of roadmap.

- To describe the work plan (timeline) for project implementation which clearly shows ramp-up reduction periods.
- To propose a process to create a policy implementation roadmap that leverages feedback from stakeholder engagement and that is consistent with the international climate action planning framework.
- To identify longer-term actions beyond the 2030 timeframe as well as high-priority actions that need to be accelerated in the near-term in order to advance on realistic trajectory to achieve carbon neutrality.
- To propose a methodology for prioritizing based on evidence such as their emissions reduction potential (identified in the 5.2.3 Technical Assessment) and social, environmental and economic benefits.
- To integrate climate adaptation into plans to achieve carbon neutrality.
- To identify opportunities to integrate relevant policies and plans and optimize climate adaptation co-benefits.
- To provide key milestones or leading indicators that can be used for policy planning, such as the rate of necessary building energy retrofits and electric vehicle adoption.
- To provide recommendations for financing the transition to carbon neutrality (Fig. 17).
- To be familiar with and leverage the latest research and best practices for achieving rapid 3Ds (decarbonization, digitalization and decentralization) including the Carbon Neutral Cities Alliance's Game Changers, C40's Focused Acceleration Report and Long-Term Low-Carbon Development Strategy.

Box 10

Carbon Neutral City Initiatives

Example 1: Washington DC, USA

- "Sustainable DC 2.0 Plan" includes DC carbon neutrality strategies.
- Climate action plan.
- Carbon neutrality strategies.

- The district's solar initiatives are working to increase renewable energy use, reduce low-income energy bills and expand green job training.
- Clean energy DC.
- Climate ready DC plan.
- Annual greenhouse gas inventory.
- Carbon-free DC.

Example 2: Berlin, Germany

- The Berlin Energy Turnaround Act (2016).
- Berlin Energy and Climate Programme 2030 (2018).
- BEK 2030.
- Berliner Stadtwerke (2014).
- The ImpulseE programme.
- The Berlin Programme for Sustainable Development (BENE).

Example 3: Melbourne, Australia

- National Carbon Offset Standard.
- Emission Reduction Plan.
- Sustainable Business Guide.

Example 4: LA County, USA

- Climate action plan.
- Sustainability plan.

Example 5: Queensland, Australia

- Queensland "the Smart State": Climate Smart State Plan.

Essential Project Deliverables for Carbon Neutral City Plan 2050

Each city has to find the most appropriate documents to be delivered to city agencies considering the local situation and respond with flexibility, but the city should consider how to incorporate the following documents into the planning process of carbon neutral cities.

Technical Assessment

- Development of tool to model GHG emissions reduction pathways.
- A summary document which describes the optimal scenarios for achieving carbon neutrality by 2050 and documentation of methodology and assumptions.

Define and Measure Carbon Neutrality

- Engage with city agencies and key stakeholders to propose a definition of carbon neutrality and recommend metrics for tracking process.
- Consult with national and international work products and subject matter experts in climate mitigation and carbon neutrality planning.

Policy Road Map

- Written policy implementation roadmap template, draft and final, including the identification of highest priority near-term actions.
- Digital materials and presentations for engagement activities that are publicly accessible and web- and mobile-friendly.

Co-benefits of Carbon Neutral City Plan 2050

The city's vision for carbon neutrality is strongly framed within the contexts of co-benefits, meaning that reduced GHG emissions will not be the only benefit derived from the carbon neutrality plan 2050, but it provides a high degree of local sustainability, urban resilience and disaster risk reduction benefits along with global climate mitigation benefits in a holistic way. In addition, the following benefits can be expected:

- Localization of the SDGS and Paris Climate Agreement.
- Local implementation of the NDC and LEDS.
- Cleaner, healthier, greener and safer urban environment.
- "Smarter" and "wiser" resource management.
- Rapid digital and disruptive technical transformation to provide connectivity in the untact era.
- Greater equity and inclusivity.
- Carbon financing and funding to reboot cities' economy from COVID-19 pandemic crisis under the Green Deal for green investment and/or stimulus plan.

- Global and regional competitiveness of city brands.
- Localization of green jobs, goods and services and job training.
- Increased sense of belongings to carbon-free cities and communities.

Evaluation of Plans, Implementation and Performance: A New MRV Approach

Appendix D can be used to measure the standards of the "carbon neutral city 2050 plan" as a checklist which uses scaling techniques for assessing the carbon neutrality plans, implementation and performance. This appendix offers the readers practical indicators for preparing the carbon city plan 2050 within the broad framework of climate smart cities and their implementation and performance. The indicators provided herein can be improved by the appropriate pilot testing and applications.

Appendix A
Comparison of the Selected Carbon Neutral City 2050 Plans

See Table A.1.

Table A.1 Comparative analysis of the selected carbon neutral city 2050 plans

City	Climate Target	Action	Initiative
Washington, DC	• **2006: Baseline** • **2032: 50% reductions** • **2050: Target year-89% reductions** • **Tree canopy cover: 40% (2032)**	• Solar initiatives – To increase renewable energy use – To reduce low-income energy bills – To expand green job training • Reducing energy use • Increasing the vibrancy and walkability of neighbourhoods • To publish energy and water performance benchmarking data from the largest buildings i • Actions for government, local businesses, and residents	• Sustainable DC plan • DC carbon neutrality strategies • Climate action plan

(continued)

K.-G. Kim and M. Thioye (eds.), *Planning Climate Smart and Wise Cities*, The Urban Book Series, https://doi.org/10.1007/978-3-030-80165-6

Table A.1 (continued)

City	Climate Target	Action	Initiative
Melbourne city	• 2002: Baseline • 2020 BAU: Target year – 100% reductions + carbon offsetting = zero net	• Carbon offset projects to absorb GHG emissions: To participate in the national emissions trading scheme – To invest in native forest regeneration – To invest in indigeneous fire management practices • Projects to reduce GHG emissions	• National carbon offset standard • Emission reduction plan • Sustainable business guide
Berlin city	• 1990: Baseline • 2020: 40% • 2030: 60% • Beyond 2050: 95% by the middle of the century	• Greener energy supply – Phasing out coal-based power and heat generation by 2030 – 100% renewable energy in the future • Energy-related building refurbishment • More school education on climate protection and clean energy: Projects in schools and nursery schools to anchor the issue of climate protection • Climate-friendly behaviour for its citizens (cutting energy use and replacing fossil fuels with renewables • Greater efforts in adapting to the effects of climate change	• Berlin energy Turnaround Act 2016: To make Berlin climate targets legally binding: • Berlin energy and climate programme 2030 (approved in 2018) • BECK 2030 which protection agreements with large public utility companies in Berlin and the residential real estate industry as well as the successful model of energy- saving partnerships • Berlin energy day • Annual “Berlin saves energy action week” • “Berlin energy saving champion” competition

Appendix B
A Methodological Guide to Using Carbon Offsets for Carbon Sequestration Outside Cities

Basic steps associated with use of carbon offsets for carbon sequestration are as follows:

Step 1: Project development, validation and registration.

- Describe the environmental setting of the project area in terms of carbon sink and storage.
- Estimate carbon stock of the area of interest for investment.
- Develop baseline scenarios of carbon sequestration.
- Develop project scenarios.
- Calculate additionality if carbon sequestration would not have occurred in the absence of the project and a market for offset credits.
- Validate and register the additionality (Table B.1).

Step 2: Project implementation and offset credit issuance.

- Invest directly in an offset project.
- Make contract for offset credit delivery with an offset project developer.

Step 3: Offset credit transfer.

- Purchase in an one-off transaction directly with project developer.
- Purchase from an offset credit broker (who has previously purchased credits from a project developer or exchange).
- Purchase via an exchange.

Step 4: Offset credit retirement.

- Purchase offset credits from a retailer (who retires credits on your behalf).

K.-G. Kim and M. Thioye (eds.), *Planning Climate Smart and Wise Cities*, The Urban Book Series, https://doi.org/10.1007/978-3-030-80165-6

Table B.1 Examples of major carbon offset programs

"Compliance" carbon offset programs (run by governmental bodies)	Geographic coverage	Label used for offset credits
Clean development mechanism (CDM)[6]	Low- and middle-income countries	Certified emission reduction (CER)
California compliance offset program	United States	Air resources board offset credit (ARBOC)
Joint implementation (JI)[7]	High income countries	Emission reduction unit (ERU)
Regional greenhouse gas initiative (RGGI)	Northeast United States	RGGI CO_2 offset allowance (ROA)
Alberta emission offset program (AEOP)	Alberta, Canada	Alberta emissions offset credit (AEOC)
"Voluntary" carbon offset programs (run by NGOs)	Geographic coverage	Label used for offset credits
American carbon registry	United States, some international	Emission reduction tone (ERT)
Climate action reserve (CAR)	United States, Mexico	Climate reserve tonne (CRT)
The gold standard	International	Verified emission reduction (VER)
Paln vivo	International	Plan vivo certificate (PVC)
The verified carbon standard	International	Verified carbon unit (VCU)

Source Broekhoff et al. (2019) Securing climate benefit: a guide to using carbon offsets, GHG management institute, Stockholm environment institute, p 9

Step 5: Carbon balance sheet.

- Include the purchased offset credits in a carbon balance sheet to achieve a net carbon footprint of zero.

Step 6: Carbon sequestration benefits.

- Consider the expected carbon sequestration benefits: While carbon offsets do not immediately benefit local residents, they can be an important driver of global GHG emissions reduction.
- Integrate the carbon sequestration benefits with a carbon neutral city plan 2050.

Appendix C
Digital Transformation to Carbon Emissions Reduction Tracking, Accounting and Trading: An Integration Platform Modeling Approach

Blockchain or distributed ledger technologies (DLTs) have received wide attention in recent years as an innovative solution to provide a reliable, incorruptible, decentralized database that allows transactions between individuals who do not know and therefore do not trust each other. In recent years, blockchain technology has been tested in various international development projects, e.g., for financial services such as the transfer of remittances, peer-to-peer electricity trading, insurance and the management of land titles and identities. However, currently, there is very little reliable information available about the impacts and lessons learned from these applications (Burg J, Murphy C, Petroud JP 2018). The blockchain could be the missing link that brings consumers, businesses and investors together on climate change (Fig. 16).

Kim (2018) has designed four IoT/blockchain platform models for new market mechanisms (NMMs) and presented them at the UNFCCC COP24, Katocvice, Poland. There are several variations of the Kim blockchain model proposed for carbon financing for various types of projects.

Blockchain model A for carbon financing (Fig. C.1) shows a GHG matrix for describing the relationship between typical measures and their potential ultimate impacts on climate smart city development activities. The basic philosophy of the analysis is that a meaningful use of blockchain technology must ultimately be a tool for determination of direct and indirect impacts on GHG emission reductions at the city scale.

Blockchain model B involves the three steps for tracking of emission reductions of each CPA, GHG reduction measures and aggregation of each emission reduction by the measures (Fig. C.2).

In Blockchain model C (Fig. C.3), each block illustrates different digital functional layers of climate markets which are connected each other. Blockchain technology can synthesize and support the transaction of all types of GHG emission-related data (e.g., climate measures, additionality) in a shared visualized database.

K.-G. Kim and M. Thioye (eds.), *Planning Climate Smart and Wise Cities*, The Urban Book Series, https://doi.org/10.1007/978-3-030-80165-6

Urban CDM Project Elements which reduce GHG emissions		Urban development activities which impact climate					
		Land use planning	Transport	Energy generation	Building energy	Solid waste	Waste water and water efficiency
Measures	Transport (e.g. Transit oriented development)	Major GHG reduction	Major GHG reduction	Major GHG reduction	Minor GHG reduction	Its magnitude cannot be determined	Its magnitude cannot be determined
	Building Energy (e.g. Mandatory building energy code)	Major GHG reduction	Minor GHG reduction	Major GHG reduction	Major GHG reduction	Minor GHG reduction	Minor GHG reduction
	Land Use Planning (e.g. Infill/densification)	Major GHG reduction	Major GHG reduction	Major GHG reduction	Minor GHG reduction	Minor GHG reduction	Minor GHG reduction
	Energy Generation (e.g. District heating & cooling)	Major GHG reduction	Minor GHG reduction	Major GHG reduction	Major GHG reduction	Minor GHG reduction	Its magnitude cannot be determined
	Solid Waste (e.g. Waste transport vehicle routing reduce km travelled)	Major GHG reduction	Major GHG reduction	Major GHG reduction	Minor GHG reduction	Major GHG reduction	Its magnitude cannot be determined
	Waste water and water efficiency (e.g. Wastewater switching treatment and optimization)	Minor GHG reduction	Minor GHG reduction	Major GHG reduction	Major GHG reduction	Its magnitude cannot be determined	Major GHG reduction

Fig. C.1 Blockchain platform model A for carbon financing

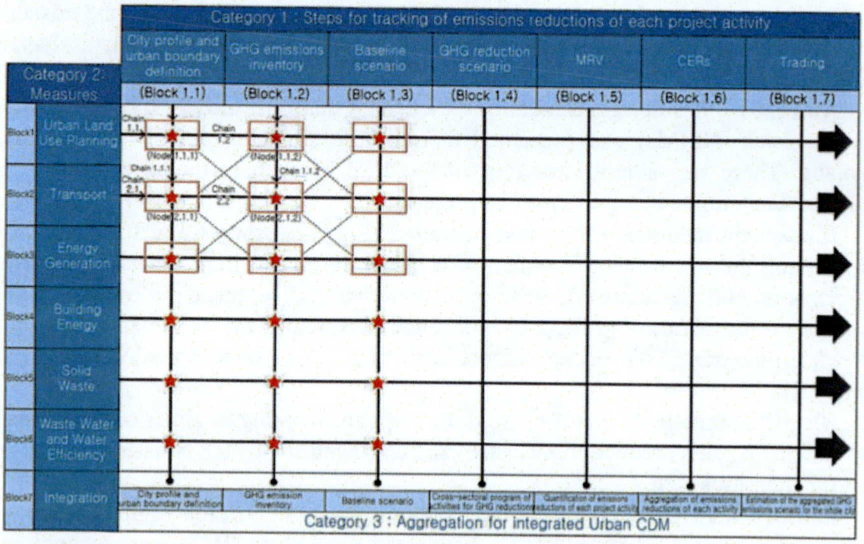

Fig. C.2 Blockchain platform model B for carbon financing

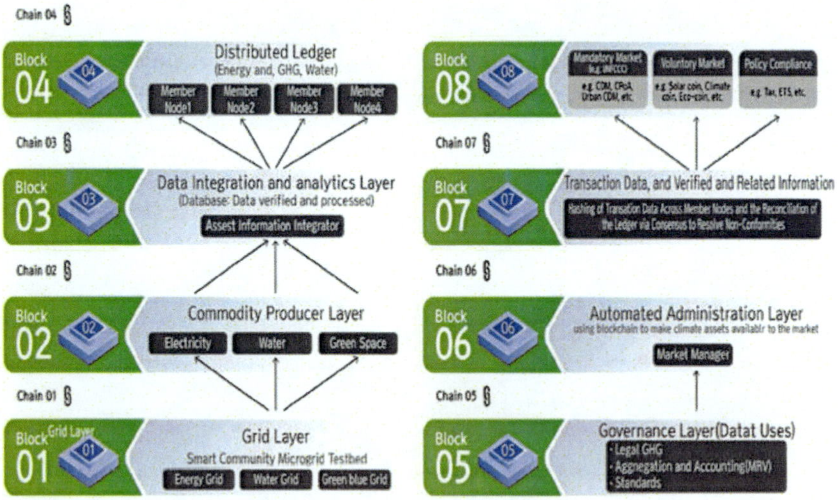

Fig. C.3 Blockchain platform model C for carbon financing

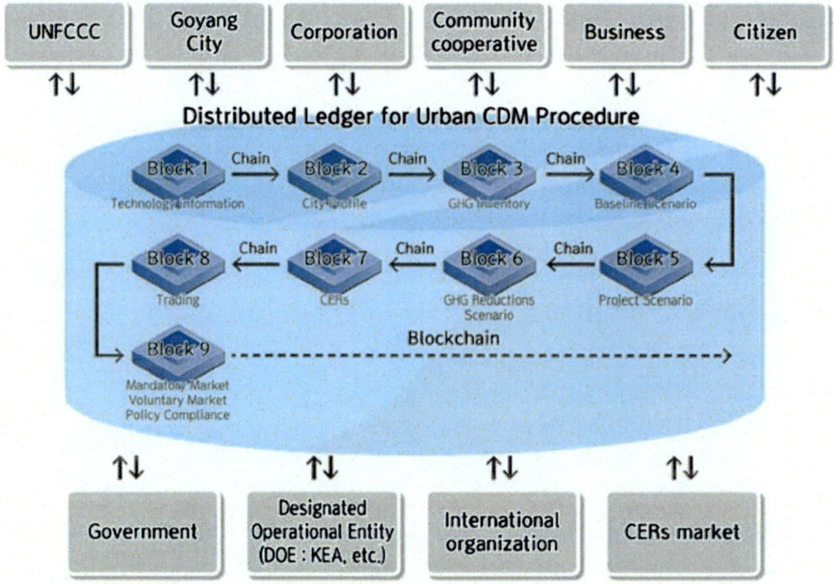

Fig. C.4 Blockchain platform model D for carbon financing

Blockchain model D shows how ledgers for urban methodology procedure can be distributed to stakeholders with the power of blockchain for the transparency tracking and control of urban carbon financing implementation (Fig. C.4).

Appendix D
Indicators of the Carbon Neutral City Plan 2050—Evolutionary Paths of Carbon Neutral City Projects

The objectives of the Appendix D are to help develop and reach out for a set of indicators (1) judging the standards of carbon neutrality plans, implementation and performance of the cities, (2) identifying challenges and opportunities and (3)creating power of making a carbon neutral city dream come alive. Key indicators have been emerged as follows (Table D.1):

Table D.1 Proposed domains, associated sub-domains and indicators for the carbon neutral city plan 2050

Domains	Sub-domains	Indicators
Overarching assessment	• Principles	• Gradual target principle • Social-spatial-digital integration platform principle • Land tenure security principle • Carbon offset principle • Digital transformation principle
Technical assessment		
Methodologies	• Strong sustainability • Weak sustainability	• Net-zero emissions • Emissions below or equivalent to the base year • Carbon emissions per capita
City profile	• Climate • Energy • Nature	• Temperature increase • Energy demand and supply • Forest, water, soil, wetland, etc.
Climate actions	• Climate targets	• Base year • 2030 • 2050

(continued)

K.-G. Kim and M. Thioye (eds.), *Planning Climate Smart and Wise Cities*, The Urban Book Series, https://doi.org/10.1007/978-3-030-80165-6

Table D.1 (continued)

Domains	Sub-domains	Indicators
	• Emissions reduction	• Land use planning – Land use patterns – Zoning – Smart platform models • Spatial planning – Urban layout – Density – Untact public spaces – Urban conversion of former industrial sites – Social-spatial integration • Urban energy – Integrated energy infrastructure platforms – Renewable energy – Smart grids – Waste management – Water management • Transportation – Transport flow and interconnections – Public transport and non-motorized transport options • Buildings – Smart building platform – Building renovations – IoT smart homes • Green space and other natural infrastructure – Integrated system for a continuous system – X number of trees – % tree canopy cover • Water – Integrated water resources management – Smart water grid platform • Industry – Smart factory platform – Smart and clean industrial activities – Climate-friendly products • Waste – Smart waste treatment platform – Taking the entire cycle of waste • Urban farming – Growing food – Food services and ICT equipment • Standards and code of practices – Standards for energy efficiency • Code of practices for climate smart cities
	• Decarbonizing the energy supply	• Solar PV • Wind • CHP, etc.

(continued)

Table D.1 (continued)

Domains	Sub-domains	Indicators
	• Carbon offsetting	• Technology • Carbon sequestration • Types of carbon offsets
Organizational assessment	• To set up climate change/ environmental policy unit	• Interdepartmental links • Cooperation with relevant authorities
	• Multi-stakeholder partnerships	• X number of stakeholders • Types of stakeholders – Private sector – Developers – Property owners – Local/community governments – The public (residents) • Self-governing – Purchasing green energy
	• Behavioral change of people	• Energy consumption • Water consumption
Pathways to carbon neutrality	• Roll-out of building energy program	• Retrofit program • Energy performance
	• Mapping carbon neutrality process	• Assessment of scenarios and assumptions • Mandatory measures • Integration into urban planning process • Grid-energy supply systems • Carbon compensation mechanisms
Implementation roadmap	• Work plan	• Ramp-up reduction periods
	• Policy implementation roadmap	• Feedback from stakeholders
	• Trajectory to achieve carbon neutrality	• Near-term actions • Medium-term actions • Longer-term actions
	• Key milestones	• Policy planning • Rate of building energy retrofits
	• Financing	• Urban CDM • ETS • Green bond • Energy production funds • Carbon offsets • Green cards
	• Best practices	• Carbon neutral cities alliance's game changers • C40's focused acceleration report • Long-term low-emission development strategy

(continued)

Table D.1 (continued)

Domains	Sub-domains	Indicators
Co-benefits	• Social benefits	• Equity and inclusivity • Sense of belongings
	• Environmental benefits	• Localization of the SDGs and paris climate agreement • Local implementation of the NDCs and LEDS • Cleaner, healthier greener and safer urban environment • Smarter and wiser resource management
	• Economic benefits	• Carbon financing and carbon funding • Competitiveness of city brands • Green investment • Green jobs and job training
	• Adaptation	• Extreme temperature • Drought • Flood • Typhoon • Sea level rise • Poor harvest of crops • Disease pandemic
	• Technical benefits	• Technical transformation to provide connectivity in the untact era
Digital dataset	• Digital data platform	• Blockchain/DLTs • 5G • Digital twin • VR, AR, MR • Cloud (virtual server) • AI application (super computer) • Big data • GIS/GPS
	• Data algorithm (solution procedure and methods)	• Emissions prediction (baseline scenarios) • Neutrality prediction (project scenarios) • Additionality calculation • Carbon emissions reduction tracking, accounting and trading • Carbon balance sheet analysis (carbon budgeting)

References

Broekhoff D, et al (2019) Securing climate benefits: a guide to using carbon offsets, GHG management institute, stockholm environment institute. p 9, 15, 20
Burg J, Murphy C, Petroud JP (2018)
City of Cape town (2020) The city of Cape Town' carbon neutral 2050 commitment. p 28
City of Melbourne (2009) Zero net emissions by 2020: update 2008, p 28, 29, 36
C40 Cities Climate Leadership Group, New York City (2019) Defining carbon neutrality for cities and managing residual emissions: cities' perspective and guidance. p 55
Global Protocol for Community–Scale Greenhouse Gas Emissions (GPC) http://www.wri.org/news/release-launch-first-greenhouse-gas-accounting-tool-chinese-cities
Kim KG (2018) Blockchain-based climate and urban CDM initiative, Goyang/CCC/UNFCCC, presented at the side event, UNFCCC COP24, Katovice, Poland, December 10th, 2018
New York City, Mayor's Office of Long-Term Planning and Sustainability (2007) Inventory of New York City greenhouse gas emissions. p 8, 15
New York City (2017) 1.5 °C aligning New York city with the Paris climate agreement, Executive Order 26, #ONNYC, p 7
Thioye M (2021) New advanced GHG accounting system for cities enabled by DLT. In: Kim KG, Thioye M (eds) Planning climate smart cities. Heidelberg, Springer (in progress)
UNDESA (2021) UNDESA voice march 2021, Making nature count. Accessed on March 2, 2021
UNFCCC (26 Feb 2021) "Climate commitments not on track to meet Paris agreement goals" as NDC synthesis repot is published, Article
Washington DC (undated) Carbon-free DC, unpaged
Water Research Foundation, CSIRO (2010) Integrated urban water management planning manual. p 31
World Bank (2010) A city-wide approach to carbon finance. The World Bank, Washington, DC, p 66
Washing DC, CNCA (2017) carbonneutralcities.org/cities/Washington-DC/
Washington DC (2018) Notice of funding availability and request for application (RFA), RFA # 2018-1907-USA, Department of Energy and Environment, District of Columbia
Zheng Y (2018) Urban computing. The MIT Press, Cambridge, MA, p 89, 313

K.-G. Kim and M. Thioye (eds.), *Planning Climate Smart and Wise Cities*, The Urban Book Series, https://doi.org/10.1007/978-3-030-80165-6

Index

K.-G. Kim and M. Thioye (eds.), *Planning Climate Smart and Wise Cities*, The Urban Book Series, https://doi.org/10.1007/978-3-030-80165-6